The Plant Engineer's Guide to
to
Industrial Electric Motors

The Plant Engineer's Guide to Industrial Electric Motors

by
Richard L. Nailen, P.E.

Barks Publications, Inc.

Copyright 1985 Barks Publications, Inc.
400 N. Michigan Ave., Chicago, IL 60611-4198

Portions of this book first appeared in *Electrical Apparatus* magazine, a Barks Publication.

All rights reserved. No part of this book may be reproduced without the written consent of the publisher. Brief passages may be quoted as part of a review, published technical paper, or magazine article provided proper credit is given the publisher and author.

First Edition, 1985

Art by Cindi Trevorrow Schrage and Richard Nailen. Jacket design by Henry Neu. Typography by Media Graphics Corporation, Chicago. Printed in Chelsea, Michigan, by BookCrafters, Inc. Edited and designed by Kevin N. Jones. Photos that appear in this book without credit given to a source were taken from *Electrical Apparatus* files.

National Electrical Code® and **NEC**®, terms used throughout this book, are Registered Trademarks of the National Fire Protection Agency, Inc., Quincy, Massachusetts.

ISBN 0-943876-01-X

*To Jack Shulman,
who first thought it was possible*

Table of Contents

Preface   x

1 Motors and Standards   1

National electrical equipment standards; converting motor specifications from English to metric units; service factor and other nameplate markings; the right way to prepare motor specifications.

2 Efficiency and Power Factor   41

The truth about efficiency and power factor; the role of power factor in motor design; stray load loss; how motors are tested; capacitors and power factor controllers.

3 Coping with Starting Conditions   91

Motor torque, inertia, and acceleration heating; typical driven machine characteristics; types of reduced-voltage starting.

4 Matching the Motor to the Power System   147

Protecting motors against overheating on starting; voltage unbalance; transient voltage; kinds of surge suppression; protecting the motor from power system irregularities.

5 The Motor and Its Environment   189

NEMA's enclosure definitions; protecting the motor against contamination; ventilation system design; dealing with explosive atmospheres; maintaining safe levels of motor noise.

6 Insulation and Windings   243

Types of coils; choosing, applying, and testing insulation material; measuring winding temperature; the selection and placement of temperature sensors in windings.

7 Bearings and Lubrication   309

Bearing standards; shielded versus sealed bearings; bearing speed limits; predicting bearing life; choosing a lubricant; the use and maintenance of lubrication systems; pressure lubrication; oil versus grease.

8 More about Optional Accessories   351

Speed sensors; anti-plugging switches; friction braking; controlling, maintaining, and repairing brakes; alternative braking schemes; terminal box standards and sizing.

9 Installing Motors Properly   **433**

Motor bases; baseplate vibration tuning; natural resonance frequency; the free baseplate; coupling selection; aligning driving and driven shafts; grounding practices.

Bibliography   **483**

Index   **489**

Preface

IN MY MORE THAN 30 YEARS of dealing with motor application, certain questions or misunderstandings have arisen over and over again. Sometimes the answers to these questions could be extracted from beneath layers of mathematics in out-of-print textbooks—if the textbooks happened to be on the shelf when needed. Often an old magazine article or a paper presented at some engineering meeting would be helpful—if it could be found.

Questions from users or consultants, ambiguities, or contradictions written into equipment specifications, and motor failures caused by abuse or misuse all indicate a need which this book may fill. This book is an attempt to bring together in one place as many of the answers as possible.

Much of the book's content has appeared from time to time, in somewhat different form, in the pages of *Electrical Apparatus* magazine. Other portions have been presented at technical conferences, workshops, or seminars. Some portions have never been published before.

This book concerns only polyphase squirrel-cage motors. No single volume could go into the necessary depth for all types of motors. Hence, there's nothing here about fractional horsepower appliance drives, wound-rotor machines, or d-c or synchronous motors.

Of course, some material in the chapters on accessories, bearings, enclosures, and installation could apply to any electrical machine. The polyphase cage motor is, however, by far the most common industrial driver.

Many texts or handbooks use basic principles to explain how such machines convert electrical energy into mechanical power. These books are of value to the serious student of electrical theory or the machine designer. But for those concerned mainly with how to properly select, describe, install, and protect industrial motors, something more is needed.

Such readers are likely to be plant engineers and operators, specification writers, consultants, machinery builders, and contractors. It is assumed that they will know, at least in a general way, what a "squirrel cage" is and what makes it work. It is also assumed that they will be somewhat familiar with basic motor terminology.

This book is not a maintenance or repair handbook. Service shop operators, especially those who undertake motor redesign or drive design rather than repair alone, may find parts of the book useful. But most of what they need to know about motor rewinding, connecting, or rebuilding is readily available elsewhere.

Richard L. Nailen

1 Motors and Standards

Application engineering

"ENGINEERING" IS NOTHING more than planning based on knowledge instead of guesswork. In this sense, everyone in service, maintenance, and technical sales work is his own engineer every day. Proper application of electrical motors doesn't take a special diploma, but it does take some fundamental knowledge, a suspicious mind, and a lot of common sense.

In choosing the best motor to do any job, we must consider three essentials of what is called "application engineering." The first is "matching the motor to the load." Motor rating and performance should be selected to suit the behavior of whatever machinery the motor is to drive. This is the most important—and the most complex—of the three areas to be considered.

The second is "matching the motor to its environment." The motor must be designed and built in such a way that it will not be destroyed by its surroundings—by heat, moisture, dirt, or vibration. Conversely, it must not in turn inflict damage, such as through noise or vibration.

The third aspect of application engineering is "matching the motor to its power system." Starting methods; protection of the system from the consequences of motor failure, as well as protection of the motor from systems problems such as voltage surges; and power factor improvement—all these must be taken into account in the choice of the motor itself, as well as in the selection of its accessories.

In dealing with these conditions, the person in the field will sometimes be stumped no matter how much he or she has learned. Suppose you have to refer to the motor design engineer at the factory. Will you know enough to ask the right questions so your problem will be clearly understood? Will you understand the answers so you can get the story to your customer?

As an example of how confusing it can be to try to judge the status of an application from inaccurate data, consider the plant superintendent who had a ventilating fan in need of a boost in output to satisfy the U.S. Occupational Safety & Health Administration's (OSHA) requirements. It was driven by a very old 150-hp motor, rated 440 volts, 178 amps, 1170 rpm. Someone checked the voltage in all three phases and got 440. He measured the amps in all three lines and got 148.

Now, if they are correct, these numbers tell us right away that the motor was only putting out about 125 hp. But then the speed was checked and allegedly found to be only 1120 rpm, which by itself would indicate a load of about 300 hp. The data just didn't fit together.

Asked to comment on all this, an application engineer can only begin by asking a lot of questions in return, such as, How was line current measured? How was speed measured? Do we know the motor has never been rewound, perhaps with changed performance? Is the motor overheating? The answers to those questions will probably lead to other questions.

Some plant operating people become upset about so many questions. Being people of action, they prefer doing something to thinking about it. There are times, of course, when thinking before doing becomes necessary.

In a Midwestern gypsum plant it was recently reported that whenever any of the plant's motors fails, it is replaced by a motor of the next largest horsepower. As Winston Churchill once remarked, "That is, at any rate, a policy." But such a policy takes no account of starting duty, cyclic loading, ventilation failure, environmental hazards, or the many other factors that may be destroying motors whatever their horsepower rating. Nor does it allow for the reduction in plant power factor which is the price for operating numbers of motors well below their rated output.

What is 'standard'?

Let's say a few words here about industry standards and their role in application engineering. We all know that certain horsepowers or voltages are "standard." When we speak of a "standard" motor, which particular standard are we talking about? Is every manufacturer's "standard" product identical in all respects? How binding is the standard?

The best way to answer these questions is by reviewing what a standard is, how it is issued, and what the various standards-making organizations are. Let's begin with national standards.

National standards, which govern the practices of most U.S. industries, fall into two categories. A few have acquired legal status—that is, some agency is legally empowered to enforce them by court action. Until April 1981, the entire National Electrical Code (NEC) was in that category, having been "adopted by reference" in OSHA regulations. Even before OSHA existed, many state and local governments had written the NEC into their own laws. Documents such as the NEC may not originate with any branch of government. Others, such as OSHA rules, were written by the federal government.

A second category of national standard includes the practices adopted by nationwide trade organizations to which all or most equipment manufacturers have agreed. For electric motors, the most important such document is MG (Motor & Generator) Standard No. 1, written by the National Electrical Manufacturers Association (NEMA). MG Standard No. 1 is actually a compilation of individual standards, some only a paragraph within a single volume. However, users, consultants, and designers invariably speak of "MG1" as a single document, normally invoked as a unit in motor specifications. Other standards, covering many principles of rotating machine design and test, have been issued by the Institute of Electrical & Electronics Engineers (IEEE).

What is a standard? One authority offers this definition: "[A standard is] a document setting forth requirements normally dictated by customary practices in industry, science, or technology." In other words, a standard is what is most often acceptable for the usual application.

The phrases "normally dictated" and "customary practices" may not seem very strict, but that is fully in keeping with the nature of our standards system. It is a "voluntary" or "consensus" system. That is, except for those few documents that have legal standing, industry standards are voluntarily adopted and voluntarily complied with. Anyone is free to manufacture or use a product not in conformity with such standards. If such a standard forms a condition of a contract between a buyer and a seller, compliance may become enforceable as any other contract terms would, but otherwise non-compliance doesn't constitute a violation of law.

All the individuals preparing a standard need not agree completely on the inclusion and wording of each provision, but they must reach a consensus. The procedures set up to compromise or resolve negative votes will depend on the organization involved.

In the electrical machinery field, the first standards-making group was the IEEE (then known as the American Institute of Electrical Engineers, or AIEE), responding to the need for common, standardized features in the manufacture and use of equipment for commercial and residential electrification. The first AIEE standards committee was formed in 1890. Six years later, at the first National Conference of Electrical Rules, the AIEE delegate was elected conference president. The group laid the groundwork for what shortly afterward became the National Electrical Code (NEC).

In 1899, the organization we now call IEEE began publishing its own standards. Today totaling more than 300, these standards deal with tests, specification writing, ways of rating insulation systems, and other matters pertinent to all sorts of electrical equipment and components. Some are only a page or two in length. Others are much longer. The *Standard Dictionary of Electrical & Electronics Terms* (IEEE 100) is a 716-page book.

The IEEE has issued three kinds of standards. Originally there were only "basic" standards—for example, definitions of electrical units such as the Henry. Then came "technical" standards, describing the behavior of certain types of apparatus, such as transformers. By 1926, "manufacturing" standards were being written, sometimes even involving dimensional interchangeability (examples: IEEE 386 on separable connectors, or IEEE 566 for design of power plant control rooms). Usually, however, IEEE standards (unlike common NEMA practice) don't deal with "commercial sizes, ratings, usage requirements or performance . . . [of] specific devices."

The mass production of weapons and vehicles during World War I created a need for broader industrial standardization. Already experienced in such documentation, the IEEE naturally led the way, along with several other organizations, in the 1919 formation of the "American Engineering Standards Committee." A few years later the Committee became the American Standards

Association, re-named the United States of America Standards Institute from 1966 to 1969, and since 1969 called the American National Standards Institute (ANSI).

Despite its name, ANSI isn't affiliated with any government agency, nor does it normally write standards (except on rare occasions when a need is not being met by any other group). Rather, it coordinates the activities of such standards-making bodies as IEEE or NEMA, adopting their documents as ANSI standards whenever:

(1) They neither duplicate nor conflict with another document already adopted by ANSI.

(2) Procedures of the issuing group—such as proper balance in the makeup of the task force or committee that drafted the standard—avoid discrimination against specific products or suppliers.

For example, NEMA MG1 and IEEE 100 are two of the many widely applicable standards that ANSI has adopted.

In the *Standard Handbook for Electrical Engineers*, former ANSI managing director S.I. Sherr described the organization this way: "[ANSI is] a federation including organizations, companies, and help from government bodies plus individual experts. It does not develop standards. Either the affiliated or organizational member companies do this, or it is done through an ANSI committee operating under ANSI rules but organized and administered by the organization itself." Such committees must avoid a preponderance of manufacturer, user, or organized labor representatives. ANSI is concerned with many non-electrical standards—graphic arts, acoustics, consumer products, and so forth.

To persons who work with hazardous or "classified" area electrical installations, one of the more confusing relationships in the standards field has involved the NEC, the insurance industry, and Underwriters Laboratories (UL). For example, users speak of a motor enclosure or application as being compatible with "NBFU" rules. These initials once denoted "National Board of Fire Underwriters." The inclusion of the word "underwriters"—known as an insurance industry term—is the source of the confusion. Here are some clarifying facts:

(1) Founded in 1896, the National Fire Protection Association (NFPA) is a non-profit organization with no ties to insurance firms. It issues a great many standards, some of them adopted as law by local governments, dealing with fire protection—fire escapes, exits and stairways, structural materials, flammable liquid storage, and so on. In the electrical industry, the major NFPA standard is the NEC. Thus, that code has no connection with any group of insurance firms or underwriters.

(2) UL was founded in 1894. It, too, is a private, non-profit organization chartered for "public service through testing for public safety." UL publishes research bulletins as well as more than 400 "standards for safety" covering almost every kind of electrical product, especially those for home use. UL, like the NFPA, is neither part of nor sponsored by insurance interests.

(3) The NBFU no longer exists. The main purpose of the agency, set up within the insurance industry and later succeeded by the Insurance Services Office (and still later by the American Insurance Association), has been to evaluate community fire protection services. It rates municipal fire departments, water supply, and other civic factors that affect local fire protection—and therefore local fire insurance rates. There is no connection between the ISO and either UL or the NFPA. (In standards work, ISO also signifies "International Standards Organization," which since 1947 has been responsible for non-electrical standardization outside the United States.)

The electrical repair business, as well as motor manufacturing, is concerned with UL. The "label service" inspection and certification procedures of UL define "explosion proof" motor capability (which we shall explore further in Chapter 5). To meet these standards, explosion proof motor rebuilding must also be performed in shops certified by UL for such work.

On the other hand, although its Article 500 basically defines the various hazardous atmospheres, the NEC has little relevance to either the manufacture or the repair of motors. The code was written primarily to provide safety in electrical "installations"—the interconnecting wiring between power source and load—rather than to prescribe internal features of motors and other equipment.

As an example of misunderstandings about NEC provisions, consider lead cable selection and connection within a motor terminal box. Although the NEC offers many rules governing connection spacing, choice of cable, and the grouping of cables within a conduit or raceway, nothing in the entire NEC chapter on "Wiring Methods" applies to the interior of any motor's factory-installed terminal box. (We'll see this again in Chapters 4 and 8.) This is made clear by this 1984 version of NEC Article 300-1(b):

"The provisions of this article are not intended to apply to the conductors which form an integral part of equipment, such as motors..."

That illustrates the need to be sure what a standard really does say, and of the applicability of its provisions, before assuming that "build per that standard" will result in the product you want.

Another often misunderstood standard is the OSHA noise regulation. We'll consider this standard in more detail in Chapter 5, but this much bears repeating: The OSHA noise regulation does not stipulate a maximum noise output for any motor. The rule governs only the allowable sound level and employee exposure to sound in the work place, whatever the sources of noise may be. Requiring an 85 decibel sound level in a shop, for example, does not mean that all motors installed there must test 85 dB or meet any other particular limit. The nature of the plant environment, and the other noise sources, must also be considered. The user must decide, from acoustical theory or testing, what to require of the motor; that limit will normally be lower than the OSHA limit for the location.

The United States' best known group of electrical machinery standards, those contained in NEMA MG1, is what is commonly meant by the term "standard motor." Their last complete revision in 1978 has been followed by numerous detail changes. How were these requirements arrived at, approved, and then revised?

The process can be long and complex. Committees and subcommittees, formed from recognized experts in various fields, may take years to complete some standards, making sure that all relevant technology is considered.

The stated objectives in MG1 are to "assist users in the proper selection and application" of machines, and to "eliminate misunderstandings between the manufacturer and the purchaser and to assist the purchaser in selecting and obtaining the proper product for his particular need."

A few basic questions about this 300-page publication should be answered here:

(1) Is everything in the standards "standard"? That is, if a user invokes NEMA MG1 when buying an electric motor, will the motor necessarily include all the features discussed in the standard?

The answer is: not necessarily. There are three "classes" of information published within MG1:

(a) "NEMA Standard." Such provisions, labelled as such paragraph by paragraph, have been approved by at least 90% of those members eligible to vote on them. They form the basic core of the document, defining such things as enclosures, foot mounting dimensions, and torque limits.

(b) "Suggested Standard for Future Design." This information deals with new features, perhaps not yet in wide commercial use, but which "suggest a sound engineering approach to future development" and have received at least two-thirds approval. Example: MG 1-14.42 concerning V-belt pulley dimensions for integral horsepower motors, which was a "Suggested Standard" from 1965 until it became a full Standard in 1969. In 1980, a revised definition of a "guarded" motor enclosure, MG1-1.25, was adopted as a Suggested Standard for Future Design.

(c) "Authorized Engineering Information." This is usually application advice that does not lend itself to precise "standard" terminology, but suggests to the user how certain operating conditions will affect his drive. Example: a portion of MG1-12.43B, describing the need to consider motor voltage and frequency variations during starting.

Categories (b) and (c) allow alternatives or merely provide background. No user should assume that any "standard" motor will suit all the application conditions discussed in those portions of MG1 unless the manufacturer has been given specific details (such as the amount of possible voltage unbalance).

From time to time, category (b) items are revised to category (a). Or an (a) item may be converted to a (c), as happened to MG1-12.51 in 1980 after 11 years of "Standard" status.

(2) Isn't MG1 biased in favor of manufacturers?

NEMA is indeed a manufacturers' organization. Their dues support its activities. Standards content, however, is dictated by what the market demands, by what public utilities may require, and by the actions of such other bodies as the IEEE. In dealing with motor testing, for instance, MG1 usually refers to applicable IEEE standards. Furthermore, the adoption of MG1 by ANSI indicates broad industry support.

(3) How is motor performance defined?

Although agreement has been reached within NEMA about voltage ratings, dimensions, and torques, the efficiency or power factor which to many people denote the "goodness" of a machine are not dealt with in NEMA MG1. These are matters for negotiation between user and supplier.

One reason for this exclusion is that such performance characteristics do not affect a motor's ability to drive its load. Also, NEMA standards must not be written in such a way as to inhibit innovation or constitute restraint of trade, even indirectly. They are not purchase specifications.

Thus, a common cause of motor misapplication is the failure to realize that MG1 cannot cover every contingency. Table 1-I shows what these standards do not cover, and knowing that is as important as knowing what is covered.

The three main categories of information in NEMA MG1 are definitions (such as enclosure types), dimensions, and performance (mostly torques and inrush current). Along with those basics, NEMA sets noise limits for the integral horsepower sizes, service factors, temperature ratings, and allowable "standard" load inertia.

The three commonly used NEMA standard motor design types are defined in terms of the relationship between motor torque, inrush current, and full load slip. These are briefly described in Table 1-II.

NEMA horsepower ranges

Keep in mind the horsepower ranges to which these NEMA design definitions apply. If someone asks for a 300-hp "NEMA Design D" motor, or a 500-hp "NEMA C," you have to stop right there. There are no such motors. (See Figure 1-1.)

Find out what specific characteristics the customer wants, and why. If he wants "200% locked rotor torque," for instance, fine. But just asking for "NEMA C" won't get it for him. Some users have described as "NEMA D" a

1. Torques or inrush current for multispeed motors.
2. The *shape* of the speed-torque curve for any motor.
3. Temperature rise at rated load for any motor with a 1.15 service factor (standard for 200 hp and below). For any large motor, temperature rise is defined only at rated load, because no service factor capability is standard.
4. Any NEMA "design letters" or the associated motor characteristics for any rating above 500 hp, regardless of speed.
5. The number of allowable starts per hour, or per day, for any large motor accelerating any load.
6. Minimum efficiency, or power factor, for any motor.
7. Maximum motor noise limits that must be met for frames larger than 445 (though "suggested" figures are stated).
8. Nature and size of any connectors used to join motor load cables to the supply circuit.
9. Any of these enclosure descriptions: "weatherproof," "dust tight," "weather protected 1½," or "watertight."

Table 1-I. What NEMA MG1 does *not* define.

NEMA design	Locked torque	Breakdown torque	Locked current	Slip	Maximum horsepower defined	Uses
B	"Normal" as in Fig. 1-3	"High" as in Fig. 1-3	"Normal"	Less than 5%	500	Compressors, pumps, fans, grinders, machine tools, unloaded conveyors
C	High (200-250%)	Normal (190-200%)	Normal	Less than 5%	200*	Loaded conveyors, pulverizers, piston pumps
D	High (275%)	None	Normal	High 5-13%	150	Cranes, punch presses, tapping machines

*Minimum NEMA C hp = 3; below that, NEMA B & C torques and inrush are identical.

Table 1-II. General operating characteristics of NEMA standard motors.

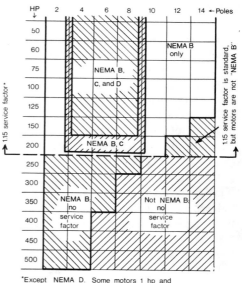

Figure 1-1: As this chart shows, NEMA design letter assignments and service factor practices are governed neither by frame size nor by horsepower alone.

*Except NEMA D. Some motors 1 hp and below have 1.25 to 1.4 service factor.

4% to 6% slip press drive motor with a high (300%) breakdown torque. What they were looking for, it turned out, was a motor-speed torque curve like that of Figure 1-2a. But a true NEMA D characteristic has no breakdown torque point at all; it looks like Figure 1-2b. The two have similarities, but an apple isn't a pineapple just because their names are similar.

If there is one provision of NEMA MG1 that has caused more confusion than any other, it is the definition of the boundary line between "integral horsepower" and "large" motors. The problem is that motor physical size and motor horsepower do not go hand in hand. Many users, for example, are convinced that a motor above 250 hp is always "large" in NEMA terms, and that any machine of, say, 100 hp is necessarily "small." It just isn't so. Motor

Motors and Standards

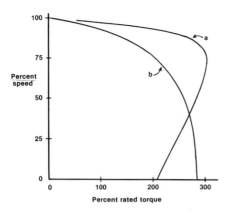

Figure 1-2: High slip punch press motors are not all alike.

speed enters into it as well as horsepower. For example, a 100-hp, 4-pole motor is definitely "integral horsepower." But at 14 poles, 514 rpm, it is a "large" motor instead.

These size distinctions are useful in judging what "standard" torques to expect. Look at Figure 1-3. Above 500 hp, all "standard" motors need have only 60% locked rotor torque and 175% breakdown torque; don't expect to get anything like the 150% locked torque that is standard for a 10 hp rating.

But when it comes to service factor, NEMA says that 200 hp is the only dividing line. Below that, standard motors have a service factor; above that, they don't, and no distinction is made involving frame size.

The torque values given by NEMA MG1 are minimum figures. Any motor built to that standard will produce at least that much torque. To ensure that

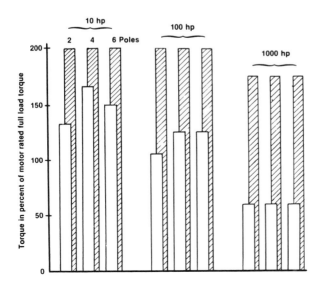

Figure 1-3: NEMA standard locked rotor torque (unshaded bars) and breakdown torque (shaded bars) for typical Design B motors (though design letter B does not apply above 500 hp). Note the tendency of torque, especially at locked rotor, to drop off as motor rating goes up.

they do, most such machines will exceed the standard. Consider a 25-hp, 4-pole NEMA Design B machine that is supposed to have at least 150% locked rotor torque. "Brand X" electric motor may actually provide 155%; Brand Y, using somewhat different parts chosen for a better fit in another supplier's overall or manufacturing process, may give 170%. Either motor meets the NEMA standard.

If a user had good results with a Brand Y motor, he might suppose that any other make of "standard" motor would do as well. But assume that, unknown to him, his driven machine actually needs 165% torque to start it rolling. Put on a Brand X motor—just as "standard" as the other—and the drive will be in trouble. This can't be blamed on the standards or on the manufacturer. The user is not relieved of his responsibility to know what his application requires.

The buyer should also make clear which standards he expects to be met. Job specifications or contracts often include a stipulation that motors conform to "any applicable state or local codes, regulations, or ordinances." Some of these may conflict with national standards—or even with one another. Hence, an installation may be completed in good faith, only to be rejected by a local inspector based on his interpretation of a law with which the contractor was unfamiliar. The only protection is either to be sure you read the laws beforehand or to get an exemption in the contract. Assuming that a "standard motor" will suffice is an unfair burden to place on any industry standard.

How important are international standards? Motor users (and manufacturers) in the United States have not always paid much attention to standards that originated elsewhere. This is changing, however. International trade agreements, along with growing foreign competition in the domestic electric motor market, have made it increasingly important for U.S. industry to be aware of world-wide standard practices.

Ironically, the chief international standards-making organization had its origin in the United States. Back in 1904, after the IEEE had started and the National Bureau of Standards had been formed, an International Electrical Congress was held in St. Louis. One of its two main committees later evolved into today's International Electrotechnical Commission (IEC). Based in Geneva, Switzerland, and supported by 44 member nations, the IEC writes world-wide standards throughout the electrical field. Standards work is carried out through 78 technical committees. The United States participates in this work through a U.S. National Committee (USNC), which was first set up by the IEEE but since 1977 has been a formal unit of ANSI.

Metrication

The greatest difference between international and domestic electric motor standards is the overseas use of metric dimensions and ratings (see below). But there are many other differences as well. For example, test methods, tolerances on motor losses, and enclosure definitions are treated in Europe quite differently from the way they are treated in the United States. In hazardous areas, Europeans may choose the "increased safety" motor—one designed and tested for a low rate of stalled temperature rise. (See Chapter 5.) No such machines are offered in the U.S. The "phase segregated" terminal box, designed to withstand the arcing and blast pressure produced by an internal cable fault, is another European standard feature not matched in U.S. practice. (See Chapter 8.)

In the future, we are likely to see increasing conformity between some of these international standards and our own, if U.S. electrical manufacturers are to compete in the world market. Meanwhile, codes and standards catalogs, ordering information, and other pertinent data may be obtained from the principal standards-making bodies listed at the top of page 12.

Let's look more closely at metrication. You realize that 100 horsepower is equivalent to 75 kilowatts. You know how many kilometers per gallon your car gets on the highway. And you are aware that one inch contains 25.4 millimeters. So now you're ready for the "metric motor," right?

Not quite. There's much more involved than just some new units of measure. Before we in the electric motor field can lay down the yardstick and pick up the meter stick, there are some other things we should know.

First, the good news.... Metrication offers little prospect of major changes in frame size, basic construction, or mounting dimensions for NEMA standard motors in this country.

Whatever does happen may be a few years off yet. At one time, it was widely expected that the next NEMA "rerate," perhaps to Class F insulation temperature as standard, would occur by 1980. A conversion to metric dimensions was thought likely to take place at the same time.

However, two developments intervened. One was the energy crunch. Renewed emphasis on motor performance means users can expect a concession to the widely-held belief that smaller, hotter motors are less efficient. This isn't strictly true, of course. As has often been pointed out, efficiency has remained relatively unchanged for many years. But power factor has certainly suffered, which can mean a growth in cost penalties as utility rates escalate. Better materials and cooling systems do permit motor redesigns on larger parts to boost efficiency, rather than allowing it to remain the same, and restore power factor in the process. Hence the next rerate will probably stick with Class B rise; no service factor.

The second development is the growing agreement between representatives of NEMA and the international standards bodies that have adopted metrication for motors in many nations. There is no NEMA "metric motor standard," nor does any timetable exist for the issuance of any such document. But NEMA has prepared a "guide for the development of metric standards" for motors, which incorporates many of the IEC standards now being used in Europe.

When a standard is prepared under this guide, it will undoubtedly express only minor dimensional differences between today's standard line and the European product. Only customer demand for those future metric designs,

Where to write for motor standards

Institute of Electrical & Electronics Engineers (IEEE)
345 E. 47th St., New York, N.Y. 10017
(212) 644-7900

American National Standards Institute (ANSI)
1430 Broadway, New York, N.Y. 10018
(212) 354-3300

National Fire Protection Association (NFPA)
Batterymarch Park, Quincy, Mass. 02269
(617) 328-9290

National Electrical Manufacturers Association (NEMA)
2102 L St., N.W., Washington, D.C. 20037
(202) 457-8400

---

principally to fit into the international market, will induce motor manufacturers to issue such standards.

How output will change

Now for the bad news. The metric motor will not be just a conversion of, for example, 40 hp to 30 kilowatts, by restamping the nameplate, or a change of 7 inches in shaft centerline height to 178 mm. The output power level will actually change.

To understand this, consider the nature of a standard line of induction motors. More than 40 years ago, U.S. manufacturers agreed to standardize on certain horsepower steps as a basis for production economy. If every motor were designed to a specific horsepower, such as 27.5 this time and 31 the next, we could not have today's mass production or manufacturer interchangeability from which all users benefit.

But what should these logical horsepower steps be? We all know what was chosen. The steps are not uniform from one to the next. For example, the next standard step above 1.5 hp is 2 hp—only 33% larger. But next above 3 hp is 5 hp—67% larger. (Because this jump is so great, one manufacturer a few years back proposed a 4 hp rating as standard, but nothing came of it.) On the other hand, from 7½ to 10 hp is back to a 33% jump; 100 to 125 hp is only a 25% increase.

Internationally, the normal method of handling any such progression in size steps is to use a series of what are called "preferred numbers." These are numbers derived from a consistent, logical mathematical relationship. Between 1949 and 1954, the ISO, with U.S. membership, produced Recommendation R17, calling for the standard use of a preferred number series called the

Renard, or "R" series, which was originated a century ago by Captain Charles Renard of the French Army.

Putting it as simply as possible, Renard's "R5" series contains successive numbers separated by the 5th root of 10. Beginning with the number 1, the second number in the series becomes 1 times $\sqrt[5]{10}$, or 1.5849. The third number would be 1 times $(\sqrt[5]{10})^2$, or 2.5119. Expanding this series and rounding off the answers, we get:

$$1, 1.6, 2.5, 4.0, 6.3, 10, 16, 25, 40, 63, 100, \text{etc.}$$

Of course, the first number in such a series need not be 1. It could be 3, or 270—or anything else—but the next number will be 1.6 times as large; the third, 2.5 times as large, and so on.

What makes this such a great idea? First, it's easy to extend the same basic principle into a compatible succession of numbers having smaller steps between each. If we change the multiplier to $\sqrt[10]{10}$, for instance, we get what is called the "R10" series, thus:

$$1 \text{ times } \sqrt[10]{10} = 1.2589$$
$$1 \text{ times } (\sqrt[10]{10})^2 = 1.5849 \text{ etc.}$$

An R20 and R40 series also fits nicely into the scheme, resulting in numbers as shown in Table 1-III. Because their products and quotients are also preferred numbers, calculations involving series of this sort are made easier. Handling them with logarithms is especially simplified.

R5 Series	R10 Series	R20 Series	R40 Series
1.00	1.00	1.00	1.00
			1.06
		1.12	1.12
			1.18
	1.25	1.25	1.25
			1.32
		1.40	1.40
			1.50
1.60	1.60	1.60	1.60
			1.70
		1.80	1.80
			1.90
	2.00	2.00	2.00
			2.12
		2.24	2.24
			2.36
2.50	2.50	2.50	2.50
			2.65
		2.80	2.80
			3.00
	3.15	3.15	3.15
			3.35
		3.55	3.55
			3.75
4.00	4.00	4.00	4.00
			4.25
		4.50	4.50
			4.75
	5.00	5.00	5.00
			5.30
		5.80	5.60
			6.00
6.30	6.30	6.30	6.30
			6.70
		7.10	7.10
			7.50
	8.00	8.00	8.00
			8.50
		9.00	9.00
			9.50
10.00	10.00	10.00	10.00

Table 1-III. Basic Renard preferred numbers.

Keep this in mind so far: The R series concept has nothing whatever to do with the metric system as such. It is independent of measuring units. It could be applied just as easily to feet, inches, and pounds as to any other scale units.

However, preferred numbers are not intended to cover only linear measurements. They express logical size progression for all quantities associated with machinery—including motor power output ratings.

The ISO R17 (and R3) preferred number recommendations are just that—recommendations. The ISO documents themselves state that "For each individual country the only valid standard is the national standard of that country." But the ISO approach is wholly consistent with an American national standard dating back to 1936: ANSI Z17.1. This document says:

"Preferred numbers are series of numbers selected to be used for standardization purposes in preference to any other numbers. Their use will lead to simplified practice and they should, therefore, be employed whenever possible for individual sizes and ratings, or for a series thereof, in applications similar to the following:

(1) Important or characteristic linear dimensions, such as diameters and lengths.

(2) Areas, volumes, weights, and capacities.

(3) Ratings of machinery and apparatus in hp, kw, kilovolt-amperes, voltage, current, . . . etc."

Whether expressed in horsepower or kilowatts, however, today's NEMA standard steps in motor rating do not follow a preferred number Renard series. Here, then, is the only real conflict between international metric standards and present U.S. practice. A true "metric motor" line would use motor kilowatt ratings following an R series progression. A comparison between such values (included in NEMA's proposed guide for metric standards) and the present ratings appears below in Table 1-IV.

Present standard horsepower rating	Equivalent of present standard in kilowatts	Proposed NEMA kilowatt rating for metric motor
1	.75	.8
1½	1.1	1.1
2	1.5	1.6
3	2.2	2.5
5	3.7	4.0
7½	5.5	5.6
10	7.5	8.0
15	11.0	11.2
20	15.0	16.0
25	18.5	20.0
30	22.0	25.0
40	30.0	32.0
50	37.0	40.0
60	45.0	50.0
75	55.0	63.0
100	75.0	80.0
125	90.0	100.0
150	110.0	125.0
200	150.0	160.0

Table 1-IV. Note how the values in the right hand column follow the progression of R10 and R20 series numbers as shown in Table 1-III (previous page).

How will the performance of the metric motors compare with what we see now? Here are some of the more important items:

Likely to remain unchanged:

(1) Insulation temperature ratings and nameplate temperature rise;

(2) Torque characteristics (though there are lower percent locked rotor torques allowed by international standard—for the equivalent of NEMA Design B, for example—U.S. metric motors will probably retain today's higher values);

(3) Operation on 60 Hz frequency only.

Expected to change:

(1) Locked rotor current limits. Whereas a 100 hp NEMA B motor today, on a kilowatt output basis, has less than 8 locked rotor kva per rated kilowatt, the proposed metric guide permits 8.7. (This does not mean, however, that manufacturers will suddenly change designs just to raise inrush.)

(2) Service factor of 1.15 for the smaller ratings will no longer be standard.

Efficiency, power factor, and the characteristic shapes of speed-torque curves will not be affected except insofar as power ratings go up or down to match the new kilowatt levels and to suit the inevitable demand for energy-saving design.

Certain differences between NEMA proposals and the IEC remain under consideration. For example, IEC standards allow three frame lengths per diameter—short, medium, and long. NEMA has stuck with only two. European enclosure definitions—the nature and extent of splash protection, for example—are quite different from enclosure definitions in the United States.

The purely dimensional differences seem to be of little importance. Tables 1-V and 1-VI illustrate that little change is needed to reconcile IEC metric dimensions with existing NEMA standards.

NEMA Frame	Shaft height inches	IEC Frame	Shaft height, IEC mm.	inches
180	4.50	112	112	4.41
210	5.25	132	132	5.20
250	6.25	160	160	6.30
280	7.00	180	180	7.09
320	8.00	200	200	7.87
360	9.00	225	225	8.86
400	10.00	250	250	9.84
440	11.00	280	280	11.02

Table 1-V. A comparison of motor shaft heights—IEC versus NEMA. Note how the IEC uses frame number designations that are the same as the shaft height in millimeters.

Dimension	NEMA value inches	mm.	IEC metric value inches	mm.
Shaft diameter U	2.375	60.3	2.36	60
Shaft height D	9.00	228.6	8.86	225
Shaft length N-W	5.88	149.4	5.51	140
Foot hole sp. 2F	112.24	310.8	12.25	311
Foot hole sp. 2E	114.00	355.6	14.00	356
Foot shaft BA	5.88	149.4	5.875	149
Foot hole H	0.656	16.7	0.748	19

Table 1-VI. A comparison of some significant dimensions between IEC 225 frame metric motor and present NEMA standard 365T.

Fortunately, the NEMA standard shaft heights adopted years ago are very nearly equal to R40 series preferred numbers. For example, the 250 frame shaft height is now 6.25 inches. The 210 frame, the next size smaller, is 5.25 inches. The ratio 6.25/5.25 is 1.19; the nearest R40 preferred number ratio would be 1.18. An attractive feature of any preferred number dimensioning is the existence of a conversion factor between different systems of measurement which is itself a preferred number. This is very nearly true with the centimeter versus inch systems of linear measure, the conversion factor 2.54 being within 2% of the R5 series preferred number 2.50. As a result, there is essential compatibility between present shaft heights and the metric proposals.

This means that tomorrow's metric motor is actually available today. One U.S. manufacturer has stated this position:

"Given a conventional hp and frame, we will convert this to kw and place it into a frame that is the metric equivalent. Mounting will be matched up or deviations pointed out. Usually foot holes and shaft diameters are no problem; shaft height may require customer use of shims. . . . Delivery is usually two weeks longer; adders vary, normally 20% to existing conventional list prices."

Foot mounting holes can be opened up. Shafts can be turned to the slightly different metric size and metric keyways provided with special cutters. This means that new casting patterns and intensive retooling will not be needed when a metric standard does take effect in the United States.

Will that happen? Undoubtedly. The metric system was first recognized by the U.S. Government a century ago. More recently, on December 23, 1975, former President Ford signed the Metric Conversion Bill which will inevitably bring about the official adoption of metric standards throughout the nation. It's a question of when, not if. NEMA can propose what and how, but it cannot dictate timing—this will develop from the needs of the marketplace.

As further background for the changeover to the metric system, those in the American electrical industry should be aware that whereas there is only one "meter," and only one value for many other metric units, there is more than one system of measurement using such units. Remember that motor application must deal with work and energy, time, force, heat, and sound, as well as just with distances, areas, and mass. Hence a U.S./International accord on metric motor standards had to include agreement on which physical system of units would be its basis.

That system is called, in French, the *Systeme International d'Unites* (SI). Briefly, there have been two other metric unit systems in common use. One is the so-called "cgs," or centimeter-gram-second system. Those three units for length, mass, and time are "primary" units—the unit of force in the cgs system is a "derived" unit which depends on the values of the primary units.

A second metric system is the "metric technical" system based on the kilogram force as a primary unit, not a derived one. Other primary units are the meter, kilogram mass, and the second. In the SI system, primary units are kilogram mass, meter, and second, with force measured in derived units called newtons. Table 1-VII gives the SI units most common for electrical work.

BASIC UNITS

Quantity	Name	Symbol
Mass	kilogram	kg
Length	meter	m
Current	ampere	A

DERIVED UNITS

Quantity	Name	Symbol
Force	newton	N
Pressure	pascal	Pa
Work, energy, or heat	joule	J
Power	watt	W

Table 1-VII. SI metric units useful in electrical work.

The distinction between a kilogram "force" (or pound force, for that matter) and a kilogram "mass" is tricky and crucial. Whole new school curricula are being contemplated to get this across to new generations of students, technical and non-technical. In the SI system, kilogram force—as the "weight" of an object—does not exist. The term "kilogram" denotes only the mass of the object. Weight, as the downward force the object exerts on a scale platform, is expressed (like all other forces) only in newtons.

Table 1-VIII summarizes some of the conversions involved in dealing with motor application quantities using the SI system. Already, U.S. motor manufacturers are encountering load torques and inertia in these relatively unfamiliar terms. This underscores what we said in the beginning of this chapter: There will be much more to the proper selection and use of the "metric motor" than just a nameplate conversion from horsepower to kilowatts.

Service factor

On any motor designer's list of performance characteristics that seem least understood by users, "service factor" must rank near the top. Manufacturers often see service factor requests in terms of load versus temperature rise specifications, which produces needlessly costly motors or contain built-in contradictions which must be resolved before a motor can be built.

Some users assume that service factor is a kind of general "extra" in motor construction or performance, allowing substantial abuse or permitting the "stretching" of whatever performance capability an application may demand. But the NEMA definition is much narrower. In Standard MG1-1.43, it reads: "The service factor of an alternating-current motor is a multiplier which, when applied to the rated horsepower, indicates a permissible horsepower loading which may be carried under the conditions specified for the service factor...." In other words, multiplying nameplate horsepower by the service factor tells how much you can overload the motor—but not necessarily what that overload will mean to motor performance. And that overload is allowable

1. **General conversion factors:**

   Length: 1 ft = .3048 meter
   Mass: 1 lb = .454 kilogram
   Heat: 1 BTU = 1055 joules (watt-seconds)
   Force: 1 lb = 4.45 Newtons
   Power: 1 hp = .746 kilowatt (746 watts)

2. **Specific relationships:**

   A.
	Present unit	Metric (SI) unit
Torque	lb-ft	Newton-meter

   Multiply lb-ft by 1.356 to get Newton-meters; multiply Newton-meters by .738 to get lb-ft.

   Today, one may also encounter torque expressed in "kilogram-meters." This is now properly written kilogram force-meter, or kgf-m, and is not an SI unit. Multiply lb-ft by .1383 to get kgf-m; multiply kgf-m by 7.233 to get lb-ft.

   Rarely, the unit "kp" (kilopond) is seen. It is the same as kgf and was used briefly to distinguish the quantity from kilogram *mass*.

   B.
	Present unit	Metric (SI) unit
Inertia of rotating parts	lb-ft^2("WK^2" or "WR^2")	kg-m^2 ("GD^2")

   "K" and "R" mean "radius of gyration," in feet. "D" is the *diameter* of gyration, in meters.

   Multiply WK^2 by .168 to get GD^2 or by .0425 to get kg-m^2. Multiply GD^2 by 5.925, or kg-m^2 by 23.6, to get lb-ft^2.

   C.
	Present unit	Metric (SI) unit
Pressure	lb-ft	Newton-meters

   Multiply lbs/in^2 by 6895 to get pascals; multiply pascals by .000143 to get lbs/in^2.

   These last two conversion factors illustrate the awkward size of some SI units. The kilonewton/m^2, or 1000 pascals, is of more manageable size.

   D. Volts, amperes, ohms, watts, time (in seconds), and temperature (in degrees C) remain unchanged.

Table 1-VIII. Metric conversions useful in motor application.

only if frequency, voltage, and ambient temperature remain as defined by the nameplate.

Several consequences of this are often overlooked. One is that both efficiency and power factor—especially the former—may drop considerably as motor output rises from rated load to service factor load. Another is that the temperature rise allowed by NEMA at the service factor horsepower does exceed normal insulation limits. (See Table 1-IX.) The result of that will be discussed later.

	Insulation class			
	A*	B	F	H
Open or TEFC motors, no service factor; rise at rated load	60	80	105	125
All motors with 1.15 service factor; rise at 115% load	70	90	115	135$^\Delta$

*Of historical interest only. Class A insulation is no longer used in industry.
$^\Delta$Not NEMA standard, but common industry practice.

Table 1-IX. NEMA standard temperature rises in degrees centigrade for various insulation systems.

What relationship exists between winding temperature at rated load and at service factor load? It isn't the same for all motors. Nor do all machines of the same type run at exactly the same rated load temperature. A uniform standard can be set up, defining service factor in terms of allowable temperature, only by assuming a load/temperature relationship that is reasonable for the average motor.

Years ago, when Class A insulation was standard, it was concluded that a 10°C added rise in temperature could be expected from a 15% increase in horsepower beyond the nameplate value. Hence, the old open motor with a 40°C thermometer rise was expected to reach 50°C rise when operated at 115% load. That 10°C increase was allowable for Class A insulation, so the motor was said to have a service factor of 1.15.

That same 10°C difference was carried over into Class B insulation system usage, so that the two temperature rises (rated load and 115% load) became 80°C and 90°C respectively. However, Class B insulation also permitted windings to operate safely 20°C higher than Class A. Therefore, that added margin was commonly accepted in industry as also permitting a 1.15 service factor (though that did not become a NEMA standard practice).

Actually, it is only common sense to expect that the added winding temperature under increased load would be some percentage of the original figure, rather than a fixed difference such as 10 degrees. At the least, then, one ought to expect service factor load in a Class B machine to produce about (50-40)/40, or 25% higher rise — which would mean going from 80°C up to 100. That figure is too high for reasonable insulation life, of course, the limit being 90°C. So the allowable rise at rated horsepower for such a machine, having 1.15 service factor, must be no more than 75% of 90, or about 67°C.

Because of the compounding effect of increased resistance in the copper when temperature goes up, the variation is really greater than this for Class B systems, and still greater for Class F. A large number of tests on various sizes, speeds, and enclosures shows that the average increase in winding rise for a 15% increase in horsepower is over 30%, with extremes ranging from 14% to 55%

Table 1-X shows what the NEMA standard is today. Note that a 1.15 service factor (once automatic for any open motor) is no longer standard above 200 hp. Keep in mind, too, that all rises are now measured by the resistance

Motor hp	Synchronous speed, rpm			
	3600	1800	100	900
½	1.25	1.25	1.25	1.15
¾	1.25	1.25	1.15	1.15
1	1.25	1.15	1.15	1.15
1½-200	1.15	1.15	1.15	1.15
Over 200	No service factor			

Table 1-X. NEMA standard service factors (NEMA Standard MG1-12.47). When a motor for which service factor is not standard is built with 1.15 service factor (at 250 hp, for example), industry practice is to assign a temperature rise rating in accordance with the 1½ to 200 hp standard and the values of Table 1-IX.

method; the old thermometer or thermocouple values have been deleted.

The motor nameplate always shows the rise corresponding to the highest horsepower output permitted. If the service factor is 1.15, for example, the nameplate rise will be that corresponding to 115% of rated load. If two different Class B insulated motors have service factors of 1.15 and 1.25 respectively, each will carry the same 90°C rise on its nameplate. The second motor must be larger than the first, however, to dissipate its greater heat loss at 125% load, rather than only 115%.

All the objectives for motor service factor (see below) are valid. In applying service factored motors, however, one should recognize several limitations in obtaining these benefits. For example, more horsepower capability does not necessarily mean any more locked or accelerating torque; it means only that at full speed the motor can supply higher shaft horsepower without exceeding a stated rise.

Some motors will produce higher accelerating torque with a service factor, simply because they are larger and magnetically "stronger" than others. But they are also likely to have higher inrush current, which can increase the system voltage drop on starting so as to cancel out the added torque.

If operated continuously at service factor load, will a motor last as long as a motor without service factor?

No. It's clear from NEMA standards that at service factor output a winding is permitted to run 10°C hotter than the normal insulation system limit. That will cut theoretical insulation life approximately in half.

The useful life span of most industrial motors is not limited by theoretical insulation life, of course; most units fail for other reasons, and most units would exceed that theoretical life anyway.

But it's unwise to expect nearly as long a service from any motor running all the time at the service factor load. As one rotating machinery engineer has said, "They shouldn't really call it service factor anyway. They ought to call it an 'emergency' or 'standby' rating for occasional short-time use, as is common with a-c generators." If only part of the service factor capability is used, though, especially with a Class F winding, the expected life may become much longer than for a motor with no service factor. (See Figure 1-4.)

A motor's service factor is intended for one or more of the following:
(1) To permit operation on unbalanced system voltages (up to about 3% unbalance), or at undervoltage. These can be particular problems on well pumps in residential service areas. (But beware of the loss of accelerating torque when voltage is low, service factor or not.)
(2) To carry a known amount of horsepower overload that may be present occasionally.
(3) To allow for a likely overload in the future because of a process change or increase in production rate.
(4) To allow for uncertainty in predicting the actual load required.
(5) To lengthen insulation life by dictating operation well below winding temperature limits.
(6) To compensate for occasional ambient temperature peaks above the normal 40° C limit.

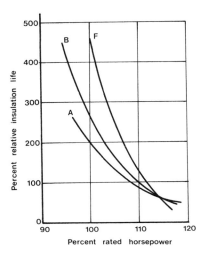

Figure 1-4: The relative insulation life of motors having a 1.15 service factor, insulated with either Class A, Class B, or Class F systems. All these windings have roughly the same theoretical life at continuous service factor loading; but at higher loads, the higher the insulation temperature rating, the longer the life.

Does the service factor allow the motor to handle intermittent high peak overloads?

Not necessarily. Just as locked rotor or accelerating torque may not be boosted by a service factor rating, neither is breakdown torque. Periodic torque peaks far above rated output are often demanded by so-called "duty cycle drives" like that of Figure 1-5. Basic motor size is fixed by the "effective," or "root-mean-square" horsepower. This motor must then have enough breakdown torque to ride through the periodic peaks without stalling. The alternative would be a much larger and probably more expensive motor, which would run considerably underloaded most of the time and give relatively poor performance. The existence of service factor wouldn't add anything to either design. This is also true for many drives involving frequent starting or speed changing.

Why has NEMA eliminated service factor as standard above 200 hp?

As already stated, insulation life is reduced under service factor operation. The more nearly continuous such operation is, the closer to 50% the life reduction becomes compared to the life of a standard machine without service factor. The larger the motor, the more expensive and critical its application, and the more time required for repair or replacement, the less desirable such reduced life will be.

Figure 1-5: An example of peak loading in a duty cycle. A 50- or 60-hp motor would handle the periodic 100-hp peaks. A 40-hp motor with a 1.15 service factor would be adequate for the effective, or "root-mean-square" (RMS) horsepower throughout the cycle but would probably not have enough breakdown torque to carry the 100-hp peaks. A special, high torque 40-hp design could do the job, but in size and cost it would essentially be a 50- or 60-hp machine anyway.

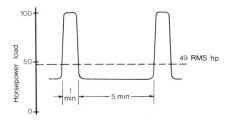

Moreover, Figure 1-6 shows that as motor size increases, the gap from one standard horsepower rating to the next steadily diminishes. This means that the next larger standard rating can usually be selected if the user needs 10% or 15% more horsepower, permitting the added load to be supplied with no theoretical sacrifice of winding life. There is also no sacrifice in accelerating torque. Finally, large drives are more likely than smaller ones to be "custom-designed" with better definitions of their exact load and operating conditions.

Below 250 hp—admittedly an arbitrary cutoff point—the situation is different. Precise load data is harder to get, so more motor margin is needed to match output uncertainty. Smaller motors are "general purpose"—more likely to be shifted from one drive to another. And in the smaller sizes, as Figure 1-6 shows, the steps between successive horsepower ratings are larger (as much as 60%), so that without service factor capability a slight overload may have to be supplied by a greatly oversized standard rating.

There is no need, though, for a great variety of different service factors, making it all the more difficult for manufacturers to offer an economical product with a short delivery time. Since 1960, the 1.20 service factor has been deleted from the NEMA standard where it formerly applied to 1½- and 2-hp sizes. For integral horsepower ratings above 1 hp, only 1.15 is now standard.

Some larger motors are ordered with 1.25 service factor. This is a reasonable specialty, typical for Class B insulated machines capable of 90°C rise at service factor load, but it is designed for the old Class A rise of 50°C at rated load. "Oddball" service factors of 1.05 or 1.175 should be avoided. They represent impractical hair splitting. (Can a user really be sure he needs 2½% more horsepower than a 1.15 service factor would provide?)

Figure 1-6: "Percent difference" stands for the percent difference in output between base horsepower motor at 1.15 service factor load and the next highest base horsepower rating. Matching a motor to its load by going to the next largest standard horsepower rating instead of using a 1.15 service factor becomes easier as motor size increases. The graph shows how the closer spacing of standard horsepowers in larger sizes brings this about. For example, a 345-hp load can be handled by a 300-hp, 1.15 service factor machine. But the preferred alternative is a 350-hp standard unit, only 1½% above the 345-hp requirement. (Actually, a 400-hp motor without service factor may offer higher efficiency at the 345-hp load than either of these alternatives.)

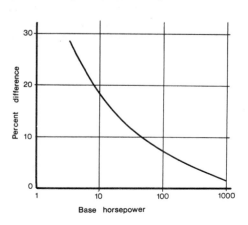

How service factor is being specified

Let's consider some of the common ways service factor (though not always recognized as such) is being specified for industrial motors today. Each has its application pros and cons which should be recognized:

(1) Specifying no service factor as such, but calling for unusually low temperature rise at rated load—such as 60°C or 70°C for a Class B motor instead of the normal 80°C.

The goal here is increased winding life at rated load. However, remember that an unusually cool motor will be oversized, more expensive, less readily replaceable, and will probably cost more to run. The oversized, more costly "energy efficient" motors on the market today will cost less to run but will not necessarily have lower temperature rises.

(2) Specifying temperature rise at rated load even though a service factor is requested. This contradicts NEMA practice. As already explained, although it's still common to assume that a 15% overload corresponds to 10°C higher rise, that is no longer true. Hence, a motor ordered as "80°C rise at rated load with 1.15 service factor" would undoubtedly exceed 90°C rise at 115%. On the other hand, calling for a totally enclosed fan cooled motor rating of 55°C or 60°C rise at rated load, with a 1.15 service factor, results in a needlessly oversized machine that will run quite cool at 115% load. (See Figure 1-7.)

(3) Specifying temperature rises at both rated and service factor loads—usually 10 degrees apart, but not always. One example is the ordering of an 80°C standard motor to be wound with Class F insulation for 115 degrees at service factor. One limit or the other must govern; there is no point in specifying both.

(4) Specifying a nameplate temperature rise at a given service factor, when the insulation class used permits a higher service factor.

A user may interject here, "Yes, I know the rating I've asked for gives me a built-in service factor higher than the nameplate shows. But I don't want it shown. If they see it, my operating people will eventually load up to it and I will lose the cushion the unused capacity gives me."

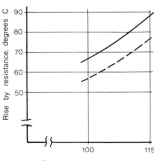

Figure 1-7: These curves are typical for TEFC motors, 75 to 1000 hp. They show why the modern practice of specifying temperature rise at service factor load may permit a motor smaller than for the old method of specifying only rise at rated load. If designed for 90° rise at service factor (solid line), the motor will run too hot for a 60° rise at rated load (the old Class B standard for 1.15 service factor TEFC designs). Meeting the 60° limit requires a larger machine (dashed line) that will be much cooler than necessary at service factor load.

That sounds logical. However, the user may benefit less than he expects. If he has only a 10-degree cushion, for example, it will probably give him only a 5% to 7% margin in loading—certainly not 15%. If he foresees a problem with voltage unbalance or regulation, how can he tell if his margin is large enough?

Although it's true that using a "cool" motor having no service factor probably allows rated-load operation at better power factor than using the next larger horsepower size and running it underloaded, efficiency may be more important than power factor. Chances are, any industrial motor running 10% to 25% below its nameplate output will have higher efficiency than at rated load—and higher than a service factor design running at overload. (See Figure 1-8.)

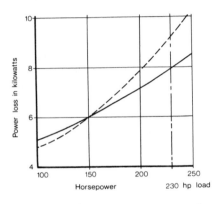

Figure 1-8: An example of efficiency comparison involving two motors, one with service factor (dashed line) and one without (solid line). If loaded to 230 hp, a 200-hp motor with a 1.15 service factor will have a shorter winding life and cost more to operate than a more expensive, 250-hp motor having no service factor.

Besides, few motors roast out just because of sustained running at too high a temperature. Extending insulation life beyond 5 or 10 years up to 30 or 40 makes little sense when frequent starts, destructive voltage surges, bearing failure, or neglected maintenance will probably wipe out the winding long before that time. If an industrial process, or an entire plant, has a project life of only ten years, or if the motor is only used a few hours a day, what are the advantages of decades of theoretical winding life?

A designer always has in the back of his mind that an express or implied service factor inherently gives a motor overload capacity. Despite what the user may say today, the designer must assume that someday that capacity will be wanted. This is especially true for service center redesigns. For years, the service industry has sold Class F rewinds for longer winding life and/or "uprating" a motor to carry more load. That capacity wouldn't be a selling point if it couldn't be used. So, any combination of insulation class and rated rise that produces "service factor" should be allowed for, as an example, in sizing lead cables to carry maximum current. That won't happen if everybody closes his eyes to the existence of the design margin being provided.

As one way of dealing with these options, some engineers have used load curves like those in Figure 1-9. These incorporate present standards and past practice involving insulation limits and motor enclosures into several possible

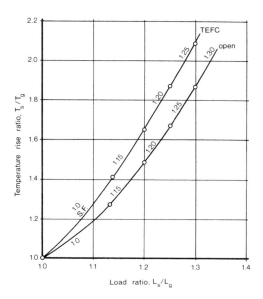

Figure 1-9: Curves like these have been used to relate frequently specified temperature rises and service factors, depending upon motor type and insulation class.

service factors. (Again, however, the 1.15 value is the only recognized standard.)

Is service factor on the way out for future standard designs? Certainly it will become less readily available. International standards do not allow for it. Should they become part of U.S. practice, along with metrication, the successive gaps between standard horsepower (kilowatt) ratings will become more uniform, easing the problem of load-matching. Meanwhile, as long as motor service factor remains a common option, those involved with motor selection and application will benefit from a better understanding of what service factor means and how it is used.

Nameplate markings

Service factor is one of the important motor characteristics that will appear on a standard motor nameplate. Anyone selecting or using motors should be familiar with what nameplate markings include and how they should be interpreted.

Who decides what's marked on a-c motor nameplates? NEMA Standard MG1-1978 sets the basic requirements. All plates must include horsepower rating, frame size, voltage, phase, frequency, full load amperes, inrush code letter, and time rating. Some expression of winding temperature must also appear. In those frame sizes for which a NEMA "design letter" dictates certain torque values, the nameplate displays that letter. You may see many other items stamped on a "standard" plate. Most are either requested by the user or are a manufacturer's option. Common examples: insulation system trade name, special service factor, and elevation above sea level.

Although most of the standard markings are well understood throughout industry, there is uncertainty about some of them. This is particularly true of a marking that appeared around 1978, the efficiency rating, which was generated by nationwide concern about energy conservation.

What does the efficiency rating mean? Efficiency may be expressed on a motor nameplate in several ways. Unavoidable tolerance variations in material and manufacture prevent the assignment of one full load efficiency value to all motors of a particular design. Hence the NEMA standard assigns ranges, or bands of efficiency, for nameplate use. These were initially expressed by the series of "index letters" shown in Table 1-XI. The letter appropriate for a particular design was stamped on the motor nameplate. (See Figure 1-10.)

As NEMA MG1-12.53b states, "Variations in materials, manufacturing processes, and tests result in motor-to-motor efficiency variations for a given motor design; the full-load efficiency for a large population of motors of a single design is not a unique efficiency but rather a band of efficiency."

That means that only statistical theory—the mathematics of probability—can derive the nameplate value for a given design, based on numerous tests. The nameplate index letter expressed "a value of nominal efficiency to be expected from a large population of motors." How large? The standard did not say, nor does it now.

Each index letter also defined a "minimum" efficiency. When tested by the accepted NEMA/IEEE dynamometer method, any motor of that particular design should show at least that full load efficiency. However, not all motor manufacturers took the same approach. One, after statistical analysis, chose a minimum figure associated with a 95% confidence level. This meant at least 95% of production units had to meet the minimum. It also meant that up to 5% of them might not. Other suppliers chose a minimum low enough to be met or exceeded by all units.

Table 1-XI

Index letter	Percent efficiency Nominal	Minimum
A	—	95.0
B	95.0	94.1
C	94.1	93.0
D	93.0	91.7
E	91.7	90.2
F	90.2	88.5
G	88.5	86.5
H	86.5	84.0
K	84.0	81.5
L	81.5	78.5
M	78.5	75.5
N	75.5	72.0
P	72.0	68.0
R	68.0	64.0
S	64.0	59.5
T	59.5	55.0
U	55.0	50.5
V	50.5	46.0
W	—	46.0

Figure 1-10: How the efficiency index letter appeared on a typical small motor nameplate.

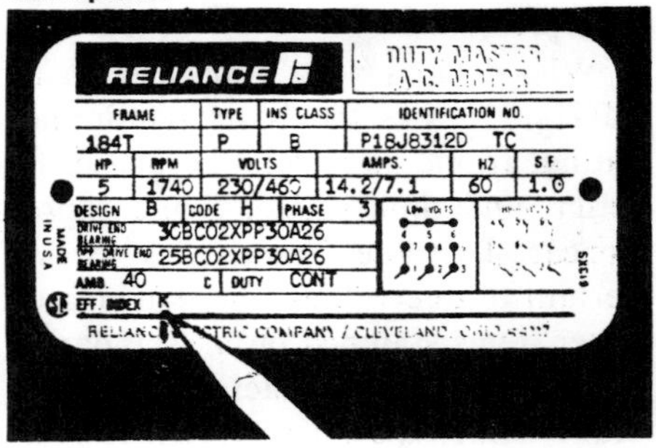

"Nominal" efficiency has been variously defined, but it is always a kind of average expected figure which "most" motors will meet. One manufacturer considers nominal efficiency as a mean value (Figure 1-11), so selected that half of all units will be above the mean, the other half, below.

Such a nominal efficiency can change as the tested "population" grows—as more units are built and tested—but the index letter system introduced some added uncertainty. Suppose tests of a design showed 87.5% nominal efficiency. The index letter G could not be used, because 87.5 is a full percentage point below the nominal level defined by that letter. Hence, index letter H had to be used, with its minimum efficiency of only 84.0%. The system thus permitted labelling a particular motor tested at only, say, 84.1% efficiency with a "nominal" nameplate value of no more than 86.5%, although the actual nominal value for the design was 87.5%. The letters, in short, didn't inform the user very precisely.

The spread could be even wider. Published data on one supplier's 5-hp, 1800-rpm motor showed more than five percentage points separating minimum and "maximum" efficiency; the nominal would fall somewhere in between.

Manufacturers were not obliged to use the index letters. As an alternative, the standard allowed nameplate use of actual efficiency numbers, and this was often done.

In 1980, the NEMA table was revised to furnish finer steps, or gradations, in efficiency. The index letter system was discontinued. Instead, actual nominal efficiencies were prescribed for nameplate use. The standard specifies that the highest nominal efficiency the manufacturer may put on the plate is that value in the table nearest to but no higher than the average of his tests (or calculations correlated with tests) on a "significant quantity" of motors. The minimum efficiency he has tested must be at least the minimum table value corresponding to that nominal. (This procedure, of course, cannot be followed exactly for a brand-new design, so that the values first selected are likely to change as more units are built, as already mentioned.) Table 1-XII gives a representative list of these numbers.

Because motor cost comparisons today depend so often on efficiency, be sure you understand the basis of the efficiency rating for all units involved in such a comparison—whether letters or numbers appear on the nameplates.

Occasionally, a nameplate will also display a nominal power factor. (See Figure 1-12.) High power factor is certainly of value, because it reduces amperes drawn by the motor from the line. That can permit lower transformer or switchgear ratings. Line losses are lower; wire sizes can be reduced. Hence, both operating and capital costs can be lowered.

However, this saving is highly variable and indirect. It may or may not be achieved, depending on conditions outside the motor itself. Higher efficiency, on the other hand, directly reduces out-of-pocket motor energy cost. So despite its concern for nameplate efficiencies, NEMA has not treated power factor the same way.

Figure 1-11: Manufacturing variations will cause a group of otherwise identical motors to exhibit a range or band of actual efficiency. If the group is large enough, the efficiency values will be distributed something like this: a few low and a few high values, with the majority clustered around a "mean," or "nominal" value.

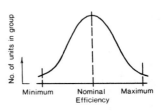

Figure 1-12: An early "high efficiency" motor nameplate, on which the manufacturer chose to use the nominal efficiency value rather than an index letter. Also shown is the nominal power factor.

```
SPARTAN™ MOTOR
MODEL      F44EJ415 0630
TYPE       CJ5B           FRAME         445 TD
VOLTS      460            °C AMB. INS.CL.  40  F
FRT. BRG.  216 S          EXT. BRG.     318 S
SERV. FACT. 1.00          OPER. INSTR
PHASE  3   HZ  60         CODE  F    WDGS.
H.P.       150
R.P.M.     1780
AMPS       172
NOM. EFF.  .940
NOM. P.F.  .871
MIN. AIR VEL. FT/MIN.
DUTY       CONT                         NEMA DESIGN  B

5NP0020                       Litton    LOUIS ALLIS
```

Table 1-XII	Efficiency, percent	
	Nominal	Minimum
	95.0	94.1
	94.5	93.6
	94.1	93.0
	93.6	92.4
	93.0	91.7
	92.4	91.0
	91.7	90.2
	91.0	89.5
	90.2	88.5
	89.5	87.5
	88.5	86.5
	87.5	85.5
	86.5	84.0
	85.5	82.5
	82.5	80.0

Whatever the motor nameplate shows, the indicated performance can be expected only under these conditions:

(1) Rated horsepower output. A motor's performance at any service factor overload won't match its performance at rated load. Or, as is true of a great many drives, if the motor rating has been conservatively chosen so that much of the time it runs well below full load, both efficiency and power factor (especially the latter) may suffer.

(2) Rated voltage. NEMA standards allow voltage to rise or fall 10%, but at these values the motor will not exhibit nameplate performance.

Do not look for nameplate efficiency ratings on motors above 250 hp, or on many units over 150 hp. The studies supporting this marking practice showed that motor energy use was most significant at 125 hp and below. Both "energy efficient" motors and nameplate efficiency markings are available to some extent well above that size, but not from all suppliers or in all ratings.

The presence of an efficiency marking on the nameplate, incidentally, does not mean that the motor is the latest "high efficiency" design. Standard machines up through 125 hp are now supposed to be so marked regardless of their design. That allows easy comparison between various makes or lines, new or old.

Remember this, too: The concept of "bands" of efficiency or power factor can lead to apparent inconsistencies between the values of those performance items and the nameplate current, especially in catalog data. For any 3-phase motor at rated load, this relationship must be true:

$$\text{Amperes} = \frac{(432)(\text{HP})}{(\text{volts})(\text{eff.})(\text{p.f.})}$$

for example, consider a 75-hp motor, 460 volts; actual efficiency .93, power factor .87. Amperes on the nameplate should then be:

$$\frac{(432)(75)}{(460)(.93)(.87)} = 87$$

If either efficiency or power factor numbers indicate a possible range or band of values, the amperes would also have to fall within some band. But there is no standard covering this, and the nameplate can show only a single number. So when choosing control or supply equipment ratings, use the nameplate amperes rather than what is calculated from the above formula.

Published motor data from one manufacturer contained these figures for a 10-hp, 6-pole machine: nominal efficiency, .85; nominal power factor, .75; full load amperes, 14.0 at 460 volts. But when the same efficiency and power factor values are inserted, the formula results in a calculated current of 14.7 amperes. Does the apparent discrepancy come from unresolved test variations? As this design undergoes more tests, will the conflicting numbers be reconciled? What will change, efficiency, power factor, full load amperes, or all three? No standard prescribes mutual consistency, as in Figure 1-13, among the current, efficiency, and power factor on a nameplate as well as in the

Figure 1-13: Although not identified as "nominal" or otherwise, these efficiency and power factor values are consistent with the nameplate amperes. The method used is mathematically correct, but the "141.3" ampere value implies greater accuracy than could normally be measured. For application purposes, either "141" or "142" would work as well.

catalog. Nor can this be checked unless all three items appear on the plate.

Another bit of useful information never seen on motor nameplates years ago but commonly encountered today is the expected motor noise level. (See Figure 1-14.) This has resulted from OSHA and other regulations controlling the industrial noise environment. Nameplate values of sound level are taken from tests on the motor running unloaded.

Although such figures permit useful comparison between motors, they cannot be used to measure an installation against OSHA limits. Nor can they indicate what the motor sound level will be under full load. The OSHA regulations, related to hours per day of worker exposure to noise, govern only the work place environment itself. They do not set limits for any individual motor or any other piece of machinery.

Nameplate sound level cannot be "verified" in the field. Because their surroundings are different, a motor noise test in a user's plant cannot match what was determined previously by a test on the manufacturer's floor. There are too many differences in background noise and in the nearby structures or surfaces from which sound can reflect. (See Chapter 5.)

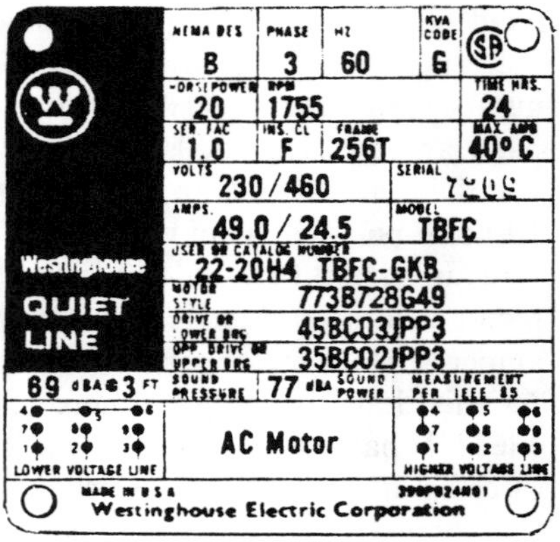

Figure 1-14: This motor, a "quiet-operating" design, has the sound level marked on its nameplate.

Still another change in nameplate information occurred some years ago in the handling of winding temperature. For any size machine, NEMA now allows two nameplate alternatives:

(1) Temperature rise by resistance (based on a 40°C maximum ambient).

(2) Insulation system designation (Class B, F, etc.), plus maximum allowable ambient temperature, omitting the rise entirely.

The first alternative is normal for "large" motors. The second is common for the "integral horsepower" motors below 250 hp.

When interpreting nameplate temperatures, remember:

(1) Rise is measured by the resistance method only (unless the machine is equipped with embedded detectors and the user specifies rise by detector). Standards no longer permit rating in terms of thermometer measurement, nor do they provide specific "hot spot allowances."

(2) If there is a service factor, the nameplate must show it. If a nameplate temperature rise appears, it should correspond to the service factor horsepower. No rise will be shown at rated load, nor is there any way to assume what that figure might be.

Users may request all sorts of special, unusual nameplate temperature markings. For example, you may sometimes see two temperature limits for two different loads. If the ambient is 50° or 60° rather than the standard 40°, it will appear on the nameplate, usually causing a corresponding reduction in the allowable winding rise (although an alternative is to retain the normal rise and wind with higher temperature insulation).

The motor "time rating" is another nameplate item that has various interpretations. A "continuous-rated" machine will run at rated conditions for an "indefinitely long period of time." Can such a motor be overloaded, even if the nameplate carries no service factor? To answer that, find out if the drive actually does operate around the clock at 100% of rated horsepower. Many do not. If the motor is shut off half the time or runs for hours at only half its rated load, it can probably safely carry considerable overload on occasion, regardless of what the nameplate shows.

An "intermittent-rated" motor is designed to stay within nameplate temperature under rated load when run no longer than the time stamped on the plate—usually ¼, ½, or 1 hour. How soon can this be repeated, and how often? Published standards provide no specific answer. Normally, though, the motor must be allowed to cool to within a few degrees of ambient temperature before carrying full load again. It cannot simply be shut off for a few minutes, then cycled through the same process over and over again indefinitely. If in doubt, consult the motor manufacturer about the proposed application.

Consider next a common item found on every induction motor nameplate: the kVA code letter. Does a particular letter imply a certain percentage of inrush current? No. Although users often assume, for example, that "kVA Code F" and "600% inrush" are synonymous, they are not. For any given horsepower rating, full load amperes varies widely depending on the number of motor poles. At low speed, power factor is low, so that such a motor has much higher full load current than a high speed machine of the same horsepower. That's because the magnetizing portion of that current goes up sharply when power factor drops.

NEMA code letters are calculated only from the number of locked rotor amperes and the horsepower. This is independent of either rpm or full load current. But the percent inrush depends on both locked rotor and full load amperes, and will therefore vary with the motor speed.

Let's look at two 125-hp, 460-volt designs—one 2-pole, the other 12-pole. The NEMA Design B inrush is 908 amperes for either one. Each tests about 850 locked rotor amperes. Here's how percentage inrush compares with kVA code for the two designs:

	2 pole	12 pole
Full load amperes	142	207
Percent inrush	600	410
KVA code letter	F	F

Although the code letter is the same for both, percent inrush is quite different. A 600% inrush may often be Code F; it may just as often represent Code G.

Actually, neither code letters nor percentages by themselves are generally useful in applying motors to power systems. Transformers, cables, and other components are rated for actual amperes, and the system designer must know, for example, "600% of what?" before he can make his selections. Table 1-XIII shows the relationship between code letters and actual ampere ranges for standard motor voltages.

Although rated speed is a motor nameplate item that has always seemed simple enough, it has assumed new importance with the advent of "high efficiency" designs. Such machines usually have a higher rpm than their older counterparts. That's because the rotor cage resistance has been lowered to reduce losses. This also reduces full load slip, resulting in higher full load speed.

But many fan and pump loads require shaft horsepower that varies as the cube of the speed. In such a drive, if you replace the existing motor with a new high efficiency unit, it will drive the load at a higher speed and therefore have

		Locked rotor amperes per hp	
Rated voltage⇒	460 v	2300 v	4000 v
KVA code letter			
D	5.02-5.62	1.01-1.120	.578-.650
E	5.63-6.25	1.121-1.248	.651-.721
F	6.26-7.01	1.249-1.398	.722-.809
G	7.02-7.89	1.399-1.570	.810-.910
H	7.90-8.88	1.571-1.773	.911-1.025
J	8.89-9.99	1.774-2.000	1.026-1.104
K	10.00-11.26	2.001-2.244	1.105-1.300
L	11.27-12.53	2.245-2.499	1.301-1.444

Table 1-XIII

to put out more horsepower. That will add an unexpected extra power cost to the drive despite the higher motor efficiency. (See Chapter 2.)

Be sure you are aware of the manufacturer's nameplating practice. Actual rpm is often rounded off to the nearest 5 when stamping the plate. Although the figure appearing on the plate may be "1775," the actual full load rpm for the design may range anywhere from 1773 to 1778.

Possible variations

Don't assume motors of widely different outputs, at the same polarity, will run at the same full load rpm. Figure 1-15 illustrates the sort of variation you may encounter. Clearly, a large error in matching motor horsepower to actual load demand can result from assuming that a 250-hp motor has the same "1750" full load speed as a 5-hp machine—yet such assumptions are common.

NEMA standards do not dictate bearing identity on nameplates. This convenience is universally provided, however, making bearing replacement data readily available to the user. Not everyone identifies bearings the same way, though. Compare Figures 1-10 and 1-12. An often-used method for integral horsepower motors is the Anti-Friction Bearing Manufacturers Association's (AFBMA) universal "identification code." It may include a dozen letters and numbers defining five basic bearing design and construction features. Among the items covered are: fits & tolerances, seals & shields, cage material, rings or races, and lubrication.

If you order a replacement bearing either from the motor manufacturer in terms of complete nameplate data or through an AFBMA supplier by identification code, the bearing should be right. But otherwise it may not be.

Bearing load, proper lubrication, operating temperature, the shaft or housing fit—all require the use of one specific bearing. The larger the bearing or the higher its speed, the sooner a small difference—such as in the internal clearance or "fitup"—may cause a catastrophic failure.

There's more to proper bearing choice than simply asking for an "electric

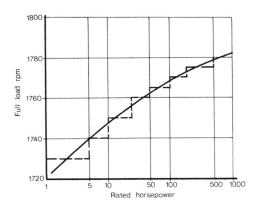

Figure 1-15: This graph shows how full load speed will typically vary with motor horsepower for 4-pole ratings. The dashed line expresses the "nearest 5 rpm" values often used on nameplates.

Taking into account the NEMA allowable 10% voltage fluctuations adds further variation in speed. For example, lowering voltage 10% reduces 1775 rpm to 1770; raising voltage 10% increases 1765 rpm to 1771.

motor grade" part rather than an "ordinary" bearing. This old misunderstanding dies hard. No such category as "electric motor grade" exists in U.S. bearing industry standards. In no way are bearings for electric motors designed or manufactured differently from bearings for any other use. The important thing is that the replacement bearing be of the same design originally used in the motor. So proper identification includes more than just a basic size number.

Be careful, then, when working from bearing nameplate numbers not in the AFBMA format. For example, the nameplate may identify a bearing as "214K." The full AFBMA code is "70BCO2X3." The 214K designation covers the correct part from only one bearing supplier; another manufacturer identifies the same item as "214S." Are they interchangeable? In this instance, yes, but that isn't apparent from the motor nameplate marking alone.

Besides the "rating" nameplate defining the motor's basic behavior, machines may carry other nameplates governing accessory equipment or application. Always look for these, and be sure you understand them. There isn't room on any one plate for everything; some large motors may have a dozen other plates to provide instructions for lifting, lubrication, accessory circuit wiring, and so on. (See Figure 1-16.)

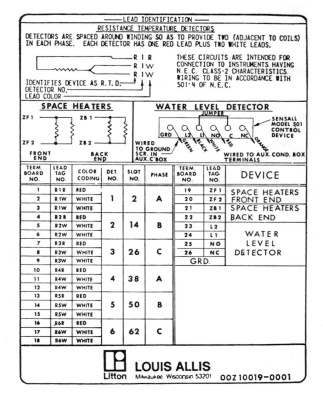

Figure 1-16: This accessory nameplate for a large motor explains the proper connections for the space heater, winding temperature detector, and air-to-water heat exchanger leak detector circuits.

These additional plates may also display safety warnings required by user practice or by local codes. An example appears in Figure 1-17. Inside this large motor conduit box are 4000-volt bare bus terminals. Such uninsulated connections, quickly taken apart and put back together for winding tests, are quite safe when internal clearances are ample. But the user requested the warning plate, using an OSHA danger sign format, "just in case."

Figure 1-17: A special type of auxiliary nameplate used at the customer's request.

Published motor standards have nothing to say about such warning or other auxiliary plates, except that certain wording is required by both Underwriters Laboratories and the Canadian Standards Association on motors labelled for "classified" (hazardous) areas. All manufacturers don't use the same auxiliary nameplate marking practices.

No matter how complete they may appear to be, nameplates can supply only limited information. They are not substitutes for instruction manuals or outline drawings. But a motor nameplate can answer many application questions—if you know how to read all of it.

Motor Specifications

Having decided which "standard" or "special" motor features are needed for an application, the user generally writes some sort of specification to describe his needs to possible suppliers. Whether simple or complex, the specification will determine whether or not the application is successful. It must be prepared carefully.

Completeness, conciseness, clarity—these three essentials of good specification writing can mean the difference between the right motor and the wrong one. Consider these points in building those three C's into your specs:

First, make sure it's all there. Claimed one purchasing expert, "Long-winded narrative descriptions are not needed on bid requests for capital equipment." Others are praising the computer as a tool to produce simplified specs for industrial apparatus, using standard phrases called "boilerplate."

But properly specifying industrial motors is an exacting task requiring strong technical background. Job requirements often mandate some "long-winded narration." Courts have held that "Engineering specs are 'warranted' in the sense that they must be skillfully prepared and must meet professional standards. Designers and engineers who fail to meet these requirements can be held liable for negligence—and damages."

A spec should be dated, signed, and, above all, titled or numbered for identification in later correspondence. Each revision should be clearly identified. Insist that the entire spec be forwarded to the motor supplier when the driver is ordered. Often the motor manufacturer receives only a handful of pages, sometimes unidentified, from the pump or fan builder. He doesn't know what might be missing.

Consider the spec that called for temperature detectors in each bearing throughout the drive, including the motor. This special requirement appeared only in the pump portion of the spec, which was not forwarded to the motor supplier. He would have omitted any motor bearing temperature detectors had he not later managed to acquire the missing pages. Don't let this happen to your drive; make sure the motor designer sees the whole document.

Second, be consistent, and say it only once. Too much content, as well as too little, leads to misunderstanding. Automation—substituting "cut and paste" copying for thoughtful editing—breeds documents full of extraneous material. Modern word-processing equipment may accentuate this. Said one consulting engineer, "Instead of editing to shorten them, the writer merely adds new paragraphs. This is the problem with any kind of automation. . . . You tend to keep those big chunks of verbiage in and add to them."

In theory, redundancy may add safety, but in practice it often adds confusion. Try to avoid basing the same requirement on two different criteria. An example is motor temperature rise. Some specs call for a rise at nameplate horsepower, typically 60°C to 80°C. The same specs then stipulate a maximum rise at service factor load, usually 90°C. One temperature or the other ought to be called for—but not both, as we saw earlier in this chapter.

Motors and Standards 37

To put it another way: When conflicting criteria govern the same parameter, apply only the most restrictive one. If a consultant is involved, both he and the user should call for the same thing. One power plant spec gave motor starting voltage as 75% of the rated value. But the consultant's spec for the same project said 85%. Which was correct?

Double check to be sure performance, temperature rating, and accessories are specified the same on a data sheet or purchase requisition as they appear in the body of the spec.

Third, are your standards standard? Mentioned earlier in this chapter was the "catch-all" motor spec provision, which read something like this: "[A motor spec] must conform to all applicable state and local codes, ordinances, and regulations." The motor manufacturer may not even know which agencies have jobsite jurisdiction.

Here's an example of how extreme this can get. A machinery manufacturer sold a unit for export to Australia. The motor supplier quoted assuming that no special features were needed. Only after receiving the order did he learn that an entire set of British and Australian standards applied—none of which he had, and some of which required parts and design procedures he had never heard of. Months passed before the overseas documents could be obtained, reviewed, and clarified.

Remember, too, that such documents as the NEC are subject to on-the-scene interpretation. How can a motor designer be expected to know that Chief Electrical Inspector A, in the City of B, has always interpreted Code Section C in a certain way, so that the motor turns out to have the wrong size conduit entrance or some other unexpected deficiency?

By all means, demand compliance with applicable standards. But try to get copies of the pertinent passages to all bidders; at least let them know where they can get their own copies.

As brought out earlier, make sure the standards cited really do apply. The statement "Torques to be NEMA Design B" often appears in large motor specs, although—as we have seen—NEMA Design B is non-existent above 500 hp. If you need some specific torque or starting current value for a larger machine, state the numbers.

Another thing to avoid is reference to "AIEE" or "ASA" standards. Both organizations ceased to exist years ago, replaced by IEEE and ANSI respectively. True, document content is more important than the issuing organization's current title. But the reader of a spec naturally wonders when he sees long-obsolete titles how up-to-date the rest of the text really is. Moreover, some standards may have changed in detail. Performance requirements that were once standard may no longer be the same—do you want conformance per the old version, or the new one?

It is also confusing when a spec begins by asserting that motors "must conform to latest issue of NEMA standards," but ensuing passages impose motor requirements contradicting those standards. A paragraph may require the motor to get a "routine NEMA standard test consisting of the following," but there follows a list of items far outside the scope of NEMA routine testing—such as performance curves or polarization index. Some of the specific tests called for may not be defined in any way, such as an "impedance test"—a term not appearing in any industry standard.

Fourth, is this requirement necessary? Many pump driver specs, for instance, call for 100% minimum locked rotor torque. As shown in Figure 1-3, NEMA standard locked torque for large machines is only 60%. Most motor designs, especially the 2-pole and 4-pole units suited to centrifugal pumps, will have 70% to 90%, which is usually more than enough.

The only purpose in having 100%—a value often necessitating special motor design, at higher cost—is to ensure safe acceleration of the load. But inasmuch as fan or pump breakaway torque, at zero speed, seldom exceeds 20%, asking for motor torque five times that much amounts to overkill.

Any starting problem such a drive may encounter is not at locked rotor but at about 3/4 speed, as we'll see in Chapter 3. A motor with 100% to 120% locked rotor torque can have just as much difficulty starting a pump (especially if the voltage sags during acceleration) as a motor with 50% locked torque.

Fifth, can you get what you're asking for? Authorities contend that specifications should be so written that "they are not only understood, but . . . also cannot be misunderstood." It is not the unreasonable spec but the unintelligible one that causes the most confusion, delay, and needless expense in processing motor orders. Although a motor manufacturer may propose alternatives or no-bid a seemingly unrealistic job, he must usually take at face value whatever technical requirements a spec imposes.

Inaccurate terminology

But some frequently encountered technical provisions in motor specs involve basic contradictions or express misunderstanding of motor behavior, such as:

(1) "Weatherproof" construction—a term not defined by motor industry standards. (See "Weather-protected," NEMA MG1-1.25H; Chapter 5.)

(2) "Pull-in" torque for an induction motor. This term has meaning only for a synchronous machine.

(3) Asking for stator temperature sensors to protect against severe starting duty. Many large machines will suffer rotor damage during starting or stalling before the stator overheats. (See Chapter 3.)

(4) Requesting a specific BIL (Basic Impulse Level) for motor windings. No impulse levels or impulse test procedures are defined for rotating machines.

(5) Asking for starting duty in terms of "allowable starts per hour" for large motors. Successive starts beyond the first two must involve lengthy cooling periods, and such machines are more typically rated in terms of starts per 24-hour day.

(6) Requesting a "motor heating curve." This can mean a variety of different plots, for different purposes: safe operating time versus current (Chapter

3) for stalling or short-time running overload protection; temperature versus load at full speed; or a curve of temperature versus time at various loads. All three versions have been requested at one time or another.

Finally, be specific; avoid vagueness. Fuzzy phrases that don't belong in any motor spec include these:

(1) "Motor starting current shall be the minimum within the standard design limits used." How is a designer to interpret that? Starting current can be varied 30% or more in a given motor size, in one or more of these ways:

(a) Change in magnetic flux density.
(b) Change in rotor/stator slot shapes.
(c) Change in rotor cage material.

Each such change involves "standard design limits." Each also affects other motor design and performance features—efficiency, power factor, full load rpm, torques, temperature, and cost. There is no way to set some "minimum" below which any "standard design limit" will be violated. All that can be expected is that inrush can be reduced to an extent that will keep all other parameters (including cost) within their permissible ranges, whatever these ranges may be.

Consider such words as "severe," "frequent," "heavy," "rigid," "normal," "extreme," and the like. What can they really signify to an apparatus designer? Unless you can put what you want in specific numerical terms, is it important at all? Not every condition can be easily quantified, but just as a motor designer views the suitable choice of parts as a professional challenge, so, too, should the specification writer regard the definition of the application requirements. Lord Kelvin, the eminent British physicist, once said, in effect, "If you cannot measure it in numbers, you do not know what you are talking about."

As an example, from a 4000-hp motor spec: "Insulation must be adequate to withstand switching or other surges well in excess of normal turn-to-turn voltages." What does "well in excess" mean? What is "normal"? This is not hair-splitting; there are no industry standards for turn-to-turn motor voltages. The wording looks good at first glance, but what will it accomplish?

Here's another: "All motors shall be adequate to start and accelerate their driven equipment . . . at about 90 percent of rated voltage." Will any two people interpret "about" the same way? Either the figure is 90% or it isn't; if it's something else, say so.

And finally: "Motor windings shall be adequately braced for full voltage starting and shall withstand the stresses caused by an immediate transfer of the power supply from one source to another." How long a time span is "immediate"—five cycles? One cycle? One second? The answer can have a large effect on motor design and cost.

While emphasizing the concrete over the abstract, avoid going to the extreme of trying to tell the supplier every detail of motor design and construction. Specifications should describe the task the equipment must perform—not the way in which it should be designed. Engineering solutions to the same problem will vary with different engineers.

If you know that a manufacturer is likely to use in his product a feature you can expect trouble with, either choose another supplier or explain the problem so he can choose a logical alternative. But don't try to write all the motor construction materials and processes into a bid specification. Describe function, not design.

So much for the Three C's of motor specs. We might close this chapter with three more, from a publication in the consulting field: "The courts have made it very clear that there are three basic ingredients to a satisfactory job: the engineer must carefully prepare plans and specifications; the contractor must faithfully adhere to them during the course of construction; and there must be rigorous inspection during the progress of the work. To accomplish these objectives, the engineer and contractor must cooperate, coordinate, and communicate."

2 Efficiency and Power Factor

UNJUSTLY ACCUSED of being energy wasters, electric motors in recent years have fallen victim to a bad press. During the period of acute fuel shortages, calls for motor redesigns were issued by some authorities, who cited figures showing that 64% of our electrical energy is used up by motors. Although fuel supplies are once again adequate, industry has come to be more aware of the potential for power savings through energy efficient electric motors.

Actually, however, motors are converters, not users, of energy. A motor transforms nearly all the electrical energy supplied to it into useful work. Only the transformer, with no moving parts, is a more efficient electrical machine.

This misunderstanding regarding motor efficiency is reflected in comments about how motor efficiency has varied over the years. Some motor users have been unfavorably impressed by the higher motor frame surface temperatures of newer machines. Aware that "rerate" programs have packed more horsepower into smaller frame sizes, they are sure that the result has been a decline in efficiency.

But that supposition isn't supported by the facts. Certainly some smaller horsepower ratings have gradually declined in efficiency over a long period of time. But, as Figure 2-1 shows, the drop was not great. It was power factor, not efficiency, that significantly decreased.

Although low power factor has an indirect effect on plant facilities cost and the utility bill, capacitors can correct that. On the other hand, efficiency has a

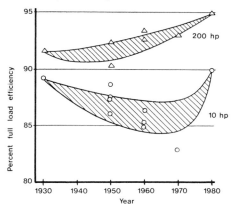

Figure 2-1: How "standard" 4-pole open motor published efficiencies have changed over the years. Evidently, recent NEMA rerates did not suddenly lower efficiency. Note that efficiency of the larger horsepower ratings has actually risen, not fallen. Newer "premium efficiency" designs represented by the 1980 figures are a major improvement—not a return to what was common just a few years ago.

direct—and much greater—influence on drive operating cost. There are other basic misunderstandings about efficiency that hinder proper motor application.

Some misconceptions about efficiency

Probably the most persistent misconception among motor users is the belief that a hotter motor is less efficient than a cooler one. "Take us back to the old days of high efficiency, low temperature motors," they say. This started with the swing towards Class B rise 15 years ago and has intensified with the greater use of Class F rise today.

Just remember that *temperature* and wasted energy, or *heat*, are two different things. Temperature simply indicates how much heat is present within a certain space or volume of material.

Consider the interior of your home. Put a certain amount of heat into it by burning a certain amount of fuel in your furnace, and the temperature throughout the house will rise to some value we can call "T" degrees. Suppose you burn the same amount of fuel—create the same amount of heat energy—and pipe all the heat into just one closed room. The temperature in that room will soar far above "T" even though the total heat involved will be the same as before.

Now suppose the same amount of heat is spread throughout the house, but this time we open all the windows, in midwinter. The heat quickly escapes outdoors. We get an inside temperature only one-third of "T." Yet the same amount of heat was provided to begin with. Stand outside one of the open windows and you'll feel warmth—much more than if the windows were closed.

We learn two things from all this. First, the same heat (maybe even a bit less) within a small space means higher temperature. Hence, a rerated motor in a smaller frame size, using less iron and copper, may run at a much higher temperature than the old design even though heat losses and efficiency may be the same for both machines.

Second, the more easily heat can escape from the space where it is being produced, the warmer the outside of that space becomes. The easier it is for waste heat in a motor to travel from the stator and rotor to the exterior of the frame, the higher that exterior surface temperature will be. This, too, is common with rerated motors. They are designed to dissipate heat more easily. That is why they often feel much hotter on the surface than some older designs do. Surface heat shouldn't be considered a drawback unless there is a chance someone may be burned.

Getting back to your house: Put your hand on the outside of your water heater while it's operating. It won't feel very hot outside, despite lots of heat inside, because the heater is jacketed with heavy insulation to keep the heat in. In a motor, though, the whole design intent is to let the heat out. Hence, the surface of a motor is bound to be at a high temperature.

It is a fact, then, that a 105°C rise motor is not necessarily less efficient than a 40°C rise design. Sometimes it is actually more efficient, particularly

Efficiency and Power Factor 43

at high speeds and light loads, when a reduction in frame size cuts down on windage loss.

Obviously, as is being proven by the new motor lines now on the market, redesign on different parts can raise efficiency. Copper and iron can be redistributed, or more can be used, to reduce losses no matter what the temperature rise may be. Depending on changes made in the frame and the cooling system, such a motor may run either hotter or cooler than the original version.

A second misunderstanding, already touched on, is the idea that "bigger is better" where efficiency is concerned. "To raise efficiency, make the motor bigger," is the common expression.

Yes, a large motor is normally more efficient than a small one. But this holds true only if the bigger machine is being used at its higher horsepower rating. If both machines are run at the same load, the big one is often less efficient. It has more iron to be magnetized, so iron loss is higher. A larger rotor diameter means higher windage loss; bigger bearings mean higher friction loss.

A third belief that may lead users astray in judging motor performance is that "efficiency is always best at full load." They think that to get the highest possible drive efficiency, horsepower rating should be matched to actual load power.

The truth, however, is that the efficiency of any motor usually peaks at 60% to 80% of rated load. Some curves are flatter than others (Figure 2-2), but don't assume that the motor with the highest full load efficiency will save the most on power bills. You have to know what the actual load is and what the motor efficiency curve really looks like before you can choose the most economical motor for the job.

For example:
(1) 750 hp load, 1780 rpm; 2300 volts
   (a) 1000-hp motor at ¾ load: efficiency = 95.8%
   (b) 800-hp motor at 93% load: efficiency = 92.6%

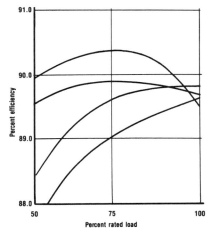

Figure 2-2: How efficiency varies with load for four typical 4-pole designs. (Not shown: how each design also varies in accelerating torque, power factor, etc.) Which is "best"? Note that the one with the highest efficiency near ¾ load has the lowest efficiency at full load. If this motor will run most of the time at 70% to 80% load, it may be a much better choice than any of the others.

Therefore, use the bigger machine, but first check the effect of its much lower power factor at this load.

(2) 350 hp load, 3560 rpm; 2300 volts
   (a) 500-hp motor at ¾ load: efficiency = 93.0%
   (b) 350-hp motor at full load: efficiency = 92.5%

This also favors the larger motor, though not by much, and it will still have lower power factor at this load.

(3) 7½ hp load, 1750 rpm; 460 volts
   (a) 10 hp-motor at ¾ load: efficiency = 82%
   (b) 7½-hp motor at full load: efficiency = 84%

This clearly favors the smaller motor, which will also have a much higher power factor at this load.

This general pattern may not hold true for other speeds, different voltages, special starting duty, and so on. Review each instance on its own merits.

Is it logical to rewind a motor just to match the peak efficiency point to an actual operating load? Some maintenance persons are considering such programs, which may be worth the cost. But simple approximations aren't too helpful here. This is a job for a design engineer.

Higher motor efficiency means a higher motor price. Will the added investment be justified? There are many ways to arrive at an answer, but they are all variations of two basic approaches:

(1) A "simple payback" analysis. This merely balances the higher initial price of the higher efficiency unit against its lower annual operating cost. Such comparisons are simple. Consider a 25-hp, 4-pole drip proof motor running 16 hours daily, 5 days a week, at a four cent power rate. Improving its efficiency by 1%, thereby saving about one-quarter kilowatt, saves 40 dollars a year. If the motor runs continuously, it's more like 80 dollars a year. Or take a 60-hp, 4-pole machine. Efficiency up 1.5% means losses down almost ¾ kW. At the four cent rate, operating 4000 hours yearly, the higher efficiency motor saves 120 dollars annually. Eventually, the accumulated power cost saving equals the new motor's price premium. This period of time is the "payback" period. User accounting practice determines the maximum allowable payback period to justify the investment. Two years is typical. If the calculated payback time falls within that range, then the second approach is used:

(2) An analysis that allows for the time value of money. Funds invested today in a new motor could instead be invested elsewhere for a known rate of return. That interest is lost if the money is spent today. Hence, available interest rates dictate a reduction or "discount" of the simple payback period. Furthermore, savings from using the new motor will change as power cost rises in the future. Will inflation get worse or better? That will affect the calculations, too.

How long the new machine will last should also be considered. Depreciation must be part of any asset valuation. The depreciation method—straight-line or otherwise—is equally applicable to any motor design, so that in itself should not influence the choice. But the length of the depreciation period may differ because the more efficient unit is likely to last longer.

All these variations are leading motor users to several different ways of expressing "efficiency evaluation" in equipment specifications. In general, the higher the motor losses quoted by a bidder, the lower his price must be for him to be considered. Here are the most common versions of loss evaluation:

(1) A dollar value for each "tenth of a point" efficiency. This method, as we shall see, is seldom realistic.

(2) Two separate dollar values, one assigned to each kilowatt of "no load loss," the other to each kilowatt of "load loss." This can be misleading because those loss descriptions do not suit motor behavior.

(3) An amount of dollars per kilowatt of total motor loss. That figure today ranges from $1,000 to $9,000, with $3,000 being a representative amount.

How are efficiency points related to kilowatts? That depends on motor rated horsepower. Some users assume that "one kilowatt loss" is equal to "one tenth of a point" of efficiency. This is similar to the common—and equally incorrect—notion that "600% inrush current" is the same as "KVA Code F."

Why not compare efficiency percentages alone, rather than kilowatts of loss? The answer is that efficiency is too broad a measure. For example, assume that cost studies dictate a value of $3,000 per kilowatt for a 1000-hp motor. If the actual loss is 41 kW at full load, motor efficiency is .948. The complexities of accurately determining efficiency by test will be discussed briefly, later in this chapter. Authorities generally agree that it's useless to quote—or attempt to test—efficiency beyond three significant figures. A value of ".9466," for example, would be rounded off to .947. Similarly, a value of .9473 or ".947286" could easily be generated by a thorough test analysis using modern computing equipment, but would still be quoted as .947.

The difference between two motors having actual efficiencies of .9466 and .9473 is .0007. If the rating is 1000 hp, that corresponds to a loss difference of about 0.6 kW. That might be evaluated at $3,000 times 0.6, or $1,800. Yet if

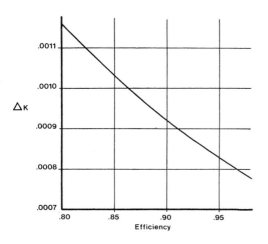

Figure 2-3: This curve shows how kW loss is related to fractions of a point in efficiency. Find the value ΔK corresponding to the nominal efficiency for the motor being considered. Then, if motor performance is evaluated at D_1 dollars per .001 efficiency: $D_1/(\Delta K)$ (rated hp) = D_2, dollars per kW loss. If the evaluation is given in D_2 dollars per kW loss: $(D_2)(\Delta K)$ (rated hp) = D_1, dollars per .001 efficiency.

the evaluation instead is made in dollars per tenth of a percent efficiency, the two motors would be considered equivalent because each has the same .947 efficiency.

One user has begun to recognize this, specifying that "bid efficiencies are to be given in 0.1% steps only. If in .01 steps, the values will be rounded down to the nearest 0.1 unit." This discourages a bidder from offering ".9466," for example, because the buyer rounds it down to .946 instead of up to .947.

Figure 2-3 shows how kilowatts loss per rated horsepower is related to a tenth of a percent in efficiency. For a 1000-hp machine, evaluations made on a kilowatt basis compare this way to evaluations made on a "percent-efficiency" basis:

Nominal efficiency	Dollar value per kW loss	Corresponding dollar value per .001 efficiency
.94	$3,000	$2,550
.95	$3,000	$2,500
.96	$3,000	$2,450

Transformers, unlike motors, operate with only two loss components. One varies with the square of the load current. The other is present at any load, light or heavy. These are:

(1) Load-dependent, variable, or "copper" loss.
(2) Constant, no-load, or "iron" loss.

Iron loss is supplied by a magnetizing current, but that can safely be neglected in considering these two basic losses.

Public utilities, accustomed to dealing with transformers, often attempt to deal with motor efficiency on the same basis. A specification may read like this: "No load losses will be evaluated at $1,150/kW. Variable losses will be evaluated at $1,080/kW."

But motor losses behave differently. Motor iron or core loss is not constant with load. In a large motor, it may drop as much as one to two kilowatts from idle to full load. The reason: Voltage across the "magnetizing branch" of the motor's equivalent circuit is what magnetizes the core. As load current goes up, the voltage drop across the stator portion of the equivalent circuit results in decreasing magnetizing branch voltage, hence lower core loss.

User specifications may similarly assign one dollar value to "core loss" and another to "copper loss." This omits stray load loss entirely. Although it resembles a "copper" loss, changing rapidly with load, it does not simply vary with the square of the line current, as would be true for transformer copper loss.

Different motor suppliers are bound to interpret such statements differently. Hence, design approaches and bids will not all be made on the same basis. If the user needs to assign different dollar values to different kinds of loss, he should clearly specify the basis on which suppliers should bid by defining what losses are to be considered fixed, as well as how the other losses are considered to vary.

Whatever motor losses may be called, how are designers reducing them? Today's energy efficient motor design is not a return to the good old days. Its performance is far superior to that of any previously available standard product. Years ago, when power might cost only 2¢ per kilowatt-hour, users could not justify the cost of such machines. But a charge of 7¢ is not uncommon today, and 15¢ to 20¢ is sometimes encountered. Those figures will continue to go up.

Widespread publicity about the new designs may give the impression that higher motor efficiency stems from technological breakthroughs stimulated by the energy crisis. That's misleading. For example, steel suppliers have not introduced any truly new kinds of lamination material. Yes, steel quality has gradually improved, but the basic grades remain what they were a generation ago. It's just that economics now supports broader use of the higher quality materials that existed but had not been cost-effective in the past.

Again, motors do not enjoy the advantage of transformer construction. Magnetic steel improvements involve the modification of the metal's crystalline "grain structure." Because much of the wasted core loss energy goes to distort that structure as the alternating magnetic field varies, two grain structure innovations can reduce core loss. One is "orienting" the crystals all in the same direction within the steel sheet. The core is then built so that direction matches the magnetic force path in the finished unit. Such "grain oriented" steel has been used in power transformers for many years. It works well because the transformer magnetic field follows straight paths along the laminations.

Unfortunately, the cylindrical electric motor core includes no such paths. Each core tooth between slots is at a different angular position, so that some teeth are in line with the direction the steel sheet was rolled in and others are at right angles to that direction. Any straight lamination path is at some point in space or time located at right angles to magnetic flux lines rather than parallel with them. Tests confirm that grain oriented steel won't help reduce motor core loss.

Another innovation, taking various forms over the past twenty-five years, has been a relatively amorphous grain structure—that is, very small symmetrical cubic grains, or (like the structure of glass) no grains at all. This would be ideal for a multi-directional magnetic field. But so far such material has either not progressed beyond a laboratory curiosity or not become cost-effective for industrial machines.

The core loss reduction in modern high efficiency motors results from other improvements, none of them really new:

(1) Reduced magnetic field strength. This usually results from longer cores or different slot shapes, leading to much lower core loss.

(2) Greater care in manufacture. Punching the laminations work-hardens the steel. Then the magnetic force cannot move the grain structure as readily, which increases energy loss. Heat-treating the laminations, to relieve punching stress, can bring the loss back down to its original level.

Pressing or grinding down the tiny burrs along the punched edges can increase the "contact resistance" between laminations, thus lowering the eddy current loss. Many other special techniques in both lamination manufacture and core assembly can yield further improvement.

(3) Thinner steel. For motor production, these three thicknesses are commercially available: 24 gauge (.025 inch thick), 26 gauge (.018), and 29 gauge (.014). For higher efficiency, No. 29 is replacing No. 26, and No. 26 is replacing No. 24. This shortens the path length for circulating eddy currents within each lamination, reducing the I^2R loss that such currents produce.

(4) Higher silicon content. Again, such material isn't new, but its higher cost has led designers to avoid it in many lines of standard motors.

All such modifications add cost. For example, using a thinner lamination—even for the same core length—requires punching and stacking more laminations. Thinner sheet in itself costs more per pound.

Each of the other four motor losses (stator winding I^2R, rotor cage I^2R, and windage, and stray load loss) is being attacked in many other ways. This can be a demanding art, because changing any one loss affects others as well. Suppose the stator slot is widened to allow more winding copper, thus lowering stator I^2R. This narrows the teeth between slots, raising the magnetic flux density so that core loss increases. The designer must find a competitive, cost-effective compromise. Overall motor design may involve scores of such tradeoffs.

Hence, no fixed rules govern the comparison between different "high efficiency" designs. Table 2-I shows what variations may exist in the market.

The effect of operating conditions on efficiency

Before replacing any standard motor with a premium efficiency unit, find out if you can save more money by modifying the driven machinery or the process controls. Go over the whole drive system from incoming power line to outgoing machine product. Many authorities have called attention to the high percentage of generated electric power which is wasted outside the motor itself. (See Figure 2-4.) So the first step towards raising drive efficiency is to lower power wastage in the driven machinery. Perhaps you can cycle the drive on and off more often. (Remember that the simplest, most economical energy-saving device ever invented is the "stop" button.) Changes in valve or damper setting, or adding microprocessor control, may help. Such measures are likely to yield a much greater saving than just replacing the motor.

A variable speed drive, with solid-state power supply, can often greatly reduce operating cost. Power loss in any fan or pump will far exceed that in the driving motor, so a more efficient operating speed always cuts energy use. Power cost will drop still more with higher efficiency in the motor itself.

However, in calculating payback on the motor investment, do not use the published performance of the high efficiency motor. That's valid only for sinusoidal voltage at commercial frequency. Internal loss in any motor will be

Efficiency and Power Factor 49

Mfr	Std. eff.	Prem. eff.	Diff.	Std. kW loss	Prem. kW	Diff.
10 hp, 4 pole						
A	.823	.895	.072	2.15	.88	1.27
B	.855	.893	.038	1.27	.89	.38
C	.853	.902	.049	1.29	.81	.48
D	.850	.902	.052	1.32	.81	.51
E	.865	.885	.020	1.17	.97	.20
F	.859	.910	.051	1.23	.74	.49
G	.850	.890	.040	1.32	.92	.40
H	.865	.900	.035	1.17	.83	.34
J	.857	.902	.045	1.24	.48	.76
			Average .045			Average .54
125 hp, 4 pole						
A	.914	.945	.031	8.8	5.5	3.3
B	.916	.950	.034	8.6	4.9	3.7
C	.921	.950	.029	6.8	4.9	1.9
D	.924	.943	.019	7.7	5.6	2.1
E	.910	.940	.030	9.2	6.0	3.2
F	.925	.945	.020	7.6	5.5	2.1
G	.918	.954	.036	8.4	4.5	3.9
H	.920	.935	.015	8.1	6.5	1.6
J	.917	.958	.041	8.5	4.1	4.4
			Average .026			Average 2.9

Table 2-I. These figures indicate the variety of published, nominal full load efficiencies available for open motors. Although standards do prescribe tests, terminology, and nameplate efficiency markings, the marketplace determines what each supplier will offer.

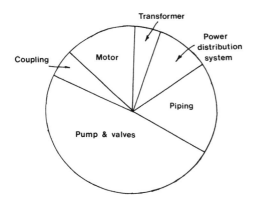

Figure 2-4: Where the total energy lost—wasted—in supplying a typical drive actually goes. Note that relatively little energy wastage occurs in the motor itself.

up to 20% higher on a variable-frequency power supply. The exact increase will depend on the type of supply (6-step inverter versus 12-step, for example). The motor manufacturer can probably estimate the change in motor efficiency if he is given enough of the power supply characteristics.

In making cost studies on any drive, be sure the drive is operating as you think it is. Are you sure of the actual horsepower required? Is it constant, or does it frequently vary? Whatever the load is, how many hours per day, or days per month, does the motor operate at that condition? Beware of evaluating motor efficiency at rated nameplate horsepower for an actual load that is quite different most of the time. It's true that a high efficiency motor will always be cheaper to run than a standard unit under the same loading. But return-on-investment calculations depend on knowing just how much cheaper.

Don't take your power system for granted. Back when greatly oversized motors ran within 40°C rise, using low-cost power, operators often cared little about maintaining rated voltage. After all, NEMA standards allowed motors to run at plus or minus 10% voltage (which is still true). Nor was voltage unbalance important, because of an ample margin in winding temperature.

Those power system conditions became more dangerous as motors became smaller and ran hotter—but only because winding life might be shortened. Now, a greater concern is the effect of poor voltage conditions on motor efficiency. When a motor price premium is justified only by high efficiency, the payback may be cancelled out by operating voltage which is unbalanced or does not match the nameplate.

For example, consider a 150-hp, 6-pole open motor with full load loss of 7.7 kW and efficiency of .935. If terminal voltage drops only 5%, the increase in I^2R—even though core loss drops somewhat—will raise total motor loss to 8.8 kW, pushing efficiency down to .927. If that 1.1 kW added loss had been evaluated at several thousand dollars, such a voltage sag could be economically disastrous.

How voltage unbalance affects motor heating has long been known. It's described in NEMA standards. "Over-motoring" or a service factor rating may compensate for the temperature but cannot make up for the extra loss caused by unbalanced voltages. A 2% unbalance may not seem severe. Typical de-rating curves, like that found in NEMA MG1, call for motor load reduction of about 5% for a 2% voltage unbalance. The reason is that motor loss goes up about 8%. What happens to efficiency? For a 500-hp machine, having 20 kW full load loss, efficiency drops to .949. An 8% boost in loss brings the total to 21.6 kW, and efficiency drops to .945. Clearly, the "slight" voltage unbalance can eat up much of the saving expected from replacing an existing motor with a higher efficiency unit.

Won't variations like that be the same for any brand of motor at the same load? No, because how efficiency varies with voltage depends on the relative magnitudes of core loss and I^2R loss, how much of the total loss is in the rotor, etc. You may find the same efficiency in two motors, one having higher core loss, the other relatively higher current-dependent losses. Any change in terminal voltage will vary the efficiencies of the two machines by different amounts—perhaps even in opposite directions.

The premium motor design tends to remain the better value. The point is that knowing how much you can expect to save—what return will pay off the premium investment—depends on knowing what actual motor performance will be. Unknown variations in power system conditions can upset that calculation.

Before replacing a motor, consider all the effects of a different driver design. For example, a new motor with lower losses will probably run at a higher full load rpm than the old one. It's to be expected that maximum efficiency requires reduction of all losses, including the rotor I^2R. When that loss goes down, full load rpm will go up proportionately.

How will that affect power consumption? Almost half of all motors in this country drive pumps or fans. The larger the motor size, the higher that percentage rises. Almost all such loads, moving some liquid or gas, require a horsepower varying as the cube of the speed. This can mean a mismatch between motor and load power ratings when the motor full load rpm changes.

Figure 2-5 illustrates this. Suppose full load rpm for a 6-pole "standard" motor is 1180, whereas the "premium efficiency" design runs at 1190. Either motor operates at whatever shaft horsepower corresponds to the intersection of its speed-torque curve with that of the driven machine—in this instance a pump. When the 1180 rpm motor is replaced by the higher efficiency unit, the new motor will therefore supply about 2½% more horsepower than the old motor.

That won't endanger the motor, but it will change the economic evaluation of its performance. An overload of 2½% represents an increase in motor loss of about 3½%. For a typical 100-hp motor, actual power output will rise to 101.5. This means about 1.5 needless output horsepower plus 200 watts higher loss, for which the power system must supply an extra 1.3 kW. If the newer machine is designed for .943 full load efficiency, its true efficiency on this pump will drop to .941 on a 100 hp basis.

This drop in efficiency may seem unimportant. However, if efficiency is valued at $2,000 per kW loss, there is a $3,500 difference between .941 and .943 efficiency. That forces a lengthening of the payback period.

Won't the extra power simply provide more useful output from the pump?

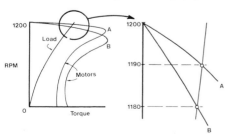

Figure 2-5: When standard 6-pole motor B is replaced with higher efficiency motor A, the operating rpm of a typical pump or fan will shift as shown. Each motor will operate at that point where its speed-torque curve intersects that of the load. Torque output of both motors differs according to $(rpm)^2$, but power output varies as the cube of the speed. Hence, motor A will see a load very nearly $(1190/1180)^3$, or 1.024 times as much as motor B.

Why should it be considered a waste? Unfortunately, when flow increases beyond the design operating point in a fluid system, most of the added energy input will be lost in the form of higher friction, throttling, or head losses.

One solution is to undersize the motor slightly. Instead of 100-hp, for example, order a 75-hp machine. More nearly matching the load horsepower requirement, the smaller motor will run at lower speed than the larger one, cancelling out the increase in shaft output dictated by a higher rpm.

Alternatives include trimming pump impellers, resetting dampers or flow control valves, and moving fan inlet vanes. These options may not be possible in some liquid or gas handling equipment, so be sure to take account of the effects of motor speed differences in all cost comparisons.

Allow for changes in other motor performance, too, such as power factor or starting current. What about accelerating torque—will it be more, less, or the same with the new motor? Do you know what the load actually requires?

Don't compare apples with oranges. One motor supplier may quote "nominal" efficiency. Don't consider that equivalent to someone else's "guaranteed" value. Perhaps when dollar evaluation is important, one should deal only in guarantees. But that's up to the user. Remember, though, that the NEMA nameplate efficiency markings do permit a range of efficiency. And, whatever its value, the nameplate figure is not guaranteed.

Avoid motor modifications that can nibble away at the efficiency improvement. Users have seldom been concerned about the effects on motor performance of an extra shaft seal, special bearing grease, a brake, a larger bearing for severe belt loading, or special paint. Whenever the application dictated such extras, they were added. No one worried about possible reductions in efficiency. But when a single kilowatt of motor loss becomes worth thousands of dollars, adding even 100 watts to the loss cannot be taken lightly.

In all these reviews, be prepared to work with a "sharper pencil," as motor designers have been doing for years, blending test and field experience with computer techniques to tighten up the design process. Users can lose all the benefit of that unless they are equally sharp in checking actual loading, controlling their power systems, and calculating cost trends.

Get all you pay for. "Premium efficiency" motors generally offer longer winding life, greater overload capacity, longer lubricant life, less sensitivity to ventilation impairment or high ambient temperature—benefits resulting from reduced internal losses. Chemical industry spokesmen, for example, have contended that standard T-frame motors broke down too often in some plants because of their lack of design "margin." Higher efficiency, more conservatively designed motors should last longer.

Finally, avoid losing that advantage—and perhaps the efficiency improvement itself—by neglecting maintenance. A proposed revision of the National Fire Protection Association's standard on Electrical Equipment Maintenance puts it this way: "Equipment which is well maintained operates more efficiently." The Committee said further, "Scheduled maintenance . . . such as removing dust and dirt from motors, cleaning of ventilating openings of equipment, and proper lubrication of rotating and other moving machinery also results in money saving and energy saving efficient operation." So, just as the motor designer must work closer to the ideal, so the user must not neglect the "premium efficiency" machine in service—if it's to prove the money-saver it was expected to be.

Test procedures

That expectation often depends on how thoroughly tested a motor design may be. Tests in the field, as well as on the factory floor, are assuming new significance.

Why perform tests on motors? One reason is to make sure a repaired machine is running properly again. But as users become more involved in motor redesign, uprating, speed-changing, and adapting motors to new applications, other types of testing may become necessary. Reworking a motor to save energy by more closely matching the load is an example—likely to be more common in the future—of the need for field testing far beyond a simple check for no-load amps or watts.

Here's a recent illustration from a large Southern textile mill. For years, it was standard practice to rewind the mill's old motors rather than replace them, sometimes more than once. But recently, said the plant engineer, "The changing economic conditions prompted us to re-examine this practice.... A number of new motors were obtained from several manufacturers and were dynamometer tested along with some of our old motors by an independent electrical shop. The results convinced us that there were enough potential savings to justify the investment in a small dynamometer...." The newer motors had higher efficiency.

So in January 1977 his firm bought its own testing equipment, for motors up to 100 hp, 2 or 4 poles. The dynamometer was a small air-cooled friction brake unit costing $2,000. Accurate 3-phase voltage, current, and power metering board cost another $2,350.

Within a year, load tests on 90 motors showed annual power waste in those old rewound units, of obsolete design, totalling as much as $7,000 annually compared to new motors. This was based on power cost of 2.5¢ per kilowatt-hour—half the price of energy in many places today.

Concluded this engineer, "Supported by the above information, we decided to start replacing our burned-out motors with new high efficiency units rather than rewinding." Most firms would be unable to justify in-house test equipment or evaluation. What is it worth? Full load testing is charged for by motor manufacturers at typical list prices ranging from $25 per motor horsepower (at 100 hp) down to $1 per hp at 5000 hp.

We can divide tests into two main categories. First, there are evaluation tests. These tell us simply whether the motor is "good" or "bad," and they include:

(1) Idle (no-load) saturation.

(2) Vibration testing.
(3) Noise measurement.

Second, we have performance tests. These show whether or not the motor is suited to a given application or meets some performance goal. They include:
(1) Losses, efficiency, and power factor under load.
(2) Locked and accelerating torque tests.
(3) Measurement of temperature rise—the "heat run."

Let's consider the methods and equipment needed to perform these tests, particularly in the second category. The basic motor industry standard for most of the above procedures is IEEE Standard 112. If you plan to get into testing that is thorough and accurate enough for customers who may place a value of $1,000 or more on each kilowatt of motor loss, and plan their drives accordingly, you will need to get that standard and become familiar with it. Obtain copies from the IEEE (see page 12 for their address and phone number). Although we will often refer here to various parts of the standard, it's much too long to go through in detail.

The idle saturation test is used to determine the core or iron loss and the friction & windage loss in a motor, as well as to evaluate the no-load or magnetizing current. Run the motor idle or uncoupled from any load, adjusting the voltage through a range from between 15% and 20% up to about 125% of the nameplate rating. At each of the eight or ten values of voltage used, read amperes and watts carefully. A single average value of amps and volts may be recorded at each point, but be sure all three phases are balanced. (See Figure 2-6.) A good test board should have switching capability to read all three phases in succession. Meters should be kept properly calibrated and be of laboratory accuracy—within ½%. Ordinary switchboard meters may be no better than 2%. Depending on the range of expected motor voltages, you will need a proper set of ammeter shunts, along with current and potential transformers.

Although the board or console setup need not be elaborate, the instrumentation is not cheap. This kind of work is no place for quick checks with clamp-on meters. How to use the data is explained in IEEE 112.

Noise tests are so specialized and can become so complex, besides being closely dependent on the shop environment, that it seems unwise for most users to attempt work in this area. Relatively inexpensive instrumentation is available, but knowing how to interpret the readings can be tough even for acoustical engineers. So we'll say no more about it here.

That brings us to performance testing. From the idle test we find two of the machine losses. Load testing gives us the other three, at whatever load is desired. This may be rated horsepower, or some percentage of it anywhere from 25 to 125.

As IEEE 112 points out, we can arrive at those losses in two basic ways: (1) by the so-called segregated loss method, in which certain measurements are combined with calculated motor equivalent circuit data, or (2) by directly loading the motor and measuring input electrical quantities and output power.

Efficiency and Power Factor 55

Figure 2-6: Non-I^2R losses are obtained from idle test in that motor input power is measured at various voltages. Small amount of power used in stator I^2R loss produced by idle or no-load current can normally be neglected. Points are plotted as shown in the graph (top). Total non-I^2R losses at rated voltage can be read from curve at rated voltage line. However, finding the friction and windage loss component is not so easy, because curve must be extended down to zero volts. This is subject to much variation in plotting. The solution is shown at bottom. Plot same kilowatts versus voltage squared instead of voltage to the first power. Most of the curve will now be a straight line, with several points close together in the low voltage range. Extending this straight line down to zero volts is easy and gives a good value for friction and windage loss. Subtracting that from total non-I^2R loss gives core loss.

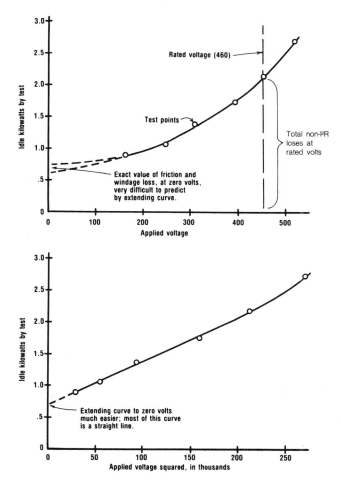

Figure 2-7: Typical small electric dynamometer test stand in operation, showing multi-channel temperature recorders at left and right.

Figure 2-8: 4250-hp 2-pole motor on test using 5000-hp water absorption dyne, left.

Most common is the dynamometer, either a straight electric type (Figure 2-7) or a water absorption type. (See Figure 2-8.) A typical small electric type on the market can handle up to 100 hp, 3600 rpm for about $1,400. Water absorption types are common in the 500- to 1250-hp range; 5000 hp is about the largest in use today, with a price tag in the $50,000 range. Secondhand equipment, especially electric, may be available for much less.

Dyne accuracy is generally in the 3% to 6% range, the lower number applying to smaller sizes. However, proper test procedure, as outlined in IEEE 112, can improve this. For best results, dyne capacity shouldn't exceed twice the rating of the motor on test.

The water absorption dyne is basically an eddy current brake. A relatively low-power d-c field coil magnetizes the dyne. When the test motor drives the dyne rotor, this field sets up a retarding torque, tending to slow the motor down. As d-c excitation is raised, braking torque increases with it.

The dyne converts all the rotational energy supplied by the test motor into heat, which must usually be carried away by cooling water. Assuming output water temperature to be limited to 140°F, such a dyne will typically require .08 gallons of water per minute per horsepower. A 1250-hp dyne test would therefore need 100 gpm of water. To adapt the dyne to a wide range of horsepower, calibrated flow restrictors and automatic valves are used to vary the flow as needed.

Unless there is some use for the hot water, the energy used to heat it (paid for by the charges for test motor kilowatthours) is entirely wasted. This type of dyne, however, is simple to control.

The electric dyne may be a separately excited d-c generator using fine field control for accurate loading adjustment. The d-c power generated during the test will usually be wasted unless a d-c to synchronous a-c motor-generator set is available to return it to the line.

Both electric and water absorption machines must have their stators trunnion-mounted to allow rotation through a small angle. This permits torque developed by the test to move the stator, through some kind of arm, against a scale to register force. Newer units may employ transducers or "load cells" instead of scales, to convert the rotational force, or torque, directly into a digital electronic display.

Another electric loading method is to use an induction machine (preferably one identical to the unit being tested) supplied from a variable frequency source. Shafts of both machines are coupled together. This is the "Method C" outlined in IEEE 112. When the driven machine is energized at a frequency slightly below 60 Hz, the test motor will be driving it above synchronous speed, causing the driven machine to generate. Power absorbed from the test motor usually is returned to the system (minus machine losses) as shown in Figure 2-9.

If it isn't feasible to change stator frequency on the drive unit to produce generator action, then you can get the same effect through overspeeding its rotor by belting the two machines together so the two rotors operate at differ-

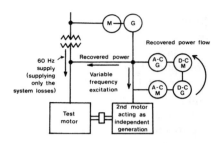

Figure 2-9: Pump-back or duplicate machine test. Variable frequency set may be d-c or wound-rotor motor driven; supplies only excitation or magnetizing kVA for 2nd machine. M-G sets in recovery loop must be able to handle full load power from test motor.

ent speeds. Such testing was common many years ago. This requires precise control of pulley diameters, however, and many motors—particularly 2-pole units—are not designed to carry belt loads.

You may also apply reduced voltage to an induction motor at normal line frequency, then use the test motor to drive its rotor in the reverse direction against the field. This has the disadvantage that the rotor in the "dyne" machine develops about twice its normal full load loss and heating, so the unit must be oversized. A further drawback is that the power output from the test motor is wasted as heat.

One of the more unusual versions of the "feed back" or generating electric dynamometer was developed by a service shop for a large rewind project for which a number of large motors had to be load-tested. Instead of supplying the loading unit with reduced frequency power to permit it to act as an induction generator, the same effect was created by slowly rotating its stator in the direction opposite to physical rotation of the test motor. This "slowed down" the machine's rotating magnetic field just as reduced frequency would have done.

This "dyne" was a secondhand induction motor, reworked so its stator could be revolved by a d-c powered chain drive. Stator electrical connections were made through slip rings and brushes. General drive arrangement appears in Figure 2-10. This is relatively economical. The d-c chain drive gives great flexibility in loading. This particular dyne was wound for use at two speeds and two voltages; speed matching to a range of test motor polarities could also be accomplished through gearing.

Such an electric dyne setup is a power saver, unlike the water absorption machine, which dumps all the test motor power input down the drain. Thus, enough kilowatthours may be saved while doing a large amount of testing to justify considerable expense for test equipment.

How much power is enough?

That brings up the important question of power supply for tests. Large sums of money have been spent equipping test departments for work that could not be performed because of an inadequate power source. Table 2-II gives some idea of what is needed.

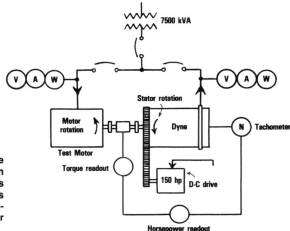

Figure 2-10: Feed back dyne arrangement using induction motor with rotatable stator as dynamometer. Large arrows (above coupled machines) indicate directions of a-c power flow during test.

HP at 1800 rpm	kW for rated load test	Full voltage kVA for locked or speed-torque test	Full voltage kW demand on locked test
50	43	350	80
100	83	650	145
500	410	3000	650
1000	790	5500	1220
5000	3900	25,500	5600

Table 2-II. Power and kVA requirements for typical motor testing. The figures will not vary much with voltage rating.

Look at the kilowatt demand that can be pulled. The utility rate structure or other regulation may permit large test loads only during off-peak periods.

For example, these figures applied to one large motor test floor:

(1) Normal plant demand: 3600 kilowatts, the average of peak demands over a four-week period.

(2) Monthly charge for this demand: $2.60 per kilowatt, or a total of $9,360.

(3) Maximum capacity of incoming lines: 8000 kVA.

(4) If a motor of about 1500 hp were to be given a locked rotor test during a peak demand period, the kilowatt demand would be about .20 power factor times 8000 kVA, or 1600 kW. That week's demand would then become 3600 + 1600, or 5200 kW.

(5) Assuming the ensuing three weeks were at the normal 3600 kW level, the total for four weeks would be 5200 + 3(3600), giving an average for the four weeks of 16,000/4, or 4000 kW. Demand charge would be $2.60 times 4000, or $10,400 instead of $9,360. Excess cost for this one test: $1,040. Tests of smaller machines would draw lighter penalties, of course.

Temperature tests take fewer kilowatts than locked tests but for a much longer time—typically from four to six hours. So demand scheduling is important there, too. (Besides, the energy is expensive. Power for a 5000-hp motor heat run lasting six hours may cost $2,500, up 400% in the past few years and likely to keep rising. Hence the economic advantage of a feed back dyne.)

Whatever the system capacity, two important power supply accessories are needed. One is a transformer of adequate capacity to sustain motor voltage at a reasonable level during locked or acceleration tests when inrush current is being drawn from the line. If only 460-volt motors are to be tested, this might not be needed. But for a range of voltages, transformer taps are required.

There also needs to be voltage control. The taps, easily changeable, or an auxiliary buck-boost autotransformer as in Table 2-III, or a series reactor may permit setting the voltage and holding it there throughout a lengthy test.

Reactors, like transformers, must be properly sized. Too small a unit can saturate and distort waveform, throwing results off. Reasonable transformer ratings are indicated in Figure 2-11.

As with idle testing, load testing requires the same capability to accurately read amps, volts, and watts—on a three-phase basis—plus speed. An electric

Rated motor HP	Transformer kVA rating, min.
50	150
100	300
500	1500
1000	2500
5000	10,000

Table 2-III. Size of supply transformer to maintain at least 85% of rated voltage during across-the-line locked rotor test. These figures are approximate in several ways, being rounded up to the nearest standard transformer ratings and dependent on the stiffness of the power system ahead of the transformer, as well as on the transformer percent impedance. Nevertheless, they can be useful as a general guide.

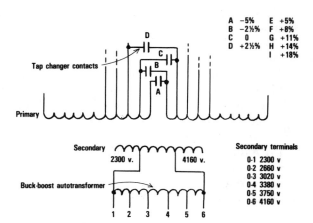

Figure 2-11: Test transformers arrangement that will provide test voltage in series with the secondary can provide further "fine tuning."

tachometer on the dyne shaft is usually the best way to measure rpm. Slip speed is used to derive rotor loss in the test motor. The average resistance between motor terminals, read by a bridge—not an ohmmeter—is used along with load current to derive stator I^2R loss. The dyne must provide one additional item needed to get total loss and efficiency: the motor output torque. Besides the scale or load cell options, this can also be obtained (as in Figure 2-10) from a torque transducer fitted to the shaft.

Stray load loss

This torque and rpm determine the horsepower being supplied to the dyne. We have already arrived at four of the five motor losses. The remaining one—stray load loss—is the difficult one.

For years, European test standards assigned this an arbitrary value, usually one percent of output on a kilowatt basis (one horsepower output equals .746 kilowatts). A recent U.S. standard for power plant motors, used by public utilities, takes the same approach. (See Figure 2-12.)

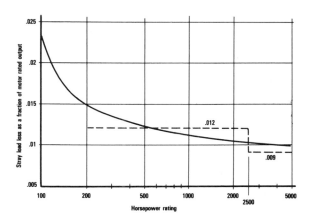

Figure 2-12: The solid curve is a reasonable plot of stray load loss versus hp for induction motors. Use it if no other data are available, primarily when segregated loss tests are performed. The dashed line shows the values adopted in 1977 by a national standard for power plant motors (ANSI C50.41). As an example, for a 500-hp motor, stray loss may be taken as 500 times .746 times .012, or 4½ kilowatts.

There are special tests, with special calculation procedures, which some engineers believe can produce more accurate stray loss values. But those methods are seldom practical in the field.

Now, at any load—determined by adjusting dyne load until torque and speed give the desired motor output—we simply add all the losses to the output in kilowatt terms, and we have the input. (That input can of course be read directly from the wattmeters but is normally a large number compared to total losses. Thus, efficiency as output divided by input becomes the ratio of two very similar but large numbers, and the result is not very accurate.)

Then we may say:

$$\text{Percent efficiency} = 100 \left(1 - \frac{\text{kW loss}}{\text{kW loss} + \text{output}}\right)$$

Use the same "input" (output + loss) figure, with amps and volts, to calculate the motor power factor at the same load.

For general performance evaluation, take readings at loads of ¼, ½, ¾, ¾, ¾, and ¾ load—from 25% to 150% of rated motor hp. Then you can plot a curve of efficiency and power factor versus load to best fit the individual data points.

Perhaps the user is interested in efficiency at only one load point. If only that point is checked, you lose accuracy, because there's no way to plot the probable curve of efficiency versus load and see how well that one point fits the curve. So to minimize inaccuracy several readings should be taken.

What if no dyne loading device is available? The IEEE 112 "segregated loss" method will be usable if you have enough motor design data. In this method, rotor I^2R must be derived by finding a value for rotor resistance, R_2. The standard way of doing this is with a locked impedance test at low frequency, normally 15 Hz. It is recognized that this frequency is still so high that the test will not accurately deal with a "deep bar" rotor as commonly used in large motors today. But it seems the best compromise.

How large a 15 Hz source is required? At this frequency the test motor impedance will be only about ¼ as great as at 60 Hz. So you would expect rated motor full load current to flow, locked rotor, at a voltage about ⅙ times ¼, or 5% of the rated value. The total 15 Hz kVA would then be only 5% of the test motor kVA rating.

However, the lowest readily available alternator voltage would probably be 440. Hence, for a 2300 volt motor test, the 15 Hz alternator rating would be 5% times 440/(2300 × .05), or at least 15% of the test motor kVA. Kilowatts supplied during the test would be somewhat lower.

Locked rotor tests, like load tests without temperature readings, take only fifteen to thirty minutes. The power demand doesn't last long. But even with a well planned power supply, the high currents needed may make it impossible to sustain reasonable voltage. Large machines may have to be given locked or acceleration tests at deliberately reduced voltage. In that case, assume that motor kVA drawn from the line varies as (voltage)$^{2.1}$. (See Chapter 3.) The minimum practical voltage for a useful locked test is about ⅓ of the rated value. Take several readings (torque, amps, and volts) from there up to as high as you can get. Plot these on logarithmic coordinate paper as in Figure 2-13, then extend the line upward to rated volts to get the proper answer.

Locked and accelerating torque

How is locked torque measured? The dyne may be arranged for locking the shaft, or at least for holding the motor down to a slow roll, which is good enough for this test. If no dyne is used, the rotor must be locked by either a

Efficiency and Power Factor 63

Figure 2-13: How to properly derive full voltage locked rotor current from a test at reduced voltages. Here are the test readings:

Volts	Amps	
900	208	(500 hp 1780 rpm
1150	265	2300 v. motor)
1400	340	

Simply ratioing the 340 amps to full voltage would give 555 amperes. From the graph, we get 570; full voltage test shows the correct figure is 572. Even larger discrepancies will apply to locked torque readings unless data are plotted as shown.

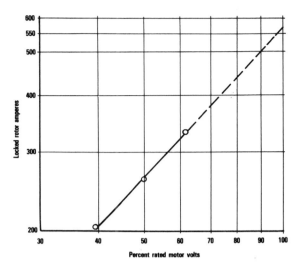

friction brake (for small motors) or a coupled torque arm or beam and scale. A 3- to 6-foot arm will usually keep scale force within reasonable limits. The beam must be designed to withstand the maximum locked rotor torque of any motor you may want to test, plus a generous safety factor. Figure 2-14 shows some typical values.

Properly fitted couplings are needed to attach this beam securely to whatever size motor shaft extension may be encountered. Some shops have used V-block clamps to tighten the arm on the shaft. But slippage and shaft damage

Figure 2-14: Assuming a typical motor has 125 percent locked rotor torque (a high figure for large machines), a test torque arm would have to carry at least these pound-feet values at rated voltage.

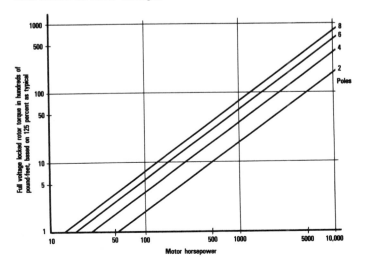

may result. A close-fitting, keyed sleeve, designed for at least as much torque as the beam itself, is preferable.

If a vertical motor is to be torque-tested, a conventional platform scale may be used to read torque only if the beam is fitted with a right-angle crank to change the direction of the transmitted force from horizontal to vertical. It may also be possible to use a transducer on an upright post instead of a scale.

There are several ways to test torque at intermediate speeds. Normally, the best way is on a dyne. Adjust the load carefully until it just matches what the motor can produce, then take readings of torque, current, and voltage as nearly simultaneously as possible, at half a dozen speeds between breakdown and locked rotor. This will generate the entire speed-torque curve for the motor.

Unfortunately, it is seldom possible to get readings between the breakdown point and about 75% speed. Motor torque changes so rapidly with rpm in this region that you can't "hold" the motor on a given speed long enough to get data.

When dyne testing isn't possible, most of the speed-torque curve below that 70% to 75% speed region can be derived in other ways. One is with the plotting equipment shown in Figure 2-15. The uncoupled motor is accelerated at a voltage chosen to bring it up to speed within five to fifteen seconds. A tachometer signal fed into the plotting circuit is electronically converted into

Figure 2-15: Speed-torque curve plotter in use during test on 2500-hp motor. This equipment costs about $4,500.

rate of change of speed with time, which is acceleration. Because torque is equal to that acceleration multiplied by a constant, the plotter can thus produce a display proportionate to torque versus speed. This displayed curve is calibrated numerically by measuring actual locked rotor torque at the same voltage. To smooth out transient disturbances during the first instants of start-up, the motor is usually started from a few rpm in the reverse direction, so torque is easily readable as it passes through zero rpm.

Such equipment is sophisticated and expensive. Another way of doing the job is by measuring or calculating the inertia of the motor rotor and then accelerating the uncoupled machine at reduced voltage. Use careful measurements or a recorder to plot speed versus time as the motor comes up to full rpm. (See Figure 2-16.) The rate of change of speed from this plot is used to calculate torque:

$$\text{Torque in lb-ft} = \frac{(\text{Inertia, lb-ft}^2)(\text{change in rpm})}{(308)(\text{time interval during which the change in rpm occurred, seconds})}$$

The smaller and more frequent the time intervals used, the more accurate the resulting speed-torque curve will be.

Many other plotting arrangements have been used, including oscilloscope circuits. Most have been worked out, built, and used by motor designers in their own plants. The components may be commercially available, but complete systems are not.

Temperature

For heat runs alone on machines (especially verticals) where no other loading method is possible, the two-frequency, or "equivalent load" test works well. This procedure, standard in many parts of the world, is included in the most recent revision of IEEE 112 as an option. Temperature rise from such a test will be a few degrees higher than from a dyne test, because the motor winding sees a distorted waveform which raises both core loss and rotor I2R.

Here's how it's done. The duplicate-machine method and the special arrangement of Figure 2-10 (page 59) both use the same principle: The dyne

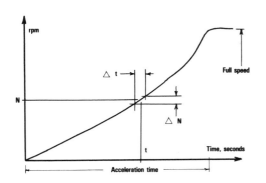

Figure 2-16: Plot of speed versus time during idle acceleration. Average torque at speed N is obtained from formula in text using speed change ΔN during the time interval Δt.

acts as an induction generator because the rotation of its magnetic field is made slower than that of its rotor, driven by the test motor. In the two-frequency test, the same effect results from mixing rated frequency with a lower frequency (usually 50 Hz) in the line feeding the test motor. The power to magnetize the machine comes from the 60 Hz line; the power to supply the losses comes from the 50 Hz source. In effect, the motor rapidly switches back and forth from motoring to generating, alternately accelerating and decelerating in response to the two frequencies. By adjusting the 50 Hz voltage, it is possible to draw rated full load current and therefore heat the motor to "full load" temperature. Losses and efficiency cannot be determined reliably from this test because of the distortion in magnetic field waveform.

The 50 Hz source puts out 15% to 20% of the test motor's rated line voltage. This lower frequency may be combined with the 60 Hz supply in two ways, as shown in Figure 2-17. At (a), a mixing transformer is used. This has the advantage of simpler windings for the 50 Hz generator; different voltage ratios, and higher voltage insulation to suit line-to-ground potential on the test motor, can be taken care of in the transformer itself.

At (b), mixing is done by passing 60 Hz line power to the test motor through the 50 Hz generator windings in series. No transformer is needed. But if the test motor is rated 4000 volts, for example, the generator windings must be insulated for at least 4 kV to ground. They must also be capable of carrying rated test motor amperes.

Theoretically, as Figure 2-18 shows, the 50 Hz supply needs to be sized for only a fraction of the test motor kVA if a transformer is used. However, too small a source may lead to instability, with hard-to-control absorption of generated power from the test motor. If a smaller generator is used, it should have fairly high inertia to remain stable—perhaps with a coupled flywheel. This may have to be determined by experiment.

The only other precaution to be taken in this test, besides using meters that will properly display RMS values of the distorted voltage and current waveforms, is to watch for damaging vibration caused by torque harmonics. Vibration is especially likely with vertical motors because they are unsupported at the top end.

For temperature measurements during any heat run, a Kelvin bridge is necessary. Use it to measure winding resistance or to monitor embedded temperature detectors if they are present. If the detectors are of the 10 ohm type, however, bridge connections should provide lead compensation. (See Chapter 6.)

Thermocouples must be installed for test if there are no detectors to indicate when temperature rise has levelled off enough to stop the test. It's handy to have a multi-channel recorder or display instrument to monitor several thermocouples at once.

Whatever the testing involved, a solid support for the test motor and any connected loads is a must. The vibration of a large coupled drive can quickly become destructive. Short evaluation tests, unloaded, may be no problem.

Efficiency and Power Factor

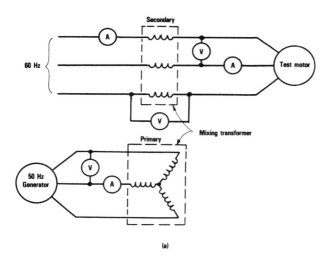

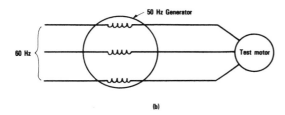

Figure 2-17: Circuits for two-frequency heat run.

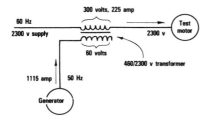

Figure 2-18: Volt-ampere relationships during typical two-frequency test of 1000-hp 2300 volt motor.

Longer tests, though, particularly with full load heating, can exaggerate slight misalignment and lead to bearing or shaft damage, if nothing worse. Without a precise, rigid foundation to which all components can be solidly clamped, proper alignment cannot be held. Floors that are strong enough may not be stiff enough.

Though by no means complete, this description of test procedures and equipment gives some idea of the complexity of accurately determining motor performance. When that performance takes on great economic importance, as is common today, anyone purchasing or applying motors must be prepared to evaluate a motor manufacturer's test capabilities as carefully as price or delivery has been compared.

The importance of power factor

Although energy-conscious motor users today are putting the spotlight on efficiency, savings in plant distribution equipment and in electric bills are just as likely to result from higher power factor.

Is it true, as one sometimes hears, that an electric motor with a higher power factor will use much less energy than one with a lower power factor, and that power factor relates to electrical energy in a way similar to efficiency? Don't you believe it. Yes, there can be great benefits in power factor improvement in a-c motor circuits, but reduced kilowatthour cost to operate the motor is not among them.

As interest grows in motor efficiency and operating cost, such misconceptions about power factor continue to crop up. These ideas can lead to wrong choices in motor selection or operation, and the resulting failure to achieve expected energy cost savings.

What do we mean by "power"? Power is defined as the rate of doing work. All work requires the expenditure of energy. (More accurately, all work requires the conversion of energy from one form into another.) Thus we can say that whenever we use energy to do work, the process involves power. This power is always "real" in the sense that real water is being pumped, real heat is being generated, or a real shaft is being turned.

In a-c electric circuits, however, the use of such terms as "real power," "watted power," "apparent power," "true power," and "reactive power" seldom aid in understanding what's really happening in the circuit. Those terms arose because from the earliest studies of alternating current it was clear that "volts times amperes" did not equal "watts" (i.e. was not the same as the rate of actual work being done in the circuit) as had always been true with direct current. The product of a-c volts and a-c amperes, therefore, was called "apparent power"; it appeared to be power, like d-c volts times d-c amperes, but really wasn't.

What, then, does "power factor" really mean? It's merely a numerical way of expressing phase difference between voltage and current in an a-c circuit. It also expresses the difference between power (watts) and the product of volts and amperes, or "volt-amperes."

Let's look again at the relationship between alternating current or voltage—the familiar "sine wave" time variation—and the "vectors" used to represent such quantities. In Figure 2-19, we see a typical a-c voltage/current relationship.

Look at the power factor angle. It tells us the phase difference between volts and amperes in that particular circuit. The larger that angle, the "worse" the power factor, meaning that the cosine of the angle becomes smaller. As the circuit power factor "improves," the angle gets smaller and its cosine gets larger.

What produces such a phase difference, and what does it mean? A phase difference is simply a time lag between the rising and falling of a cyclic voltage (or current) in one part of a circuit, and the rising and falling of another voltage (or current) in another part. Such differences result from the two circuit properties of capacitance and inductance. Because all circuits involve conductors, they all contain some capacitance, a characteristic of any two conductors separated by insulation. And because any current-carrying conductor surrounds itself with a magnetic field, all circuits contain some inductance.

As a simple way to visualize the effects of these properties, consider first the inductance, a good example being the coils in a motor winding. At the instant an alternating voltage is first applied to a coil, it is unopposed by any internally generated coil voltage or "back e.m.f." So its full value appears across the coil terminals. Current begins to flow. As the current builds up, it generates a growing magnetic field within the coil that acts to oppose the applied voltage. Because the applied voltage is cycling (alternating), the current will do likewise but always later in time than the voltage, because of the delay, or "magnetic inertia," resulting from the buildup or decay of the the coil's magnetic field.

Thus, current builds up to the peak value of its cycle after the voltage has already peaked. We say that the current "lags behind" the voltage—there is a "phase difference" with the current lagging. In a perfect inductance, the lag

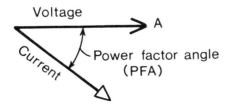

Figure 2-19: Voltage and current vectors in an a-c circuit containing inductance. These vectors rotate counterclockwise at a speed dependent on frequency. This is a useful way to visualize the rise and fall of alternating volts or amperes with time; an observer at Point A, facing the center of rotation, would see the voltage rising as the vector rotates from the position shown, falling again, then peaking in the opposite direction. The current, lagging behind the voltage, would appear to reach its corresponding peaks or zeros a little later.

would be 90° of angle for the current vector, or one-quarter cycle of the current wave. (See Figure 2-20.)

In a capacitance, the opposite is true. Current does not flow "through" the insulating material separating the plates in a capacitor. When the circuit is first closed, current must flow into the capacitor to "charge" it before a terminal voltage can develop. Hence, current into the capacitor must "lead" the voltage or "be ahead in time phase" of it—again, by 90° or one-quarter cycle. (See Figure 2-20.)

Because applied voltage alternates, or cycles back and forth in polarity according to system frequency, this process repeats itself continuously as long as the circuit is closed. The phase relationship remains unchanged, current lagging voltage in the inductive circuit and current leading voltage in the capacitive circuit.

Now consider the power or energy relationships in those two kinds of circuits. Whatever the current or voltage might be, in a perfect inductance or capacitance no energy would be lost. There would be no work done, nor any power "wasted." In real circuits, however, energy is lost in the movement of iron molecules as cores are magnetized and demagnetized. All real conductors have some resistance, so that current of any kind—whether leading or lagging the voltage—produces I^2R heat loss when flowing through any resistance. All "non-conductors," as in the capacitor dielectric, undergo some electron movement when the a-c circuit is energized. That too produces heat loss, corresponding to the behavior of a resistance.

In a resistance, no phase difference exists between current and voltage. (See Figure 2-21.) Hence, when the effects of resistance are combined with those of inductance, as in actual motor windings, the phase difference will be somewhere between 0 and 90 degrees, as in Figure 2-19.

Resistance and inductance may be connected in parallel or in series; either way, the same "current lagging" phase relationship will exist. In the parallel circuit of Figure 2-22, there can be only one common voltage E across both branches of the circuit. The currents in the two branches will therefore be out of phase with each other, as shown. In the series circuit of Figure 2-23, the current must be the same throughout, so the voltages across the two components will be out of phase. But the total current I lags the total voltage E in both circuits.

The product of volts times amperes will represent actual power—energy consumption—only for the resistive part of the circuit. That is, watts, or kilowatts. In the purely inductive (or capacitive) part of the circuit, the product does not signify energy usage at all. It is properly termed "volt-amperes reactive"—"vars," or kilovars.

Because the watt and var products are derived from current and voltage vectors separated by the power factor angle, the same current/voltage triangles of Figures 2-19, 2-22, and 2-23 can be redrawn to a different scale of units to represent a "power triangle." (See the example in Figure 2-24.) This triangle is most useful in evaluating power factor of a power system or of a

Efficiency and Power Factor　　　　　　　　　　　　　　　　　　　　　　　　71

Figure 2-20: In a perfect capacitance (near right), current I_C leads the applied voltage by 90°. In a perfect inductance, current I_L lags the voltage, again by 90°. At any given point in the alternating current cycle, therefore (far right), capacitive current and inductive current—when the two circuit elements are connected in parallel—are exactly opposite in direction, or "180° out of phase" with each other.

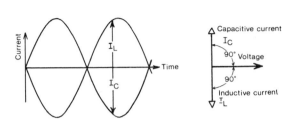

Figure 2-21: In a perfect resistance (left), there is no phase difference—no lead or lag—between current and voltage.

Figure 2-22: Top right: In a parallel circuit containing inductance L and resistance R, the same voltage exists across each. The inductive current I_L lags that voltage by 90°; the resistive current I_R is in phase with the voltage. The resulting vector diagram appears at bottom right, with I lagging E (as it does in Figure 2-19). The inductive or "reactive" current I_L is not some different kind of current; an ampere is an ampere. It is simply current that is "out of phase" with the voltage.

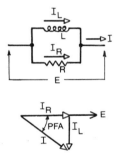

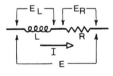

Figure 2-23: Top left: If R and L are reconnected into a series circuit, there can be only one current through both. But that current, flowing through L, must still lag behind the voltage E_L, so the vectors will now appear as below, left. Voltages E_l and E_R combine to form the total voltage E, and again I is lagging (though the individual current and voltage magnitudes will not be the same as they were in Figure 2-22).

Figure 2-24: To a different scale of units, the current/voltage triangle of Figure 2-22 (above) can be considered a "power triangle." But in fact only the watts—volts times I_R in the resistive part of the circuit—represent power. The vars appear in the inductive portion of the circuit only, and do not represent power, energy, losses, or work done within the machine.

Total volt-amperes are what would be measured at the motor terminals.

"Watts" is associated with "watted" or "working" current L_w, equivalent to the I_R of Figure 2-22. "Vars" is associated with inductive current I_L.

piece of equipment, such as a motor. As in Figure 2-19 or Figure 2-22, the power factor angle separates watts and vars; the power factor is the cosine of that angle.

Clearly, then, a lower power factor in an electrical device, such as a motor, in no way implies a waste of energy. Whatever energy transfer there is occurs only in resistive portions of the motor circuit.

Plant power factor

Another misconception is that the overall plant power factor (PPF) somehow influences the behavior of a motor, including its efficiency. One expression of that view came from the chief design engineer in a manufacturing plant. In trying out a special motor, he determined that its line current was "substantially" different from the motor manufacturer's tested value. This was his explanation for the apparent discrepancy: "Our plant power factor is approximately 72% lag, and we believe that your motor ratings were made from laboratory tests with 100% power factor input."

But that simply isn't possible. In the first place, the power factor within a 3-phase motor depends only on its internal design and on the magnitude of the applied voltage (assuming the voltage to be balanced). Remember that the motor is simply an impedance—a combination of resistance and inductance, R and X—which is paralleled with other impedances (loads) on the plant power system. Motor current will assume a phase angle (power factor angle) with respect to system voltage depending only upon its own R and X values, which vary only with shaft load. Thus, the motor power factor angle helps fix PPF, but the reverse can never be true.

In the second place, it's impossible to test a motor in a laboratory, or anyplace else, with "100% power factor input." Because a motor, like any other wound device, always has both R and X, it can never of itself operate at 100% power factor. Therefore the circuit supplying its power cannot do so either.

Similarly misleading is this quotation from a 1979 magazine article on the selection of more efficient motors. Said the author, "[Plant power factor] also has an effect on motor efficiency.... Adding capacitors to improve the plant's power factor can, at the same time, dramatically increase motor efficiency."

As we have seen, however, PPF as such does not influence current or losses in any motor running on that system. There's no way PPF can do anything for motor efficiency. Of course, if a low PPF raises total amperes in the distribution system to a point where motor terminal voltage is significantly reduced by drop in the feeders, then motor efficiency can certainly suffer. (Winding temperature will go up, too, so that nuisance tripouts, rather than lower efficiency, may become the main problem.) But it's low voltage that's the problem then, not the PPF as such.

Related to the notion that PPF influences motor behavior directly is another misconception: that motor efficiency is directly affected by the motor's own internal power factor. Several years ago, for example, an author claimed:

"Power factor relates to electrical energy in a way similar to efficiency when discussing mechanical energy.

"Because of losses resulting from friction, etc., the mechanical power delivered from a machine is never equal to the power put into it. In a similar way, we could state that it is always necessary to supply more energy to a machine than the machine can deliver in terms of work. . . . Power factor can be considered as the wasted or non-used electrical energy which must be delivered to an electrical system. . . . "

No way. Power factor neither equals nor represents energy at all, wasted or otherwise. An electric motor is an energy conversion device. It changes electrical energy into two other forms: heat and mechanical work. Friction, I^2R, and iron loss represent most of the heat. The shaft power output represents most of the mechanical work. Power factor as such does not enter into those energy conversions. The energy supplied to a low power factor machine, with the exception of a little more I^2R loss in the windings because of the higher magnetizing current. Otherwise, higher amperes do not represent greater energy demand from the power system. Current is not energy; watts represent energy.

Or consider this quotation from an electrical contractor: "A-c motors, 10 hp and down, [today are] designed to use much less energy. For example, 10-hp motors, 230 volts, single-phase, normally use 47 to 70 amps. Some that we connected in 1978 used only 39 amps." Again, the number of amperes drawn from the line does not necessarily indicate the amount of energy a motor is using up in losses. That's measured by watts, not amperes.

Redesigning

It's sometimes erroneously assumed that any motor redesigned for high efficiency will have higher power factor as well. But there is no assurance that it will. Consider how the motor manufacturer might make such a redesign. If the magnetic densities, core loss, and inrush current are relatively low, he could "strengthen" the machine by reducing the effective winding turns. Doing so lowers stator and rotor losses (fewer turns, of larger wire, rapidly decreases "R" in the I^2R losses). The result may be significantly higher efficiency—but the power factor will go down, because magnetizing current, which is almost entirely reactive, will go up.

Or he could choose to put in more copper using a larger stator slot. A larger rotor slot would lower I^2R loss in the cage. Though likely to raise efficiency in most instances, both measures would increase magnetic density and drive the power factor down.

If stray load loss is important, the air gap could be increased. This, too, means reduced power factor.

Where existing magnetic density won't permit bigger slots, core stack length could be raised to compensate for such efficiency improvements. This would permit the power factor to remain unchanged, or to rise. Non-electrical efficiency improvements, involving bearings or fans, would not measurably influence power factor one way or the other.

What then can be expected when a service center "upgrades" a motor in the field, making changes only to the winding? As would be true at the factory, efficiency is likely to go up and power factor down. Figure 2-25 shows this situation. Magnetically, the effect of upgrading by reducing turns is comparable to increasing voltage, and vice versa. If the horsepower rating is raised at the same time, which is frequently done, both efficiency and power factor may go up together. As useful output increases while losses do not (that's the reason for higher efficiency), we also may find I_W going up faster than I_L. (See Figure 2-24.) That will raise the power factor.

Wait a minute, you say. If we drop turns, doesn't the higher magnetizing current flowing through the windings raise I^2R loss so efficiency will drop? Not necessarily. In the stator, although more magnetizing amperes does raise IL, most of the "I" in stator I^2R loss is useful load current which undergoes little or no change. And the "R" value goes down rapidly, because not only are there fewer turns, but the wire size has increased as well. In the rotor, a magnetically stronger design lowers the slip, meaning lower effective rotor resistance, so the rotor I^2R decreases too.

So it's difficult to generalize. Power factor and efficiency do have some interrelationship. But changing one does not change the other automatically in the same direction or to the same extent.

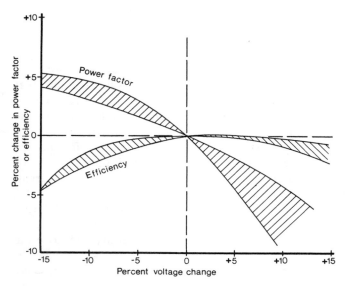

Figure 2-25: How typical motor power factor and efficiency (below 250 hp) may vary with motor terminal voltage. Adding coil turns to raise power factor has an effect similar to reducing the voltage—and will not necessarily raise efficiency. Reducing turns, equivalent to raising the voltage, will sometimes improve efficiency (an effect augmented by raising the horsepower rating) but makes power factor lower.

Capacitors

What about the power factor correction literature that says you can "reduce current demand" by connecting capacitors to the circuit? If the motor draws less current from the line because of its improved power factor, won't motor I^2R losses go down?

This, too, is a misunderstanding. When capacitors are connected to a motor feeder circuit, the current drawn by that feeder from the plant system does decrease, but the motor itself continues to demand the same current and has the same losses it had before. Some or all of the motor's magnetizing current is now being supplied from the capacitors, not from the feeder. The current flowing within the motor windings, however, hasn't changed a bit—and neither has motor efficiency.

Of course, we don't get something for nothing here. Capacitors aren't magic. How can they provide current while the feeder supplies fewer amperes than before? It's the "time phase" property of a-c circuits that provides this benefit. Current flows toward the capacitors during those portions of the cycle when the motor magnetizing current is flowing away from the winding. Later in each cycle, when magnetizing current reverses, the amperes flow from the charged capacitors to the motor.

Look at it another way. In Figure 2-20, properly matching capacitor and inductance sizes would result in reactive current into the capacitor which is always equal to and opposite in direction from the inductive current (assumed to be motor magnetizing current for this example). Thus, as far as an external power supply system is concerned, the two reactive currents cancel each other, resulting in unity power factor. If the capacitor is somewhat smaller than that, some net "inductive amperes" must still be supplied by the system to the motor, and the power factor is no longer unity, but has a small lagging value.

The capacitor, then, really isn't "manufacturing" amperes. Rather, by providing a 180° phase difference with the motor's inductive magnetizing current requirement, it is effectively neutralizing the need for that magnetizing current from the system.

If power factor improvement doesn't help cut the energy cost to run the motor, then what good is it? The answer is that the I^2R losses caused by current flow in the supply circuit will be reduced if motor current demand from that circuit is lowered. The cables, transformer windings, switches, and bus bars all have resistance, which in each instance is fixed. When the connected loads draw lower currents through these resistances because of improved load power factor, wasted energy throughout the power system will be reduced—though not within motors themselves.

There may even be a reduction in the size of system components (which are generally rated in terms of the total amperes they must carry). That extends clear back into the utility system. For this reason, most utilities offer users financial incentives to raise plant power factor.

These incentives, however, are extremely varied and complex. Some utility rate schedules contain a "power factor clause" which must be "negotiated" depending upon plant type and size. Others will exact a 1% boost in demand charge for any power factor below a limit such as .95, .90, or .80, or offer a 2% cut for any power factor above such a limit. Or, both demand and energy charges may be boosted by 10% for a power factor below .85. Many other versions exist. In some localities there may be no utility penalty at all for low PPF.

What that means is that a careful study by an engineer is the proper way to decide whether power factor should be improved or not, and if so, by what method. The choices include:

(1) Individual capacitors connected directly to each motor on the system. This can be costly and space-consuming. (See Figure 2-26.) It also requires re-examination of starter overload ratings for each unit. Moreover, for high inertia drives or motors subject to frequent circuit opening and reclosing, directly-connected capacitors may produce damaging transient voltages and torques.

(2) Capacitors only at selected motors.

(3) A capacitor bank somewhere on the plant bus. (See Figure 2-27.)

(4) Power factor "controllers" on certain feeder circuits. These sense when power factor falls outside pre-set limits, then switch capacitors in or out of the circuit to compensate for the variation. (See Figure 2-28.)

Although many handbooks, articles, and papers explain the basics of capacitor use, questions arise for which answers are not so readily available. We shall consider some of these questions one by one.

What's an easy way to figure the corrected power factor for a given capacitor size?

For those handy at manipulating the vector diagrams of power and kVA in the motor equivalent circuit, it's not difficult to get the answers with a small calculator. But the chart of Figure 2-29 offers a simple alternative. It works with any size or speed of squirrel cage machine. Here's how to use it:

Figure 2-26: Separate power factor correction capacitor mounted at every motor is seldom cost effective.

Efficiency and Power Factor   77

Figure 2-27: Typical outdoor capacitor bank used to correct plant power factor. This will have no effect on individual motor performance except insofar as it may correct a low voltage condition.

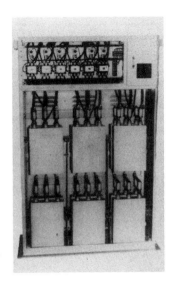

Figure 2-28: Electronic circuitry in this power factor controller (upper compartment) senses reduced PPF and switches capacitors (lower compartment) onto the system as they're needed.

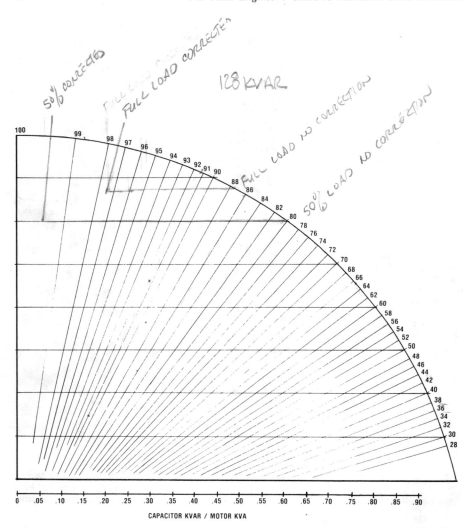

Figure 2-29: To calculate corrected power factor for a given capacitor size, find the point where the radial line, corresponding to the motor's uncorrected full load power factor, intersects the upper boundary arc of the chart. From that intersection, using the horizontal scale below the chart, lay off the horizontal distance representing the ratio between capacitor kvar and full load rated kva of the motor. The corrected power factor then corresponds to the radial line reached by the end of that distance.

Efficiency and Power Factor 79

(a) Find the point where the radial line, corresponding to the motor's uncorrected full load power factor, intersects the upper boundary arc of the chart.

(b) From that intersection, using the horizontal scale below the chart, lay off the horizontal distance representing the ratio between capacitor kvar and full load rated kva of the motor. The corrected power factor will then correspond to the radial line reached by the end of that distance.

For example, take a 500-hp, 4-pole, 2300-volt motor with uncorrected full load power factor of .875. No-load amps are 32; full load, 115. Maximum allowable capacitor size using the normal rules would be $(\sqrt{3})(32 \text{ amps})(2.3 \text{ kV})$, or 128 kvar. (Actually, 125 kvar would be the closest readily available package.) Rated full load motor kva is $(\sqrt{3})(115 \text{ amps})(2.3 \text{ kV})$, or 450. The ratio of capacitor-to-motor kva is 128 to 458, or .279. If we measure off .279 horizontal scale on the chart, starting from where .875 power factor would be (between .86 and .88), we get a corrected power factor of about .974, as in Figure 2-20.

Suppose we want to find out what this capacitor will do at only half load on the motor. First we must know, of course, what the uncorrected motor power factor and the current will be at half load: .800 and 62.5 amps, respectively. The half load motor kvar is $(\sqrt{3})(62.5)(2.3)$, or 249. Capacitor-to-motor kva ratio is 128/249, or .543. Using the chart as shown in Figure 2-30, we now find a corrected power factor of .994.

If we were to go all the way down to zero load, we would get a corrected power factor of 1.0—to be expected, because this size capacitor compensates for all the reactive kVA in the machine when it is running idle.

To find the capacitor size needed for a given power factor correction, measure off the horizontal distance between the corrected and uncorrected power factors and multiply the ratio that distance represents by the motor kVA at the uncorrected condition. Figure 2-31 shows how different capacitor sizes will affect power factor for a typical motor at various loads.

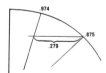

Figure 2-30: Using the chart of Figure 2-29 for full load (a) or half load (b).

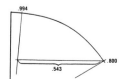

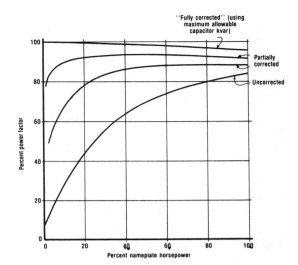

Figure 2-31: How corrected power factor varies with horsepower load for different capacitor sizes with the same motor.

Isn't it true that damaging transients will not occur if capacitor size does not exceed the no-load or idle kvar of the connected motor?

This is not always true. The rule is a handy guide for preventing overvoltage from generator action in the motor, excited by the connected capacitors. It is a conservative rule. But you must leave some latitude. Remember that motors themselves, by NEMA standards, are rated for operation at plus or minus 10% of nameplate voltage. That could mean a 21% above or below rated kvar for connected capacitors. For this reason, the air conditioning industry has suggested that capacitor size be limited to 90% of the theoretical maximum.

Of course, when voltage goes up, magnetizing kVA in the motor goes up too, but not always as rapidly as capacitor kvar. Also, manufacturing tolerances may produce actual capacitor kvar as much as 15% above rated kvar.

Beware of a drive that is likely to be shut off and then quickly restarted, as in jogging, reversing, or plug stopping. Many operating duty cycles may permit this. Normally, after a second or so of de-energized time, the trapped magnetic flux in all but the largest motor cores will dissipate, so that no generated voltage will remain at the terminals when the circuit is reclosed. If capacitors are connected to these terminals, however, they will sustain the generated voltage for two to ten times as long, or even more, depending upon capacitor size.

When reclosure does occur, oncoming line and motor voltages are likely to be out of phase with each other. The resulting transient current surge may produce torques 20 times normal. With a very light connected load having little inertia, the rotating parts may sustain that torque impulse without damage, transferring it harmlessly into an abrupt speed change. But if the load has

Efficiency and Power Factor

high inertia, rapid speed changes aren't possible. Something will give—usually the motor shaft. Fast reclosing of one 300-hp motor broke several shafts before the user realized what his problem was.

When a motor manufacturer answers the question, "What is the largest size capacitor I can safely switch with this motor?", he will not generally take account of possible rapid on-off switching, or of high load inertia. He usually isn't told of those application details. So watch out for them in applying capacitors.

One leading consultant in this field has claimed wide success for a "50-50" rule. That is, he proposes using capacitors of kvar equal to half the motor horsepower, up through 50 hp. (For a comparison with the normal rule, see Table 2-IV.) He advocates adding capacitors only to the largest size motors in a plant, to the extent needed for the total corrective kvar that is economically justified. But he warns that each motor corrected must be fully "eligible"; that is, not connected to high inertia, or subject to rapid on-off cycling.

What do I do if I can't get a high enough corrected power factor with the maximum allowable capacitance?

The upper limit to capacitor size applies only when the units are connected to the motor side of the starter—that is, when they are "switched with the motor." If they are connected to the line side of the starter instead, they are automatically separated from the motor winding whenever the contactor opens, so no torque or voltage transients or "over-excitation" can occur in the motor.

To figure savings from capacitor installation, isn't it sufficient to calculate the total reduction in system current and the resulting effect on supply circuits or power rates?

No. One other element is necessary: the time factor. Most motors do not run continuously. You may figure an impressive saving on a plant air compressor motor, for example, only to find that it runs just a few hours a month. Depending on how its running time fits in with plant demand periods monitored by utility metering, the effect on overall cost could actually be negligible. A capacitor investment is usually justified only for a substantial drive operating time. Ordinarily, for the greatest savings, the largest motors in a plant should be corrected first—but only if they run long enough to make correction worthwhile.

Motor hp (1800 rpm)	Normal max. kvar	"50-50" max. kvar
5	2	2½
10	3	5
25	6	12½
50	11	25

Table 2-IV

Will power factor correction capacitors help a motor start more easily?

Yes, but probably not much. Because capacitors help reduce line current demand, resulting in less voltage drop in the power system during starting, they will raise motor terminal voltage somewhat. Unless line drop is a major factor, however, the change may not be great. Many large motors are started with special connected capacitor banks to boost the voltage, but this requires about 20 times as many kvar as would normally be used just to improve power factor during running. Such banks must be switched out of circuit when the motor reaches full speed, because otherwise once the inrush current disappeared they would generate a potentially damaging voltage rise on the bus.

Some standard motors in the 1- to 200-hp range are now being sold with power factor correction capacitors that can be mounted alongside and cabled right to the terminal box. Isn't this logical for larger motors too?

Unfortunately, maximum capacitor physical size goes up faster than motor size, particularly above 600 volts. (Above 500 hp, most motors are 2300 volts or higher.) So it isn't feasible to mount capacitors on or at the motor itself, especially in the motor terminal box, above the NEMA frame sizes.

For example, Figures 2-32 and 2-33 show typical unit size and construction at high voltage, and the total amount of kvar normally usable for large motors. These higher voltage capacitors are bulky because they must include high voltage fuses. When used on power distribution networks, they have no fuse compartments because the fuses are provided separately. But in motor application these are part of the capacitor unit.

Moreover, capacitors need ventilation. They do not develop a great amount of heat, but there are internal losses. Maximum capacitor ambient is 55°C. At the 10% over-voltage allowable for motors, capacitor heating may go up considerably. Be sure they are not mounted too close together without room for free air circulation. Follow the manufacturer's suggestions for proper installation.

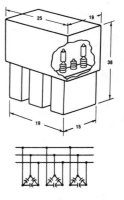

Figure 2-32: Three-phase 150 kvar 2400 volt capacitor bank, showing internal connections (below, right) and terminal fuses.

Figure 2-33: Typical single-phase 2400 volt capacitor without fuse compartment.

Given the total amount of capacitance for a particular circuit, how do we make the actual hookup? How do we divide up the total capacitance among the three phases?

Let's begin with an example quoted from one of the more popular handbooks:

"The motor is rated 700 hp, 2,300 volts.... The magnetizing current at rated voltage is 40 amp. Because power factor is low, motor-magnetizing kvar are approximately equal to $(\sqrt{3})$ EI/1000 or $\sqrt{3}$ (2300)(40)/1000, or 159 kvar.

"In microfarads the capacitance $C = 1000$ kvar/$(2\pi f \text{ kv}^2) = (1000)(159)/(2\pi(60)(2.3^2))80\mu f$. X_c, the capacitive reactance, in ohms = $\frac{1}{2\pi fC}$ or $\frac{10^6}{2\pi(60)(80)} = 33.2$ ohms.

79.7 μf   *33.15*

Thus three capacitors, each having $C = 80 \mu f$, to obtain $X_c = 33.2$ ohms, may be connected in wye at the motor terminals."

This is theoretically correct. But where do you get three such capacitors? And why do three capacitors of 80 μf each combine to produce a total capacitance of only 80 μf?

Taking the second question first: Capacitance, like other circuit elements such as resistance, does not vary with voltage. What we are really supplying to the circuit is not capacitance, however, but kvar—reactive kVA. As already mentioned, when a capacitor is operated at a voltage other than its rated value, its effective kvar will change. Here is the rule:

Capacitor kvar is directly proportionate to applied voltage squared.

Suppose we have three equal 2300 volt capacitors, each rated 50 kvar. If we connect them to a three-phase circuit in delta (as in Figure 2-34), each one

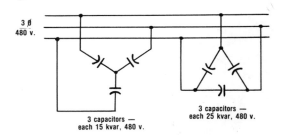

Figure 2-34

sees its rated 2300 volts, and the total circuit capacitance will be equivalent to 3 times 50, or 150 kvar.

If we now reconnect the same capacitors in wye (Figure 2-34), each one sees $2300/\sqrt{3}$, or 1328 volts. Each capacitor's effective kvar now becomes 50 times $(1328/2300)^2$, or 16.7 kvar. Now, total circuit capacitance is equivalent to 3 times 16.7, or 50 kvar—the same as the rating of one capacitor alone.

Actually, standard power capacitors are not rated at 2300 volts or at 4000, but at 2400 or 4160. Thus if we operate a 50 kvar unit at 2300 volts on a motor circuit, its actual kvar will be $(2300/2400)^2$ times 50, or 46.

What determines whether to connect delta or wye? The overall economics, availability of units, and total kvar needed. Total dollars per kvar can vary widely depending on installation. This leads back to the first question from our earlier example: What standard capacitor sizes are available? You can't get 159 kvar, for example, or 37, or 233. But you can come close. Normal unit sizes are shown in Table 2-V.

Depending upon the manufacturer, some (not necessarily all) of these standard ratings may be purchased as either three-phase or single-phase units. The former supplies its total rated kvar to the circuit; the latter units must be combined in multiples of three to give the required total.

Suppose you need a total capacitance of 140 kvar for a 4000-volt motor. One choice would be a 4160-volt, three-phase 150-kvar assembly. This would provide $(4000/4160)^2$ times 150, or 138 kvar ... close enough, especially in view of that plus 15% tolerance on kvar. If such a unit weren't readily available, three 4160-volt, 50-kvar, single-phase capacitors could be connected in a

Voltage rating	Rating kvar
240 or 480	1*,1½,2,2½,3,4,5*,6,7½*,8,10*, 12,15*,18,20*,22½,etc. in 2½ steps through 50.
	(One supplier offers 7,9,11,14,16,21,26, 30,38,45,53,60,68)
	Larger units 125,150,175,200,etc.; up through 600 may be available.
2400 or 4160	50,100,150,200 (One supplier offers 25 and 75 also)
	*Most widely offered

Table 2-V

Efficiency and Power Factor

delta bank. Alternatively, the same delta bank could use two 25 kvar units in parallel. Or, connect a wye bank of three 50 kvar capacitors to the same lines as a three-phase 100 kvar capacitor. (See Figure 2-34 for a similar example.)

Always play safe by choosing the most economical, readily available combination giving total kvar below the desired limit, rather than above. Figure 2-35 illustrates what is normally available for large motors 2300 volts or above; most of the handbook tables do not go above 500 hp.

What about motors with special connections, such as part winding, multi-speed, or dual voltage? Can power factor correction capacitors be applied to these?

Avoid such situations if you can. There are ways to handle some of them, but the risks may be great. For example, a few years ago a user connected capacitors to both windings of a two-speed machine. When one winding was energized, the closed inductance-capacitance circuit of the other winding overheated and burned out from circulating currents. (The solution is a special contactor arrangement to disconnect the capacitors from the de-energized winding.)

Recently, another user asked about connecting 230 volt capacitors of the proper size permanently across terminals of a 230/460 volt motor as shown in Figure 2-36. The winding will be magnetically unbalanced to some extent, with currents in the inner and outer phase groups not in phase with one another. There is a possibility of resonance and transient oscillations. The exact effect will depend on capacitor size and motor design. Unless you're

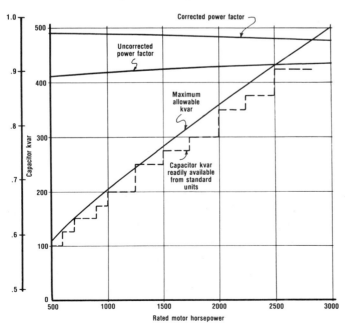

Figure 2-35: Typical power factor correction kvar that may be applied to large motors with little risk. Note that in all cases the corrected power factor is well above .95, usually more than enough.

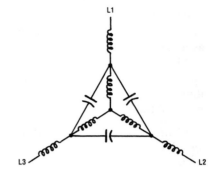

Figure 2-36: Proposed "permanent" connection of 240 volt capacitors across inner legs of 460 volt motor. Unbalance of winding, and possible resonance, is a potential source of trouble.

prepared to verify satisfactory motor performance with this connection, it is something to avoid.

Can capacitors be connected in series, rather than parallel, for more flexibility in getting certain kvar combinations?

Yes, this can be done. Voltage across each capacitor in a series grouping will be reduced in inverse proportion to its kvar rating; that is, a higher kvar capacitor (having lower impedance) will have a relatively smaller share of the total circuit voltage across its terminals. At this lowered voltage, its effective kvar will be reduced according to voltage squared.

The rule is:

$$1/\text{kvar total} = 1/(\text{kvar})_1 + 1/(\text{kvar})_2 + \text{etc.}$$

For the simplest case of two capacitors in series:

$$\text{Total kvar} = \frac{(\text{kvar})_1 (\text{kvar})_2}{(\text{kvar})_1 + (\text{kvar})_2}$$

Example: 25 and 50 kvar units in series; total kvar

$$= (25)(50)/75 = 16.7$$

Remember, though, that effective capacitor kvar varies with the applied voltage. To use these formulae, then, all capacitors involved must be rated for the same voltage to begin with—the overall line voltage.

Instead, suppose we have a 50 kvar and a 100 kvar capacitor, each rated only 2400 volts but connected in series to a 4160-volt line. The calculation of capacitor impedance, current, and terminal voltage shows that the 100-kvar unit, having the lower impedance, will operate at only 1385 terminal volts instead of its rated 2400, thus developing only $(100)(1385/2400)^2$, or 33.3 kvar. The smaller unit, with a much higher impedance, will see 2775 volts. This is well above its normal rating, and perhaps damaging if a further 10% rise occurs on the circuit. The kvar developed will be $(50)(2775/2400)^2$, or 66.6. Total kvar is then 100. But from the series capacitor formula above, we would expect only $(100)(50)/150$, or 33.3.

If both capacitors are rated 4160 volts, however, their impedances will be much larger, total current will be smaller, and the voltages across each will again be 2775 and 1385. But effective kvar will be $(100)(1385/4160)^2$, or 11.1 for the larger unit and $(50)(2775/4160)^2$, or 22.2 for the smaller unit. The total of 33.3 agrees with the formula.

A further precaution: In connecting 2400-volt capacitors in series on 4160-volt lines, you may still have 4160 volts to ground, depending on the type of power system. Capacitor bushings may not be insulated for that voltage level.

To sum up the connection question: Delta and wye groups of capacitors can be paralleled on the same motor circuit (as in Figure 2-34). Three-phase and single-phase units can be paralleled. Series combinations are possible. The important points are to maintain the same capacitance in all three phases and to match the capacitor voltage rating to the circuit line voltage.

Power factor controllers

We've all heard about another type of "power factor controller" for motors, which is—with justification—claimed to improve motor efficiency by lowering motor losses. It's the so-called "Nola device," invented by Frank Nola of the National Aeronautics & Space Administration (NASA) and now being marketed by a number of firms. "Totally stops wasted energy...up to 400 hp," says one maker of this controller, adding that the user can "plan to deduct 40% to 50% from the electric bill" for a-c induction motors at no load.

Despite the name "power factor controller," however, these devices actually control motor voltage. When motor shaft load decreases, the Nola device electronically senses the reduction in working current, the amperes supplying output torque, compared to the unchanged value of magnetizing amperes. It then uses its control circuitry to reduce the voltage applied to the motor terminals. This drops the magnetizing current as well as greatly reduces the core losses. The result? Motor power factor and efficiency for that particular load are increased. Figure 2-37 shows what happens.

Figure 2-37: How the Nola controller adjusts motor losses by changing voltage to suit the load. At full load, inductive current I_{x1} results from full voltage and magnetic flux density. Combined with the resistive or "watted" current I_{w1}, supplying losses and output, it results in a power factor angle PFA_1. When motor load is greatly reduced, the much lower I_{w2} combines with the unchanged inductive current I_{x2} for a larger power factor angle PFA_2. The controller senses that and lowers motor voltage to reduce inductive current (largely magnetizing) to the smaller value I_{x3}. At light load, most of the watted current supplies core loss, and that is greatly reduced also when voltage is lowered (I_{w3}). The power factor is thus restored to its original higher value—PFA_3 equals PFA_1.

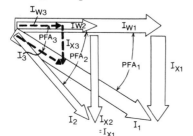

Reduced motor voltage also raises the load component of current, I_{w2} or I_{w3}, which tends to increase copper loss in the windings, so a net gain in efficiency results only when the load is quite low.

So, although this is one instance of motor power factor that apparently has an effect on efficiency, the benefit results only from an adjustment of motor voltage. For a clearer understanding of what makes one motor "better" than another, then, keep the concepts of efficiency and power factor separate from each other. There is no direct relation between them.

The theory is quite sound. Putting numbers on the results is not easy, however. Figure 2-38a shows the sort of power saving claimed by the patent, from which the graph was directly taken. For any 3-phase motor, however, and for most types of single-phase or fractional machines (such as the shaded pole, split phase, or repulsion types), the actual shape of the normal watts input versus horsepower output relationship is more like Figure 2-38b. Power saved by specific voltage reduction may drop off very quickly above very light loads.

With a little thought we see that these plots cannot be straight lines. We know that at zero load there is core (iron) and friction/windage loss in any motor, and these losses do not change much as the load is increased. Relative power input to the motor, as a percentage of output, must therefore be higher at light loads than at heavy loads.

The Nola circuit reduces motor voltage in the same basic way any other solid-state power supply functions—by "chopping" the waveform to vary its effective magnitude. Magnetization—and iron loss—decreases when the volt-

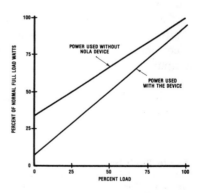

Figure 2-38a: Power savings for small motor operated from a Nola device, as illustrated in the Nola patent described in this chapter.

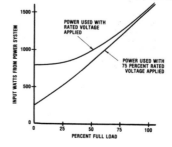

Figure 2-38b: The actual power saving for a standard 10-hp, 4-pole open motor with voltage reduced 25%.

age is lowered, but other losses in the motor will go up in two ways.

First, whatever shaft load there is will draw current which produces I^2R heating loss in the stator winding and in the rotor cage. Reducing applied voltage will force that current, and those losses, to rise. Figure 2-39 illustrates this effect. This has been taken into account in plotting Figure 2-38b, which is why there is little gain at full load where the higher winding losses almost compensate for the lowered core loss.

Second, the harmonic content of the "chopped" voltage waveform will increase motor heating still further (especially in the rotor). It is not clear from published data how great this effect might be. But in a 400-hp, 3-phase machine it can certainly not be neglected.

The expected savings will generally vanish above ⅓ to ½ load. For fractional hp motors that run idle much of the time, as in some office equipment, machine tools, and vacuum pumps, the power factor controller can pay for itself quickly. But for larger industrial polyphase motors, the device is of little benefit. A series of tests in 1980 and 1981 showed that on a 7.5-hp motor the actual input watts saved with a power factor controller was zero at half load and 260 at 5% load. On a 25-hp unit, the figures were zero and 40, respectively. For a 50-hp motor, the saving reached 175 watts at half load and 450 at quarter load.

Furthermore, energy is lost in the controller itself. For a 100-hp, 460-volt drive, this can add 4% to 6% to existing motor losses at full load. Depending upon the loading versus time relationship, this can cut expected savings in half.

Control of motor torque has presented some problems also, for two reasons:

(1) Like any electronic device that adjusts motor voltage by varying the waveform (see Chapter 3), the controller generates harmonics that not only increase certain motor losses but also reduce shaft torque.

(2) When the voltage has been lowered to suit a light load, sudden application of full load may stall the drive because controller "recovery time" to restore full voltage is too long. This has caused trouble with bowling alley equipment, for example, although the latest controller circuits provide a faster response.

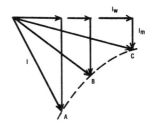

Figure 2-39: The approximate plots of current in a loaded motor show how reduction in core loss (represented by the magnetizing current I_m is balanced by increase in load current I_w—both changes being caused by a lowering of applied voltage. The motor winding heats in response to the total current I, the heat being least when the current is smallest (point B). At either A, full voltage, or C, with greatly reduced voltage, the stator copper loss will be higher. This has to be accounted for in determining overall energy saving possible through voltage reduction.

This diagram is oversimplified, but is an easy way to visualize the relationships.

In considering the application of these controllers, be prepared to deal with confusing and sometimes misleading data. For example, the curves in Figure 2-40 have been used to show how "average standard" motors waste power at light load. But what is plotted as "idle" can't be correct. By definition, efficiency of any motor running without load must be zero. Hence, the fraction of electrical input power that is wasted in such a motor must always be 100%—regardless of motor size. The lower curves on the graph make sense; the upper one does not, and was probably arrived at by confusing power factor with efficiency.

Whatever the plot looks like, remember that a power factor controller can reduce only the iron loss portion of that waste. Friction and windage loss will be unaffected, and at light load the other losses are already negligible.

An advantage sometimes claimed for the controller is a reduction of utility power factor penalty by raising motor power factor. But utilities base penalties on total plant conditions. Even a large number of small, lightly loaded motors will have little effect on that, because the actual reactive (lagging) amperes they draw will be small. Consider Table 2-VI for a larger unit. Thus, although idle power factor is only ⅛ of the full load value, total reactive current is still half that at full load. This is likely to greatly influence neither plant power factor nor the penalty status.

So economic justification for a controller of this type may not always exist. Exact calculation may be difficult. The cost for commercial versions of the Nola device has ranged all the way from $15 to $150 per horsepower, depending on the manufacturer and the motor size.

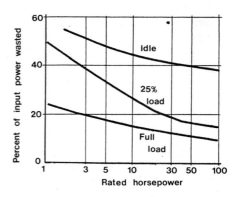

Figure 2-40: These curves have been used to show how much power lightly loaded motors may waste. But although a high percentage of total input is lost in such a motor, that total itself is far lower than it would be at full load.

Motor load	Motor power factor	Total motor amperes	Total reactive motor amperes
100%	.80	150	90
0% (idle)	.10	45	44.5

Table 2-VI

3 Coping with Starting Conditions

PERSONS WHO FEAR air travel seldom mind the flying itself. It's the takeoffs and landings that bother them. Similarly, most motor application problems arise in the drive starting or stopping process. Nature adapts readily to change, provided it doesn't occur too rapidly, and that applies to change in machine speed as well.

Selecting the proper motor requires an understanding of what's needed to start the driven machine. This has nothing to do with the "running horsepower" which the machine may demand. The most important single characteristic of any motor-driven machine, or "load," is the shaft torque which it requires. Torque, also thought of as "turning effort" or "twisting force," is measured by the product of force times distance—normally defined in terms of the number of pounds applied to cause rotation of a shaft multiplied by the radial distance from the center of rotation to the point of application of the force. (See Figure 3-1.) If you increase the "leverage" by moving the same force twice as far away from the center of rotation, you double the torque.

Clearly, then, the resistance to rotation within any machine is some amount of torque produced by an internal force. This may be a friction force—in solid or liquid material inside the machine, such as coal in a crusher, or a thick liquid such as sugar syrup inside a mixer. The force can be a "pressure head" in the casing of a pump, or the resistance to compression of a gas within a compressor.

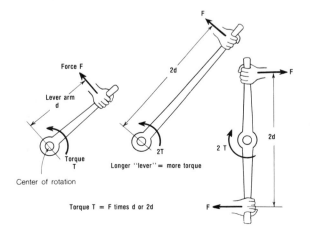

Figure 3-1: Basic turning effort relationships in rotating motion.

Torque is normally measured in units of pound-feet or pound-inches. As the metric system becomes more widely used in the United States, the units will become kilogram-meters (or "Newton-meters"), for example. But the measurement will always be in terms of force times distance. So we should avoid speaking of torque simply in "pounds," even though a 1-foot radius is commonly implied so that pounds and pound-feet become the same numerically. It still isn't right, and can mislead you, just as it can become confusing to speak of pressure in a water system as "pounds" when you really mean "pounds per square inch."

You will also often see torque expressed in "foot-pounds" rather than pound-feet. The units look the same, feet and pounds just being in reverse order. But as any engineering handbook will tell you, "foot-pounds" are units of work or energy rather than torque; 33,000 "foot-pounds" per minute equal one horsepower. It's bad enough to be uncertain of our numbers in dealing with motor applications; let's not be confused in our concepts too. Sloppy terminology leads to sloppy thinking.

Machine torque

Two different kinds of machine torque are important to drive acceleration. The first is the amount of torque it takes to start the machine rotating from its initial position of rest—the torque needed to "break it away," which is called the *breakaway torque*. As we shall see, for some kinds of loads this can be quite large—much larger than the torque needed to keep the machine running at full speed. In other loads the breakaway torque may be small, but it is important to be sure. Unless the motor can exert enough torque of its own to overcome this breakaway required by the load, the drive cannot start.

The second load torque to consider is not a single value, but the range of values required at the shaft as the drive comes up to speed. During acceleration, load torque can vary greatly with speed. A little further on, we shall see how different these variations may be depending on the type of machine in question. Again, the motor must be able to exert enough torque to overcome the load's *accelerating torque* demand at all speeds; otherwise the drive will not reach full speed but "hang up" at some intermediate rpm until the motor is tripped off the line.

How much output torque is available from standard motors for load acceleration? NEMA stipulates certain values. But remember that NEMA does not specify the shape of any motor speed-torque characteristic curve, only the torque values at locked rotor and at the breakdown point. Nor does any other published standard. Figure 3-2a shows how different two motor designs can be in their ability to furnish accelerating torque, even though they are identical in locked and breakdown values.

Very well, you may say, suppose we control the minimum or "pull up" torque by specifying it to be at least a certain minimum value. Figure 3-2b shows why this may not help either; both motors now have the same pull up

Coping with Starting Conditions

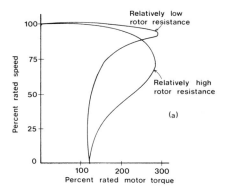

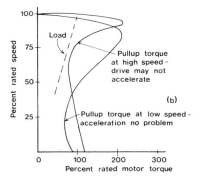

Figure 3-2: Left: Motors of widely different accelerating characteristics may have the same locked rotor and breakdown torques, as in Figure 3-2a (top). Each is identically "standard" in NEMA terms, yet they are not at all the same. In Figure 3-2b (bottom), both motors have the same minimum or "pull up" torque during acceleration, yet these designs also are quite different from each other. For proper application, there is no substitute for knowing what the entire motor speed-torque curve looks like, no matter how "standard" the motor may be.

Figure 3-3: Right: Speed-torque curve shapes typical for small motors (top), 5 to 50 hp for example; and large motors (bottom), 300 to 500 hp. The difference in shape, particularly for the Design B, can easily mean the difference between a drive that will start and one that will not.

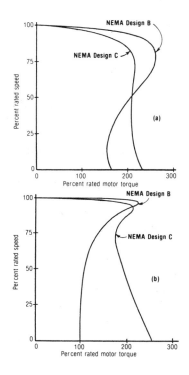

torque; one will accelerate the load, but the other will not.

If you look in reference books for curves on "standard" motors, in many you will see what is shown in Figure 3-3a. These may be fine for motors of 5 or 10 hp. But, as was pointed out earlier, locked and breakdown torques for large motors are smaller, and the curve shapes themselves are quite different, as is apparent from Figure 3-3b.

Why do these differences exist? There are many reasons which have to do with basic changes in the relative proportions of the working parts. For one thing, as motors get larger the full load slip (and the slip at breakdown torque) tends to decrease. The enclosed, heat-producing volumes of copper and iron become larger with increasing size of frame than does the exposed surface area that dissipates the heat. So large motors do not cool as well, requiring lower rotor resistance (and slip) to compensate.

This difference in full load slip—and rpm—can be important for other reasons. Pump or compressor builders often assume that a 100- or 1000-hp motor design is just like a 10-hp design, only bigger. They recommend or assume use of a 300-hp motor on a 290-hp compressor, for example, but base it on 1750 full load motor rpm, which would be reasonable for a much smaller machine. If the motor reaches "full load" at 1785 rpm, the compressor (whose input power varies as the cube of the speed) will demand 310 hp, thus overloading the motor. (See Chapter 2.)

Speed versus torque

From all this we can conclude that the speed versus torque curve of the load is the basic tool needed to apply any motor correctly. When such a curve is made available, take these three steps right away:

(1) Make sure you understand the units in which the curve is plotted. Don't just look at the numbers. If either speed or torque is plotted in percent, make sure you know percent of what. To work with the motor, you need to refer everything to motor shaft speed. If the drive is belted or geared, and the curve comes from the manufacturer of the driven machine itself who may not know anything about the rest of the drive, it's likely that the speed on the curve will be that of the driven machine shaft. Percent torque (in terms of motor full load rated torque as 100%) will be the same at any speed throughout the drive. So will horsepower, except for power lost in gears, couplings, etc. But if torque is given in pound-feet, it will differ at different shaft speeds.

(2) Make sure you know whether "100%" torque on the curve means 100% of rated motor output. Often it does not. For example, a pump builder determines that it takes 275 hp to drive his pump at full output. Someone else orders a 300-hp motor to drive it. But to the pump designer, "full load" or 100% torque isn't 300 hp torque, but only the 275 hp. A torque curve through the higher full speed value could be much harder for the motor to accelerate, particularly when a reduced voltage start cuts down drastically on available torque. (See Figure 3-4.)

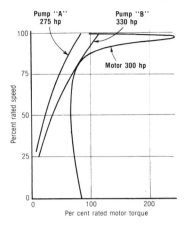

Figure 3-4: A 300-hp motor can easily accelerate the 275 hp pump load "A." Starting full 300 hp pump load might not be so easy. If, as often happens, the actual pump load were 330 hp—well within the nameplate capability of a 300-hp motor with a 1.15 service factor—we would have curve "B," and acceleration may be impossible. This is why, if load speed-torque curve is given in percent, it's necessary to be sure that you know what the percentage is based upon.

(3) For direct-coupled drives, remember the potential problem of big differences between the design speed of the load and the rated output speed of the motor (such as the 1750 versus 1785 rpm mentioned above).

(4) Check to see that the full speed torque doesn't exceed the motor capability. This is surprisingly common, usually because whoever specified the motor did not thoroughly check to see what the load required under all conditions. For example, look at the fan curve in Figure 3-5. It shows a full speed torque requirement of 35,700 pound-feet at 1190 rpm. But the specs called for a 7000-hp motor with no service factor to drive this fan. It doesn't take a motor designer to figure this out. By simply using the basic formula below relating torque to horsepower and speed, we find that the motor can deliver only 31,000 pound-feet at full speed.

$$HP = \frac{(RPM)(Torque, lb\text{-}ft)}{5250}$$

Figure 3-5: Fan load curve for "7000" hp, 1200 rpm motor application. Full speed torque required by the fan exceeds motor rating by almost 16 percent.

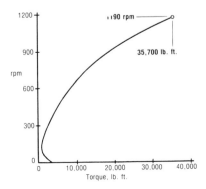

So questions were asked. It turned out that the fan only required this excessive torque when the air being blown through it was cold. The problem was that there were no data on how long the air would stay cold, or how the curve changed after the air warmed up. So an 8000-hp motor was negotiated instead.

You probably don't get involved with motors that size. Few people do, very often. The reason that example was included is to emphasize that the same miscalculations and fuzzy thinking occur when dealing with the largest drives just as with the smallest ones. And it happens the same way on 50-hp, 1.15 service factor motors, too, when the load turns out to be 65 hp.

That is one of the complaints users have about the T-frame rerated motor, of course. Maybe the old U-frame or earlier design would take the 65 hp, or 75 hp, and nobody worried about it. Nowadays, the answer is either to replace it with the next size larger or just rewind it Class F. But that's not application engineering. That's "cut and try." And Class F motors are being pushed, too; as margins become available, sooner or later they will be used up.

The time to recheck fan damper settings, install compressor bypass valves, or trim a pump impeller is before the motor has failed to carry the load, not after. This is why load torque curves should be carefully and knowledgeably reviewed when the motor is being selected.

Load torque

How can you find out what load torque is required? There are many sources of information:

(1) Look at the motor ratings, characteristics, and operating records for drives as similar as can be found.

(2) Find out what the load controls—dampers, valves, etc.—will be. Are they automatic, or subject to mechanical override? How do they work?

(3) For machinery processing bulk material—conveyors, crushers, mixers—find out if the motor must start them loaded or empty. Does material consistency change during processing, bringing torque changes with it?

(4) "If all else fails, read the instructions." How is the driven machine supposed to be operated? Who and where is the manufacturer, and what information can he provide?

It is not always easy to get these answers. But we should realize the importance of the questions, and the degree of risk if they aren't even asked.

Table 3-I indicates the range of motor torque commonly needed for various types of loads. Machines in the "normal" or "low" torque category include most fans and pumps. A recent study by the Federal Energy Administration showed that 47% of all motors in this country drive one of those two kinds of loads. Perhaps the most common is the centrifugal pump. The motor is generally 1800 or 3600 rpm, though in large sizes speeds down to 720 rpm are sometimes found.

Centrifugal pumps are so named because in operation the pump impeller

Coping with Starting Conditions 97

Type of load	Typical motor torques needed, percent of rated	
	Locked rotor	Breakdown
Rubber mill, pug mill, mixer	150-175	225-250
Centrifugal fan or compressor; dampers closed	60	200
Pulverizer	200	250
Loaded conveyor, bucket elevator	150-200	225
Crusher	80-100	250
Solid waste shredder	125-150	250-300
Centrifugal pump	60	200
High pressure piston pump	200-250	200
Mixed flow pump, closed discharge	80-100	200
Rubber or plastic extruder	100	200
Belt feeder	100	200
Hydropulper	125	200
Brick press or briquetting machine	100	200
Ball mill	150-200	225-250
Agitator	100-150	200
Rotary kiln (loaded)	250	200
Plasticator	150	250

Table 3-I. Some commonly encountered industrial machines listed alongside the motor torques usually needed for their operation.

throws liquid radially outward by centrifugal force, converting its energy of rotation into a "velocity head." If the discharge valve from the pump is open, a fixed value of driving horsepower is needed to keep the fluid flowing, and this is the power required from the driving motor. That power is quite sensitive to pump impeller speed, as noted earlier, varying as the cube of the rpm.

Assuming pump horsepower and motor rating to be the same and the discharge valve open, we would have the load curve of Figure 3-6. This sort of drive normally presents no acceleration problem, because the inertia of a centrifugal pump impeller is very low—typically only a small fraction of the inertia in the motor rotor itself. But a problem can develop quickly if reduced voltage starting is desired to lower the inrush current. A reduction of 20% to 40% in motor voltage, resulting in a drop of as much as 70% of the full voltage

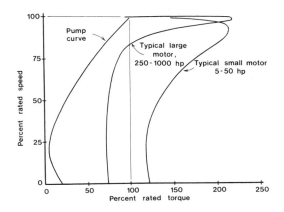

Figure 3-6: Typical open discharge centrifugal pump drive characteristics, showing once again the contrast between accelerating torque of large and small drive motors.

motor torque, may easily reduce the motor torque below what the load demands—and the drive never comes up to speed.

For this reason, many pumps must be started with the discharge valve closed. Closing the valve lowers the torque required by the pump, because now the trapped liquid cannot escape; it develops no velocity head. Do not assume, however, that the pump torque decreases by some fixed, standard amount when the valve is closed. The result may vary widely, as seen in Figure 3-7. It will depend on the type of fluid, the pump design, and the temperature. The "churning" of the fluid in a shut-off pump produces heat, which can raise pump temperature to an unsafe level. In some pumps this is avoided by a "recirculation valve" which keeps a small amount of the fluid circulating to carry away the heat. This in turn raises the shaft torque required by the pump.

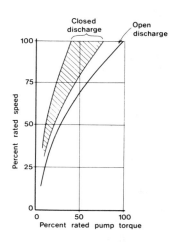

Figure 3-7: Typical speed-torque curves for high speed centrifugal pumps, showing the variation in required accelerating torque with closed discharge.

So if a "closed valve start" is specified, find out what the actual resulting speed-torque curve for the pump will be. Guessing can get you in serious trouble if there is any starting voltage reduction.

Whereas centrifugal pumps build up pressure by throwing fluid radially outward, the "axial flow" pump works like a propeller—pushing the fluid along parallel to the shaft. These normally work at a much slower speed than the centrifugal type—typically 600 rpm or less. The speed-torque curve has the same shape as the curve for the centrifugal pump.

But there is a major difference. When the centrifugal discharge valve is closed, the impellers—simply churning the fluid—require less driving torque than they do with the valve open. The axial flow pump with discharge valve closed attempts to compress the stalled fluid between the rotating propeller blades and the outlet, causing the required shaft torque to rise rather than fall. (See Figure 3-8.)

Before assuming that a closed valve start is an aid to acceleration, then, know which type of pump is involved. Fortunately, none tends to have high

Coping with Starting Conditions

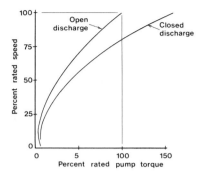

Figure 3-8: Axial flow pump speed-torque characteristics, showing how accelerating torque increases when discharge valve is closed—just the opposite as for a centrifugal pump.

inertia. The "mixed flow" pump has characteristics of both centrifugal and axial types. Another version is the screw pump, often used in pumping heavy fuel oil; this can demand very high accelerating torque to get the viscous fluid moving—to overcome the friction in the oil. (See Figure 3-9.)

To generate extremely high pressures, such as for moving solid/liquid "slurries" in pipelines, piston or plunger pumps are used. These consist of pistons in cylinders working off a crankshaft, as in a gasoline engine, with extremely high motor torque demand at low speed during starting—as much as 200% to 250% of rated torque.

The common centrifugal pump will not "stall" a motor once up to speed. However, operating conditions may change so as to overload the motor. In liquids other than water, operating temperature may increase fluid viscosity to demand more horsepower. The chemistry of the liquid may change, with the same result. Piping changes or new valve settings may boost the pressure head against which the pump must work. Careful questioning of the drive history from the pump standpoint may turn up the reasons why a driving motor has wound up in the repair shop. Only when you know what is being pumped, and something of how the pump operates, can you properly choose the driving motor.

Figure 3-9: Motor needed to accelerate the screw pump, with its much higher starting torque, must evidently be a different design from that needed for the centrifugal pump.

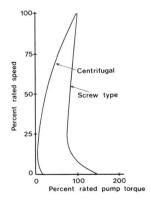

Many of these comments apply equally to fans, used to move gases just as a pump moves liquids. The centrifugal or axial flow fan operates like its pump counterparts. There is this difference, though: Air or gas temperature has a far greater effect on fan performance than liquid temperature has on a pump. Also, gases are relatively compressible, whereas liquids are not.

Hence, the fan speed-torque curves for either centrifugal or axial designs all look like Figure 3-6 (page 97). But they do change drastically with gas temperature. Air is particularly troublesome. Power plant fans, for example, may be designed for average summer conditions when outside air is warm. When a cold winter day comes along, the increased density of the chilled air greatly increases the fan accelerating torque as well as the full load horsepower. It may be necessary to partially close dampers in the system, reducing the flow, to compensate for this.

Unfortunately, the damper may not be located in the right place. One 300-hp drive had the fan and motor located a considerable distance upstream from the damper, and at startup the fan saw cold air for several minutes until the outlet duct filled and pressure built up against the closed damper, or until the process warmed the air. The difference in accelerating torque required from the fan motor is shown by Figure 3-10. It was impossible for the motor which had been purchased to get the fan started.

Three expensive alternatives faced the user: (1) buy a new motor, rated 300 hp but using 1000-hp parts, not only much more expensive but operating most of the time at poor power factor and efficiency; (2) buy a centrifugal clutch or special coupling to "unload" the motor during starting; or (3) install another upstream damper to shut the fan duct off until the air was warmed. All this could have been foreseen had someone considered the fan system characteristics, including the damper location and operation.

The final major difference between pumps and fans is that the latter tend to have very high inertia. Gases are so much lighter than liquids that to develop the desired velocity head takes a high fan blade tip speed—which means a large diameter wheel. And the spaces between blades, forming the flow area through the wheel, must be large. The result is a wheel with high mass or inertia (a property we shall discuss later).

Figure 3-10: 300-hp fan drive, required to start at 80 percent voltage. With cold air and the closed damper far downstream, the motor could not accelerate despite its high torque design. Once the air was warmed and the duct filled, load torque dropped more than 50 percent—but by then it was too late.

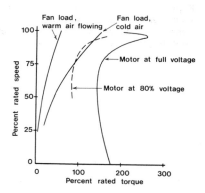

Many fan drives, therefore, mean severe accelerating duty for the motor even if motor voltage is not reduced. It may not be necessary to choose an extremely high torque motor to start a fan, but it may be necessary to provide for a great deal of acceleration heating, often with high resistance rotor bars to get the heat out of the stator winding or shorting rings.

Inertia is the second property of all motor-driven machinery that is important in assessing drive acceleration capability. Inertia is a resistance to acceleration, by virtue of the *mass* or weight of the machine, just as torque is a resistance to motion through friction *force*. Thus, inertia and torque demand are not at all the same thing. A very large, heavy mass, such as a fan wheel, may be finely balanced on well-oiled bearings so that one man can easily exert enough torque by hand to start it turning very slowly. But let the same man try to accelerate it quickly to a high speed, and its high inertia will render his efforts useless.

It is the resistance to *change in speed*—that is, to either acceleration or deceleration—that is the chief effect of inertia. Just as a massive vehicle is more difficult to steer through sharp changes in direction, so a massive rotating machine is more difficult to accelerate. To produce a given change in speed of a high inertia motor load takes far more turning effort, or torque, than for a low inertia load. Hence the importance of load inertia in motor application.

What happens when through either inertia or friction torque the load demands more accelerating capability than the motor is designed to provide? To begin with, as we have already seen, too much breakaway torque means no start. Too much accelerating torque means the motor will bring the load only up to an intermediate speed where it can go no further, as in Figure 3-4 (page 95). But suppose there is a little surplus motor torque at all points along the curve—then there should be no acceleration problem, right?

Acceleration heating

Wrong. We come now to the most important factor in motor starting capability: *acceleration heating.* Anyone who has been around motors very long realizes that some starts are easier on motors than others. But what does this really mean? We said that torque, speed, and horsepower are related by a formula, given on page 95.

The formula is just as true at some intermediate speed during acceleration as it is at full speed. Horsepower means energy, and energy means heat. Without going into all the mathematics of it, we must recognize here that the motor is producing output horsepower at the time it is accelerating. This comes from energy supplied by the motor's electrical source.

Where is this energy going? Part of it goes into the useful work of bringing the rotating parts up to speed. When we get the drive rotating, we add energy to it—the "kinetic energy" of rotation. To stop it again from that speed, we would have to remove that amount of energy, either in the heat of friction brake shoes or in the heat of bearing and air friction as it coasts to a stop by itself.

The remainder of the energy being fed into the accelerating motor goes to heat up the rotor and stator windings. From the physics involved, it can be shown that in fact exactly half the total energy required for acceleration of an inertia load goes into kinetic energy of rotation, while the other half goes into heating the motor. This relationship appears graphically in Figure 3-11.

Figure 3-11: To any convenient scale of energy units (such as foot-pounds or watt seconds), the total rectangular area represents total energy involved in a drive acceleration. Half of this goes into rotor heating in the motor itself, the other half into rotational energy of the drive. These two halves are represented by the two triangles in the figure.

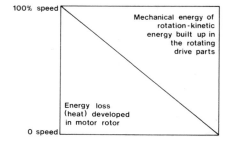

Note one most important thing about this illustration. Energy going into motor heat is not supplied at a uniform rate, but rather heating takes place very rapidly at low speeds, gradually tapering off at higher speeds.

There is something else to note about Figure 3-11. It applies strictly to what we called an "inertia load"—that is, a rotating mass which demands no torque to overcome friction while coming up to speed. As mentioned earlier, however, real machines do embody friction load (load torque) as well as inertia—some more than others. Carrying the physics a step further, we find that when a load has both inertia and torque (as in Figure 3-4), the total acceleration energy required will increase, and all the increase shows up as additional motor heat. The original motor heat energy, at any given speed, must be multiplied by the ratio:

$$T_M/(T_M - T_L)$$

in which T_M is motor torque and T_L is load torque. This gives us the relationships of Figures 3-12 and 3-13.

So now we can see the real extent of the motor application problem. We can see that a motor speed-torque curve which exceeds the load speed-torque curve at all points may still represent an unsafe start for the motor. If the $T_M/(T_M - T_L)$ ratio is high enough, the extra energy can overheat the motor dangerously even though the drive will reach full speed.

Let's look closer at what this extra energy does. First of all, it is convenient in motor design work to consider the complex electrical configuration of the actual motor in terms of a simple "equivalent circuit." Figure 3-14 illustrates such a circuit. Normally, only the motor designer is concerned with this, or with the resistance and inductance values associated with it. But it can be useful in visualizing the acceleration problem.

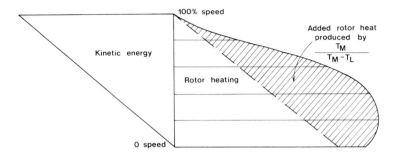

Figure 3-12: Switching the positions of the two triangles in Figure 3-11, we now see the increased rotor heat energy caused by the load torque. Each horizontal "slice" of the rotor energy triangle is increased in area by the ratio $T_M/(T_M-T_L)$ during that portion of the motor and load speed-torque curves. Rotational energy, in the lefthand triangle, remains unchanged.

Figure 3-13: These curves show how rapidly rotor heating increases during acceleration when the average load torque (T_L) becomes high compared to average motor torque T_M. These "averages" are not valid for specific application calculations, but are useful here to show trends.

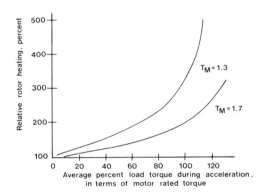

Figure 3-14: Simplified induction motor equivalent circuit; magnetizing branch neglected (S = per unit slip).

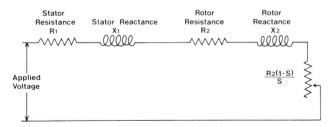

During acceleration, current through the stator resistance R_1 and rotor resistance R_2 is quite high. The motor must reach about 90% speed before the current falls as much as 30% below the locked rotor or "inrush" value. This high current produces I^2R loss, or heat, in the stator winding at a rate 30 to 40 times faster than when the motor is running at full speed. The effect of the heating is multiplied by two other factors: (1) the motor is not at full speed, so it doesn't cool itself too well; and (2) the high current produces pulsating forces on the winding end turns which are also 30 to 40 times greater than at full speed. The extra heat plus the high end turn forces can lead to premature winding failure.

These effects are also present in the rotor. But there are other conditions there which can make things even worse. First of all, the rotor circuit is made up of bars in slots joined at their ends by shorting rings. When those rings get hot, they expand. The expansion exerts large radial forces on the ends of the attached bars, tending to bend them outwards. Small cast aluminum rotors aren't so subject to this. Larger machines, however, can easily be damaged. Hence, allowable motor starting duty must limit the temperatures the shorting rings can reach during acceleration.

"Deep bar"

In most motors, rotor bar temperature is increased by another factor: the "deep bar" effect. The rotor bar currents during acceleration are at a fairly high frequency, because the rotor "sees" the relative speed of the stator magnetic field compared to physical rpm of the rotor. When that actual rpm is low, rotor frequency is high and at locked rotor will be the full line frequency. Under that condition, the effective current-carrying depth of the rotor bar becomes only about ⅜-inch for copper and half an inch for aluminum or brass, no matter how deep the bars actually are. Thus, the apparent a-c resistance of the bars becomes much higher at low speed than at full speed, raising the I^2R loss and the bar temperature just that much more. If hot enough, the bars lose strength, so that they cannot resist even slight bending force from the shorting rings for more than a fraction of normal motor life. If heated still further, the bars may actually melt.

If the rotor does get that hot, won't the stator have become hot as well, so that winding temperature sensors will trip it off the line? Not necessarily. The relationship between R_1 and R_2 in Figure 3-14 will govern which gets hottest the fastest, rotor or stator, and this relationship, or "resistance ratio," varies widely among types and sizes of motors even from the same manufacturer. There have been instances where rotor bars melted and flew out into the winding and only the resulting insulation failure took the machine off the line.

Let's get more specific now about *heat energy* versus *temperature rise*. Though closely related, the two are not the same. We have been discussing the ways in which heat is developed in windings when a motor starts; but from a given amount of heat, what is the resulting temperature rise in the motor parts? How can motor designs be chosen to minimize excessive temperature?

This is important because it is not heat as such, but the temperature, that produces the ring expansion, the loss of bar strength, or the insulation damage. The relationship between temperature rise and heat input to a mass of material, such as a set of rotor bars, is given by this sort of formula:

$$\text{Temperature rise, }°C = \frac{(K)(\text{heat input in watt-seconds})}{\text{weight of material, lbs.}}$$

The "specific heat constant" K varies with the type of material involved. For copper, it is 5.74; for aluminum, 2.5.

From the formula we can see that the larger the mass of material being heated, the lower its temperature rise will be for any given amount of heat input. So if we want to keep the rotor shorting ring temperature rise down, we make it bigger; same thing with the bars. To keep stator winding temperature rise low during acceleration, we put in more copper.

But there are limits. More copper means bigger slots, which means less iron and a higher magnetic flux density. So we may have to go to a larger motor frame size—with resulting poor performance, especially power factor, because the parts are being used at a lower horsepower than normal for the size.

Another alternative, when stator heating is the problem, and the rotor bars are cool during starting, is to increase the bar resistance by using brass or bronze instead of copper (with aluminum, there may be no alloy options). That increases the proportion of total acceleration heat that is produced in the bars; the "resistance ratio" R_1/R_2 goes down, so the stator sees less of the total heating. We get a better "balance" of heat distribution.

All that is fine from the designer's standpoint. But the application engineer in the field has no data on motor design constants from which he can make such choices. For this reason, working only from motor and load speed-torque curves as in Figure 3-2b, he *cannot* make any decision as to whether or not the motor will overheat starting that particular load. Users often ask for the motor speed-torque curve, superimpose it on their load curve, and "see" if a safe start is possible. From what we have just been discussing, however, it should be apparent that this simply can't be done without knowing the relation between heat input and temperature rise throughout the motor. Only the motor designer can do this.

Basic mechanics

Let's take a closer look at the process. The laws of motion and units of measure used in analyzing the workings of machinery are a whole course of study in themselves. We have space here only for those fundamental formulae that are most useful in applying induction motors to drives. The units are those most commonly used today. Because the metric system is now being more widely adopted, however, some conversions between metric units and "English" units will be given.

Machine drives involve two types of motion: rotation, or turning motion, as found in shafts, pulleys, fan wheels, and bearings; and translation, or linear motion, as found in conveyors, hoists, and cranes. There are many similarities between the laws governing these two types of motion, as shown in Table 3-II.

Rotation
Accelerating torque in lb-ft,
$T = (I)(\alpha) = [(WK^2)(\omega/sec)]/308$
Rotational speed in rev/min $= \omega$
Angular acceleration, or rate of change of speed, in rpm/sec $= \alpha$
Inertia, in lb-ft squared $= I$ or WK^2

Stored rotational energy or "kinetic energy," in ft-lbs,
$E_K = [(WK^2)(\omega)^2]/5910$
Rate of doing work, or horsepower $= [(T)(\omega)]/5250$

Translation
Accelerating force in pounds
$F = (M)(A)$
Velocity in ft/sec $= V$
Linear acceleration, or rate of change of velocity, in ft/sec/sec $= A$
Mass M, pound-seconds squared per foot $=$ weight in lbs/32.2
Kinetic energy E_K, in ft-lbs
$= MV^2/2 = [(Weight, lbs)(V)^2]/64.4$
Horsepower, $P = [(F)(V)]/33,000$

Table 3-II

There are many useful relationships to be derived from Table 3-II. For example:

$$\text{Rated motor full load torque in lb-ft} = \frac{(\text{Nameplate hp})(5250)}{\text{Nameplate rpm}}$$

At 1750 rpm, one horsepower is about equal to three lb-ft of torque. You need only ratio hp and rpm to estimate mentally the torque of any motor.

One of the more frequent problems that arises in application work is figuring the time required for a drive to accelerate. Again, the simple formulae above are a basis for this. Let's assume that we choose a time period during which the accelerating torque (which is the difference between the torque produced by the motor and the torque demanded by the load) is constant. Then: $T = (I)(\alpha)$ and, since the change in speed during this time is equal to the acceleration (the rate of change in speed) multiplied by the length of time in seconds,

$$\text{rpm}_{final} - \text{rpm}_{initial} = (\alpha)(\text{time})$$

If we solve for α, we get:

$$\alpha = (\text{rpm}_f) - \text{rpm}_i/\text{time})$$

Substituting that value of α in the first equation,

$$T = (I)(\text{rpm}_f - \text{rpm}_i)/\text{time},$$
$$\text{or time} = (I)(\text{rpm}_f - \text{rpm}_i)/T$$

Coping with Starting Conditions

When numerical constants are inserted to put all the units on a consistent basis (converting minutes to seconds, for instance), we find that the formula for time becomes:

$$\text{Acceleration time, seconds} = \frac{(WK^2)(rpm_f - rpm_i)}{(308)(\text{Accel. T, lb-ft})}$$

There are two important limitations to the use of this formula. The first is that it applies only to a period of time during which the accelerating torque can be considered constant. In Figure 3-15, for example, you *cannot* simply take some estimated average accelerating torque from zero to full speed and get the correct answer. The correct way is to divide up the total acceleration period into a number of segments, such that accelerating torque is fairly constant during each segment. Calculate the acceleration time for each, then add them up. For the example of Figure 3-15, here's how the right way compares with averaging:

(1) Numerical averaging of torque:

Maximum accel. torque at 94% speed = 190 lb-ft
Minimum accel. torque at 70% speed = 95 lb-ft
Average of the two = 143 lb-ft

$$\text{Accel. time} = \frac{(300 \text{ lb-ft}^2)(1800 \text{ rpm change in speed})}{(308)(143 \text{ lb-ft})} = 12.6 \text{ sec.}$$

(2) Dividing acceleration up into five segments, or five steps, each step being 1800/5 or 360 rpm:

	Step	Accel. torque	Time
1.	0-360 rpm	130	2.7
2.	360-720	118	3.0
3.	720-1080	100	3.5
4.	1080-1440	96	3.7
5.	1440-1800	170	2.1
	Total acceleration time	=	15.0 seconds

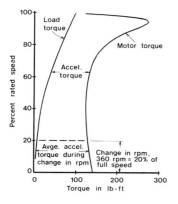

Figure 3-15: Typical load and motor speed-torque curves showing data used to calculate drive accelerating time as well as motor heating. Total system inertia = 300 lb-ft².

This "five-step" method is usually accurate enough. Dividing this example up into ten steps, each one of 180 rpm, gives an answer of 14.99 seconds, which is almost the same as before.

Because this sort of load torque curve, usual for fans or pumps, does have a basic resemblance to the motor speed-torque curve (each shows generally increasing torque with increasing speed), such applications can sometimes be "averaged" more accurately than constant torque loads like conveyors, or the "oddball" type of load such as many compressors. The five-step or *incremental time* calculation, best for any load, is the only safe way to proceed for those.

For *translation*, starting with the $F = (M)(A)$ formula and proceeding in much the same manner as for rotation, we can work out an acceleration time formula for straight-line motion:

$$\text{Acceleration time, seconds} = \frac{(\text{weight in lbs})(V_{final} - V_{initial})}{(32.2)(\text{Acceleration force F, lbs})}$$

Or, if we use velocity "V" in feet per minute instead of per second:

$$\text{Acceleration time, seconds} + \frac{(\text{weight})(fpm_f - fpm_i)}{(1932)(F)}$$

For example: Assume a planer bed and load weighing 35,000 lbs., to be accelerated to 200 ft/min during its return stroke. The accelerating force is 2400 lbs. Acceleration time is:

$$\text{Time} = \frac{(35,000)(200-0)}{(1932)(2400)} = 1.5 \text{ seconds}$$

If force F varies during the total acceleration period, divide that period up into segments during which average force may be considered constant - just as in the case of rotation.

In a rotating drive system, why does the inertia "seen" at one shaft speed differ from that which is effective at some other shaft speed in the system? The principle involved is that the stored rotational energy, or E_K, is a constant throughout the system. Suppose there are two shaft speeds involved, rpm_1 and rpm_2. At both shafts, E_{K1} and E_{K2} must be equivalent. So we can say that:

$$(I_1)(\Omega)_1^2 = (I_2)(\Omega_2)^2$$

so that

$$(I_1) = I_2(\Omega_2/\Omega_1)^2$$

which is the same as saying

$$(WK^2)_1 = (WK^2)_2 \ (rpm_2/rpm_1)^2$$

Figure 3-16 shows an example.

Suppose we have a system in translation rather than rotation—elevator, log carriage, planer, etc. How do we find the equivalent inertia at the rotating drive shaft for the load that is not rotating? Again, the stored energy throughout the system must have the same constant value, no matter what the type of motion:

E_K rotational = E_K translational

$$\text{or } \frac{(WK^2)(rpm)^2}{5910} = \frac{(weight)(V)^2}{64.4} \text{ or } \frac{(weight)(fpm)^2}{232,000}$$

$$WK^2 = \frac{(weight)(fpm)^2}{(rpm)^2}(.0254) \text{ lb-ft}^2$$

As an example, consider a motor belted to a conveyor so that 590 rpm of the motor shaft produces 400 fpm of the conveyor, which weighs 20,000 lbs. The equivalent conveyor inertia at the motor shaft is:

$$WK^2 = (20,000)(400/590)^2(.0254) = 235 \text{ lb-ft}^2$$

If there is a vertical rise or lift involved in the conveyor line, this does not affect equivalent inertia. But it does affect the driving horsepower required.

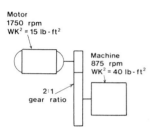

Figure 3-16: A simple drive system to show how total inertia is calculated with reference to motor shaft rpm.

In working with large flywheels such as those used for wood chippers, mine hoists, or punch presses, it is sometimes convenient to know the stored energy in terms of horsepower-seconds. The work done in bringing a system up to speed, or in slowing it down, is equal to the change in system kinetic energy. Work done is horsepower times time, or "horsepower-seconds." But one hp = 550 ft-lb of energy per second, so that:

Hp-seconds represented by a change from one system energy content to another

$$= \frac{(E_{K2}-E_{K1})}{550}$$

$$= \frac{(WK^2)(rpm_2^2-rpm_1^2)}{3,230,000}$$

For translation:

$$\text{Hp-sec.} = \frac{(weight)(fpm_2^2-fpm_1^2)}{127,600,000}$$

If there is *no load torque*, it can be shown that heat loss developed in the rotor when a motor accelerates is equal to the kinetic energy imparted to the rotating elements of the drive, and is therefore equal to E_K. This was expressed in Figure 3-11. The formula for kinetic energy given earlier was:

$$E_K = \frac{(WK^2)(\Omega)^2}{5910}$$

For a motor accelerating from rest, or zero speed, to full speed, this is more correctly expressed as:

Rotor loss in kilowatt-seconds =

$$\frac{(.00136)(WK^2)(\text{synch. rpm})^2(\text{initial slip}^2 - \text{final slip}^2)}{5910}$$

This is the basic formula the motor designer uses to calculate rotor heating. He wants it in kilowatt-seconds (or watt-seconds) because watts are his universal unit of power loss.

"Slip" is the difference between actual motor running speed and its synchronous speed. For example, a 4-pole motor with 2% slip at full load runs at 98% of 1800 rpm, or 1764 rpm; the "slip rpm" is 2% of 1800, or 36 rpm. At zero speed, or locked rotor, slip is 100%. *Per unit* slip is 1/100 of the percent slip, or 1.0 at locked rotor and .02 at full load. It is the per unit figures which are used in the above formula for rotor loss.

This basic formula tells us several interesting things. To begin with, it tells us why a two-speed motor absorbs less heat in accelerating a drive than a single-speed machine. Suppose we need to bring a load up to 1800 rpm. If we use a single-speed, 1760-rpm motor:

$$\text{Initial slip} = 100\% = 1.0 \text{ per unit}$$
$$\text{Final slip} = 40 \text{ rpm}/1800 \text{ rpm} = .022 \text{ per unit}$$

$$\text{Rotor loss} = \frac{(1800)^2}{5910}(1^2 - .022^2)(WK^2)(.00136) = .746(WK^2)\text{kW-sec.}$$

If we use a two-speed 4/8 pole motor instead, 1760/880 rpm:

Low speed initial slip = 1.0
Low speed final slip = .022 $\Big\}$ Accel. from 0 to 880 rpm

$$\text{Rotor loss} = \frac{(900)^2}{5910}(1^2 - .022^2)(WK^2)(.00136) = .186(WK^2)\text{kW-sec.}$$

High speed initial slip = .51
High speed final slip = .022 $\Big\}$ Accel. from 880 to 1760 rpm

$$\text{Rotor loss} = \frac{(1800)^2}{5910}(.51^2 - .022^2)(WK^2)(.00136) = .194(WK^2)\text{kW-sec.}$$

Total rotor loss this way = only .380 (WK^2) kW-sec., or about half that for the single-speed motor.

The second thing the formula shows is that more heat goes into the rotor during the low speed portion of an acceleration than during the high speed. For example, assume a 20% change in speed, starting from rest. From the formula, we get:

$$\text{Rotor loss} = \frac{(WK^2)(\text{synch. rpm})^2(.00136)}{5910}(1.0^2 - 0.8^2)$$
$$= (\text{a numerical constant})(0.36)$$

To reach full speed, the total rotor loss would be this same numerical constant times almost 1.0. Hence, we can see that nearly 36% of the total acceleration heat is developed in the rotor during the first 20% of the speed range.

For the same change in rpm, but accelerating from 20% speed up to 40% speed, we find that the rotor loss becomes the same numerical constant times $(.8^2 - .6^2)$, or .28. For this portion of the acceleration, only 28% of the total heat appears in the rotor. And as the drive approaches full speed, the share of rotor heating being developed continues to decrease.

The relationship makes sense, even without the mathematical background. When the rotor is barely turning at low speed, little rotational energy is being developed. Most of the energy involved is going into rotor heat. On the other hand, when the drive is almost up to full speed, the mechanical energy of rotation is large, and little is going into heat.

If we now introduce into the rotor loss equation the necessary multiplier for load torque, or $T_M/(T_M - T_L)$, we will see that a very low value of accelerating torque $T_M - T_L$ is much harder on the motor during the low end of the speed range than during the high end. ("Average" accelerating torque is no more valid for heating calculations than for accelerating time.) This is why Figure 3-17a represents a much more difficult start for the motor than Figure 3-17b.

Note the contrast here between acceleration heating and acceleration time. For any given value of T_M and T_L, acceleration time is the same from 80 to 90 percent speed as it is from 0 to 10 percent, or 30 to 40 percent. But the rate of rotor heating is not at all the same for each of those three speed increments.

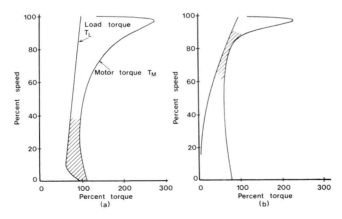

Figure 3-17: The shaded areas show where the ratio $T_M/(T_M - T_L)$ gets quite high, because $T_M - T_L$ is low. Note that in (a) $T_M - T_L$ is greater than in (b), but motor will overheat in (a) at a much lower value of load inertia than at (b) because the motor heat loss due to acceleration is much greater at a low speed than at a high speed.

This appears graphically in Figure 3-18. This rate of heat production within the rotor is, of course, not the same as the rate of temperature rise; actual temperatures within the bars will tend to go up somewhat faster than this at first because of the deep bar effect, tending to restrict the portion of bar cross-section that is being heated. Less mass of bar to absorb the heat means higher temperature. But as the drive comes up to speed, the temperature rise will rapidly taper off because heat "soaks into" the laminations and is given off to the air. The overall pattern remains, however, with temperature as well as heat input; most of the acceleration problem occurs at the lower speeds.

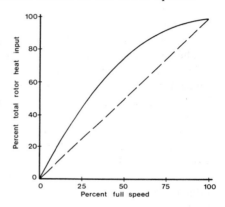

Figure 3-18: Rotor heat loss generated as a uniform function of speed during acceleration would follow the dashed straight line. Because more heat is produced at the lower speeds, however, it actually follows the solid curved line, even with zero load torque during the acceleration.

The third point to remember from all these loss equations is that *they are not functions of time*. In other words, the amount of heat produced in the rotor (and in the stator) during motor starting is exactly the same no matter what the acceleration time may be.

Of course, if acceleration time is lengthened by lowering motor voltage, the ratio $T_M/(T_M-T_L)$ will go up and heat input will increase. Or, if time is lengthened by adding more WK^2 to the load, obviously the heat loss will increase. But if there is no change in inertia and no load torque, a motor will get no hotter from a long (reduced-voltage) start than from a shorter (full-voltage) one.

The loss formula is somewhat awkward to handle as given above. If we multiply and divide the whole thing by 100^2, or 10,000, we get:

$$\text{Rotor loss} = .00231(WK^2)\left(\frac{\text{Synch. rpm}^2}{100}\right)(\text{slip}_i^2 - \text{slip}_f^2)\left(\frac{T_M}{T_M - T_L}\right)$$

in kilowatt seconds

To get the answer in watt-seconds, use 2.31 instead of .00231.

As a simple example of how this is used in duty cycle calculations, in the five step method, take the curves of Figure 3-15. Breaking out the constant numbers in the rotor loss formula just given:

$$(2.31)(300 \text{ lb-ft}^2)\left(\frac{1800}{100}\right)^2 = 225{,}000$$

For the first interval from 0 to 20% speed, $\text{slip}_i = 1.0$; $\text{slip}_f = .8$; $T_M = 135$ lb-ft, $(T_M - T_L) = 130$. Therefore, for this first interval:

Watt-seconds rotor heat $= 225{,}000\,(1^2 - .8^2)(135/130) = 84{,}200$

Second interval, 20% to 40% speed:
Watt-seconds $= 225{,}000(.8^2 - .6^2)(125/1181) = 66{,}700$

Third interval, 40% to 60% speed:
Watt-seconds $= 225{,}000(.6^2 - .4^2)(126/100) = 56{,}900$

Fourth interval, 60% to 80% speed:
Watt-seconds $= 225{,}000(.4^2 - .2^2)(143/96) = 40{,}300$

Fifth interval, 80% to 98% speed:
Watt-seconds $= 225{,}000(.2^2 - .02^2)(250/170) = 13{,}200$

Total watt-seconds $= 261{,}300$

If the average R_1/R_2 ratio during acceleration is 0.4, then an additional 0.4(261,300), or 104,500 watt-seconds of heat is produced in the stator.

High torque loads

Let's look more closely now at "high torque" loads—those that require motors designed with very high locked rotor or accelerating torques, such as the NEMA C. These drives normally process solid materials, rather than liquids or gases.

The first of them, and the only one that is not, strictly speaking, a rotating load at all, is the conveyor belt. It takes a fairly high torque to break away, to get motion started in the rollers and pulleys; from then on up to full speed the torque is constant. (See Figure 3-17.) The inertia of a loaded conveyor belt can be high, though seldom as extreme as for a large fan.

Comparing Figures 3-3b (page 93) and 3-19 shows why a conveyor can represent severe starting duty which can seldom be handled by a NEMA

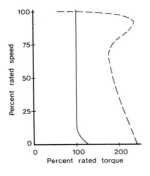

Figure 3-19: Speed-torque requirement of a loaded belt conveyor. Normal driver is a NEMA Design C motor, with speed-torque output per the dashed curve.

Design B motor. There just isn't enough motor torque available to overcome the high load requirement in the lower speeds, especially if any voltage drop occurs during starting. A high torque NEMA Design C machine, usually with a double cage rotor, is normal for conveyors.

Unfortunately, in many conveyor systems, too much motor torque is as bad as too little. Excessive torque can over-stretch and damage the belts. Whenever you are involved in motor selection for any belt conveyor, then, make sure you get from the conveyor manufacturer any limitation he has on *maximum* torque during the acceleration period. Over and over again, it turns out that this upper limit is so close to the *minimum* torque needed to accelerate the drive that no realistic motor design is possible. The solution: conveyor redesign, or a clutch.

The same type of motor is also needed for the pulverizer—most often found in power plants to produce finely powdered coal for boiler firing. Similar in their operation are many types of crushers and grinders, including ball mills used for processing paint pigments. The speed-torque curves of Figure 3-20 are typical. Again, inertia is seldom a problem, but motor torque must be unusually high.

Figure 3-20: Several types of pulverizer/mill speed-torque curves; again, the dashed curve is the NEMA C motor capability usually required.

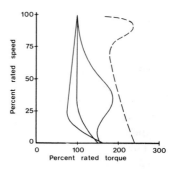

Caution: although these loads do have relatively low inertia, motor acceleration heating can be severe because of the high $T_M/(T_M-T_L)$ ratio at low speeds. The NEMA C double cage rotor develops most of this heat in the top or outermost of its two squirrel cages, which because of space limitations has much less mass or volume of bar material than would be true for a normal-torque single cage design of the same diameter. Less material plus extra heat input means very high cage temperatures. Do not expect a double cage motor to safely withstand nearly as severe a start as a single cage design. This makes it all the more important that complete application data be obtained on such drives.

Most crushing or pulverizing machinery is started "unloaded"; that is, with the bowl or chamber empty. If the machine stalls while running, the supposition is that it will be emptied manually—sometimes by tedious digging out with shovels—before the motor is restarted. But you can expect someone sooner or later to try a loaded restart, figuring that the motor is husky enough to save all that hand work. Therefore, the drive motor must be designed with enough torque (and thermal capacity) to start loaded, despite the pulverizer manufacturer's claim that this should never happen.

Mixers present similar problems, especially in the rubber industry. The dragging of paddles or rollers through a mass of viscous, semi-solid product is often an extremely high torque load for the motor. If the mixer is of the batch type, in which one lot of material is mixed, then unloaded, and the motor restarted after reloading, production rates will depend on speed of getting the processed material out and the new load going. Operators are reluctant to let the drive just coast to a stop when mixing is complete. So it's common practice to plug-stop the drive by pushing the "reverse" button while the motor is running.

Borrowing again from the physics of the induction motor, it can be proved that plug-stopping a motor puts three times as much heat into the rotor and stator as one normal start does. So this is rough duty. Most manufacturers will not warrant their motors for plug-stopping or reversing above the NEMA frame sizes because of the extreme heating and stress on parts that plug-stopping causes.

Even without plugging, mixers need high torque. Furthermore, the torque may be hard to predict. Material being processed often changes viscosity during the process, either because of temperature changes alone or through chemical reactions.

As a group, then, mixing machinery is one of the most difficult of motor applications to deal with. Fortunately, these make up only a small percentage of all motor drives.

What kind of high torque load is most likely to call for the NEMA Design D motor? The punch press drive is one of the most common applications, but in recent years scrap metal or solid waste shredding machinery has begun using motors of this type up to 6000 or even 10,000 hp. The Design D machine is intended for the type of load which imposes periodic sudden and very severe torque peaks on the driver. The recurrent strokes of the press, or the steady feeding of heavy chunks of scrap for chopping or hammering into smaller pieces, are typical examples. No motor could withstand these constant torque surges without damage or overheating, so such drives normally include a relatively high inertia flywheel—either as a separate item, or built into the rotating structure of the shredding machinery.

As already pointed out, high flywheel inertia constitutes a high resistance to any change in drive speed. When a sudden resistance to rotation appears in the form of a high torque demand from the load, considerable energy must appear from somewhere to overcome that obstacle—and in a flywheel drive that energy is supplied by the high rotational energy stored in the rotating inertia. The drive (including the motor) tends to try to slow down suddenly, but this simply allows the flywheel — which tries to prevent that change in speed—to supply most of the sudden power demand out of its own stored energy. Unlike Designs B & C, the motor itself has a speed-torque curve which permits a large slowdown without going anywhere near the stalling point.

So a NEMA Design D motor can be considered a "lazy" motor which "lets go" of the load considerably when suddenly hit by a torque peak, letting the connected flywheel step in and do most of the hard work.

Because such motors have high slip, as much as 13% at full load instead of the 1% to 3% common with Design B, they must have high resistance rotors. Their very high full load losses also mean that NEMA D motors tend to be oversized for the horsepower rating—in the larger sizes, as much as double the physical size of a NEMA B of the same horsepower.

They are therefore capable of accelerating quite high inertias, which is nicely suited to their frequent use in drives of that type. But unless the great capacity for periodic slowdown and speedup is needed, as with the shredder drive, the high losses and high cost of the NEMA D do not make it an attractive choice.

The last major load category to consider, representing about 18% of all motor-driven machinery, is the compressor or blower. These form a class by themselves because of the great variability in their speed-torque and inertia characteristics. No assumptions are possible. There is no "typical" compressor.

To realize why this is true, consider what a compressor does. It is used to pressurize some type of gas—either to move large volumes at relatively low pressure, as in a natural gas pipeline, or to put fairly small volumes under extremely high pressure, as in a nitrogen liquefaction plant.

In contrast to pumps, where there are few basic variations in the construction of fluid-moving impellers, compressors come in a bewildering variety of constructions. There are the "bladed" types, with radial or axial blades like the fans (but which may be either straight, twisted, or elaborately curved). There is the "lobed" type, in which intermeshing lobed rotor drums move the gas. There are screw compressors; sliding-vane compressors in which the fluid is moved by paddles thrown outward by centrifugal force so they form a seal against the inside of the compressor housing; and piston-type compressors that work like the hand-operated tire pump.

Many of these variations have different speed-torque requirements for acceleration. The curves vary with gas density, with gas temperature, and with the use of unloading or bypass valves of many different kinds.

Figure 3-21 is an indication of the range of compressor speed-torque curves which you may encounter. What may appear to a process plant operator as relatively minor changes in his operations can swing a compressor drive from one of these curves, representing easy starts for the motor, to one representing a severe start. Often the operator will be unaware of what he has done. The compressor manufacturers themselves, as has been repeatedly learned through negotiations with them, seldom realize the exact speed-torque curve for a particular machine because they have not tested it to find out. If you ask a compressor manufacturer for "curves," you will usually get performance curves involving pressure and flow or efficiency; many times speed-torque

Coping with Starting Conditions

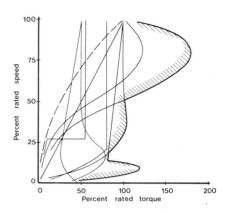

Figure 3-21: A group of actual compressor speed-torque curves, at speeds (full load) from 705 to 3560 rpm. The dashed line, for comparison, is a standard pump or fan "square law" load curve. Obviously, there is no "typical compressor curve." To handle any of these loads, a "typical" compressor motor would have to have an accelerating torque in excess of any of the values bounded by the shaded "envelope" of all the curves—an unlikely condition in any but machines of the lowest horsepower.

curves are not even available. Such curves are seldom thought of because they do not enter into compressor "performance"—and this often happens with fan or pump manufacturers as well.

Figure 3-22 is an example. Here we have four different manufacturers of the same basic product (bladed-rotor compressors, centrifugal) operating at the same speed, handling the same product (atmospheric air), and being driven by the same type of motor—but the compressor speed-torque curves are claimed to be widely different.

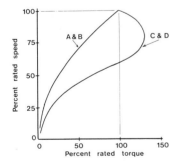

Figure 3-22: Are high speed centrifugal air compressors all alike? Manufacturers A and B say their curve applies, whereas manufacturers C and D quote something quite different. The same motor may be sought to accelerate either version.

One type of compressor which creates special problems is the reciprocating unit. This was the most common type of compressor for supplying compressed air in industrial plants until the modern "plant air package," with its high speed centrifugal machine, took over. Because of the constant cyclic buildup and exhaust of pressure in the individual cylinders, the "recip" demands a pulsating or cyclic load torque from the motor. These rapid torque fluctuations are accompanied by similar swings in motor line current. An annoying flicker of plant lighting may result. To damp out those swings, NEMA has determined that the current or torque pulsations must be limited, which in turn requires a certain amount of inertia in the drive. Just as with a punch press, the inertia or "flywheel effect" acts to smooth out the torque pulsations.

Sometimes the required flywheel is so large that a special motor design must be used.

Evaluating inertia

In diagnosing motor problems traceable to load acceleration, or in trying to apply new motors to existing machines, what can be done to evaluate load inertia? The first step, of course, is to seek information from the driven equipment manufacturer. Remember to account for two things in any drive:

(1) *Total* connected inertia. The motor must accelerate not only the driven machine itself, but all couplings, gears, or belt pulleys between motor and load as well. In some drives these "minor accessories" add more inertia than the driven machine itself.

(2) The speed to which each value of inertia is referred. In a direct-coupled drive this doesn't matter, but it does when belting or gearing is involved. (See Figure 3-16, page 109.)

If a user or machinery builder gives you this kind of information for a belted drive, check it out:

> Driven sheave pitch diameter 16.6 inches
> Driver sheave pitch diameter 5.9 inches
> Inertia of load 459 lb-ft^2 at driven shaft
> Inertia reflected to motor shaft (neglecting sheaves) = 45 lb-ft^2

Every so often, you'll find that the user did his arithmetic wrong. The drive speed ratio has to be the same as the ratio of sheave pitch diameters. Therefore, in this example, the inertia at the motor shaft has to be 459 $(5.9/16.6)^2$ which is 57.6, not 45.

Watch out, also, for this sort of data:

> Driven shaft speed 545 rpm
> Motor shaft speed 1190
> Driven sheave pitch diameter 30 inches
> Motor sheave pitch diameter 14.15 inches

Again, the speed ratio must be identical to the sheave diameter ratio. But here we see a speed ratio of 545/1190, or .458, with a sheave diameter ratio of 14.15/30, or .473. You would be surprised how often this happens, often the result of a machinery builder's having assumed the wrong motor speed, a problem referred to earlier.

In evaluating motor starting duty, what is the application engineer's role? First, he should make every effort to find out what the actual load speed-torque curve will be so he can let the designer know. Consider, from typical curves shown earlier in this chapter, how to decide whether or not a curve given by a user or equipment manufacturer is reasonable based on operating conditions.

Second, the man in the field should find out how often the user expects to start the drive. Is it started and stopped automatically, perhaps several times hourly? Are there "run-in" periods during which this can occur at infrequent intervals, or does it go on around the clock? Is the drive to be used one shift a day, or 24 hours a day seven days a week? What kind of operator monitoring will there be?

Third, the application engineer should realize how acceleration capability varies with motor size. The smaller a motor is, the more severely and the more often it can be started. There are several reasons for this. One is that as motor size decreases, the mass of material provided for the horsepower tends to increase, thus tending to lower acceleration temperature rise. Also, as Figure 3-3 (page 93) shows, motor accelerating torque tends to be much higher for smaller ratings. Finally, stator end turns in small motors are much shorter and stiffer than in large motors, and rotor bar overhang either shorter or nonexistent, so that magnetic forces and heating have less effect.

Thus, a 3-hp motor may be able to start its load several times a minute, continuously. But a 300-hp machine may have to be limited to only a few starts daily, spaced far enough apart so the motor can cool itself back down to normal temperature before another start is attempted. Typical starting frequency limits are shown in Figure 3-23.

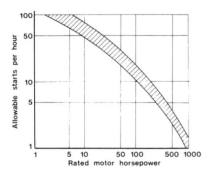

Figure 3-23: Number of starts per hour permissible for typical motors accelerating low inertia drives such as centrifugal pumps. Above the 200- to 250-hp range, the figures may apply for only one hour at a time, to be repeated infrequently.

Motor starting

Besides an understanding of load and motor characteristics, a knowledge of power system and starter behavior is important in analyzing motor acceleration. To get the motor started, we must somehow supply from 5 to 10 times its normal rated current for a short time. A power system may simply be incapable of doing this. Or it may be able to supply the current, but only at greatly reduced voltage. When high motor starting current flows through the impedance of feeder cables and transformers, voltage available at the motor either dips too low for safe starting, or to sustain the other loads connected to the system elsewhere—contactors drop out, lights go dim; other motors no longer develop enough torque to keep running.

If system voltage during motor starting cannot be kept high enough any other way, users should consider capacitors on the motor feeder circuit. Firm advice on capacitor sizing and cost is normally beyond what the motor application engineer can provide, but he should be aware of this solution. The capacitors will be much larger than those which would be used for correcting motor power factor.

For example, one 400-hp motor on the end of a "weak" power line had its starting voltage brought from 90% of rated up to 99% through the addition of 1300 kVA of shunt capacitors costing $2 per kVA. Power factor correction capacitors would only have totalled about 175 kVA.

Users are sometimes confused about starting capacitors, thinking their main purpose is to adjust power factor. This is certainly one result of the installation, but not its main function, which is to boost line voltage. Because that boost becomes quite large when the motor has reached full speed and its current demand drops off, means must usually be provided to switch starting capacitors off the line quickly when the drive has fully accelerated, or a little before. This can be done with speed switches; with timers if acceleration time is always the same; or with var-sensitive switching equipment which in effect monitors the variation in line power factor. Control or capacitor manufacturers can best advise the details.

"Reduced voltage starting" of motors to minimize the system disturbance by lowering motor inrush normally employs one of the methods outlined in Table 3-III. Although deliberately reducing motor starting voltage (and torque) may seem self defeating, it is common simply because if necessary the lost accelerating torque can often be restored through special motor design which increases torque per ampere of current. So the motor may still be capable of accelerating its load.

Actually, though usually considered in this category, neither part-winding starting nor star-delta starting are "reduced voltage" methods. Full line voltage is applied to the motor in either case, but in part winding starting it is applied to only a portion of the entire motor winding. In star-delta, the motor internal connections are changed so the full voltage is applied across more coils in series, reducing the torque developed as well as the current required.

In considering the effect of voltage reduction on a drive, it is common practice to assume that motor torque varies directly as the square of the applied voltage. This, unfortunately, is not true. It is also not true that inrush current varies directly as the applied voltage. The reason lies in the basic behavior of electromagnetic circuits containing the mixture of iron and air flux paths present within a motor. These paths "saturate," as we see in Figure 3-24. In the usual working range of a motor's magnetic field, the flux (generated by the applied voltage) is not directly proportional to current (which produces the torque). The relation is non-linear. Any change in voltage or flux, on the horizontal scale in Figure 3-24, is associated with a relatively larger change in current or torque on the vertical scale. Tests show that, in general, motor torque during acceleration (the T_M we looked at earlier) varies about as volt-

Reduced voltage starter characteristics

Feature	Autotransformer Closed transition	Primary Resistor	Primary Reactor	Part-winding	Star Delta
Smoothness of accel. (1 = smoothest)	2	1	1	4	3
Application flexibility (1 = most flexible)	1	4	5 (normally used with the highest voltages & currents)	3	2
Allowable acceleration time (typical)	30 seconds	5 seconds	15 seconds	5-15 seconds, depending on motor design	45-60 seconds, depending on motor design
Equipment	3 contactors, timer & starting device (the transformer)	2 contactors, timer & starting device (resistor)	2 contactors, timer & starting device (reactor)	2 contactors & timer	3 contactors & timer (open transition), or 4 contactors & timer & resistors (closed trans.)
Approximate cost comparison (1 = lowest cost)	3-4	2½-3	2½-3	1	2
Approximate motor cost comparison	standard squirrel cage motor				may cost ⅓ to ½ more than std. motor

Table 3-III. A few comparisons between the popular methods of reduced voltage starting. (Part-winding and star delta are not reduced voltage methods, strictly speaking, but are used for the same purpose.)

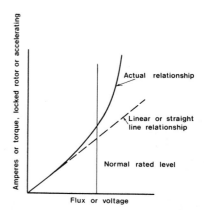

Figure 3-24: How magnetic saturation affects motor behavior. As voltage rises above rated, current (and torque) tend to go up faster than the relative voltage changes. When voltage decreases, torque (and current) decrease relatively faster than the voltage.

age to the 2.2 power, or $T_M = V^{2.2}$. Current varies as voltage to the 1.1 power, or $I = V^{1.1}$. If we apply 80% rated voltage to a motor, we don't get 64% as much torque as before; we get only 60% to 61%. Tables 3-IV and 3-V are useful in evaluating this effect for typical motors.

Concerned about loss of motor torque with reduced voltage starting, many users have begun specifying motors with higher than normal torque to begin with. But they often go about it the wrong way. One of the more common requests is for motors with "100% minimum locked rotor torque." How useless that is as a means of ensuring safe acceleration at reduced voltage is clear from the curves of Figure 3-25. (Look again at Figure 3-2, as well.) Note how the motor with 80% locked rotor torque can accelerate easily at the lower voltage, whereas the one with 100% locked rotor torque cannot.

As pointed out earlier, there is no substitute for knowing what the load torque is, and making sure that the motor designer knows it, too, as well as knowing what the motor speed-torque curve itself can be expected to look like.

When reduced voltage starters of any kind are used, there are two critical factors to consider besides reduction in motor torque:

(1) What is the relationship between nominal "starting voltage" and the actual motor *terminal* voltage?

(2) Is it necessary for the motor to reach full speed before the starter restores nominal full voltage?

It often happens that starter and motor are located some distance apart. Even when the straight line distance is small, the cabling may follow a lengthy path. One user with a group of 200-hp motors was encountering starting difficulty because of voltage drop. His power system was relatively "stiff"—that is, it was able to supply high starting current with little voltage drop. But the motors were 1300 feet from the starter—a quarter mile. Despite large cables, terminal voltage sagged considerably during motor starting.

So when investigating any such problem in an existing drive, make sure that you don't rely on voltage measurements at the control center as any indication of what's happening at the motor. And don't assume that the feeder cables are

Coping with Starting Conditions

Type of starter	Starting characteristic in % of full voltage value		
	Voltage at motor	Line current	Motor accel. torque
Full voltage	100	100	100
Autotransformer-tap:			
80%	80	64 (61)	64 (61)
65%	65	42 (39)	42 (39)
50%	50	25 (22)	25 (22)
Primary resistor	80	80 (78)	64 (61)
Part-winding	100	60-70*	40-50*
Star delta	100	33 (29)	33 (29)

* There are literally dozens of part-winding connection schemes having widely varying effects on torque and current. The values shown here are typical of commonly used connections. In scientific applications, find out from the motor manufacturer what values will apply.

Table 3-IV. You will often see tables like this in starter catalogs or application literature, but without the figures in parentheses. Because motor torque does not vary as voltage squared during starting, nor current directly as the voltage, those figures have been inserted here as reminders of the more likely true values of accelerating torque and current that will result from using the various starting methods listed.

Percent rated voltage at motor terminals	Multiply torque at rated volts by ...	Multiply amperes at rated volts by ...
95	.893	.945
90	.792	.890
85	.700	.836
80	.612	.782
75	.531	.729
70	.456	.675
65	.396	.622
60	.325	.580
55	.268	.518
50	.218	.466

Table 3-V. Use these multipliers for more realistic results when ratioing motor starting torque and current down to account for reduced terminal voltage.

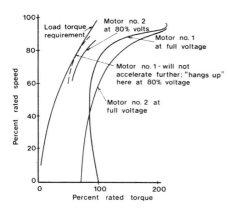

Figure 3-25: The common request for "100 percent minimum locked rotor torque" for a large motor will not ensure drive acceleration at reduced voltage, as the curve for "Motor No. 1" shows. Motor No. 2, with less than 80 percent locked torque but designed for the specific application, will do the job easily. Smaller motors have higher torques as standard, but with higher-torque loads or lower starting voltages the comparisons will be similar.

short, just because the motor sits fairly close by.

The bus voltage itself may also droop, even when a reduced voltage starter has been used to lower motor inrush. What might have been a 50% dip in voltage for an across-the-line start becomes 20%—still a serious condition that can keep the drive from accelerating.

It cannot be assumed, then, that an "80% tap" autotransformer start will produce 80% of rated voltage at the motor terminals. What can be done about it? Larger cables; circuit rerouting; bigger transformers—all of these are possible solutions. So is the substitution of a higher-torque motor at the same inrush. But none of these solutions is inexpensive. Some such steps may be impossible. The most common remedy is to simply let the motor accelerate only part-way to full speed on the reduced voltage starter and to allow the initial step of inrush to trigger system or feeder regulators.

These devices, reacting to the sudden current demand, will either change transformer taps, alter line impedance, or cut in capacitor banks to help restore the drop in voltage. They can be installed in the plant, or on the utility line. After a timed-out interval of three to five seconds, during which it doesn't matter much whether the motor reaches any particular speed, the starter switches to full voltage. The motor then sees enough voltage to get it the rest of the way up to full speed.

The curves of Figure 3-26 show how this can mean the difference between a routine acceleration and one that is impossible. Motors have been specially designed on costly, oversize parts because they were ordered for "reduced voltage starting" when, in reality, they didn't have to "start" at all, only to hang in there for a few seconds while the regulators functioned. For this reason, the application engineer needs to be fully aware of how the starting system will be used.

Those initial few seconds of greater than normal motor heating are seldom a problem. In any event, the motor designer, if properly consulted when the drive is planned, can calculate how many seconds may safely elapse before the transfer is necessary.

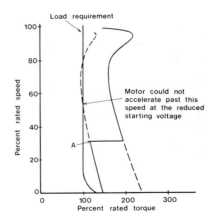

Figure 3-26: High-torque motor accelerating conveyor using 80% autotransformer start and switching to full voltage at about ⅓ speed (Point A). Note that if the drive had to reach full speed at 80% voltage, the motor would have to be designed for much higher torques. Point A can be chosen by the motor designer to be as far out in elapsed starting time as possible without producing excessive motor heating, which (already high in this low speed range) is greatly increased by the high ratio of $T_M/(T_M-T_L)$.

Since it makes little difference in this "voltage recovery" application whether the motor even starts rotating on the first step of the starter, a relatively low torque motor can be used. On the other hand, if regulators can't be of enough help, or aren't being used, then the motor must be capable of bringing the load to full speed on the reduced voltage, and a higher torque design will be needed. (See Figure 3-26.)

The question, "Does the motor have to reach full speed on the first step?" is particularly important for part-winding starting. Most motor designs are inherently unable to reach full speed under load on a part-winding start. Many cannot do it even when unloaded. This type of starting is unique in that it depends for its operation on special winding connections inside the motor itself, rather than on the behavior of the starter alone.

Literally dozens of different arrangements have been tried, many of them well-suited to commercial application. Over the years, differing emphasis on one performance item or another has caused shifts in popularity among several types of part-winding connections.

Such variations result from a basic difference between this mode of starting and the autotransformer or reactor arrangements. The latter are true "reduced voltage starting" methods. In each of them, a normal motor winding is entirely in the circuit on the "first step" (closing of the contactor which energizes the initial portion of the winding).

But because of either the motor connection itself or the starter circuit, every coil on that first step sees a lower voltage than for an across-the-line start. The reduced voltage may be 50%, 65%, or 80% of normal (for the autotransformer start), or various other figures for tapped reactors. That lowers motor starting current accordingly, a major objective of reduced-voltage starting. All winding currents and voltages, as well as the magnetic field, remain balanced and symmetrical.

Almost all part-winding schemes differ from that in three ways. First, not all the motor winding is energized on the first step. Only half, two-thirds, or some other fraction of the coils is involved. Second, the energized coils may not be uniformly distributed around the stator. The magnetic field is therefore non-uniform or high in harmonic content. That can drastically change the shape—as well as the magnitude—of the motor's speed-torque curve during acceleration. Third (though there are exceptions), those coils which are energized on the first step will see full voltage. Part-winding is really a "reduced current" starting method rather than reduced voltage.

Those first two differences between this and other starting methods make possible a wide variety of connections, the largest group in common use being the "half-winding" arrangements in which 50% of the coils are used on the first step. A 1955 IEEE paper reported tests of ten different half-winding connections for a 75-hp, 4-pole machine. Figure 3-27 shows the coil groupings for some of them; Figure 3-28 gives test results. Most such connections result in motor starting torque of about half the full-winding value; current, from 65% to 80%.

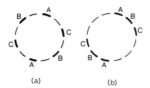

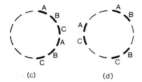

Figure 3-27: Four of the many possible coil groupings for energizing half of a 4-pole, 3-phase motor winding. Probably those in most common use today are (a) and (c). The version at (b) is also a strong contender.

Figure 3-28: Speed-torque curves showing the part-winding starting behavior of the connections in Figure 3-27. The table also shows how inrush current and motor noise on the first step compare with a linestart. Because none of these "half-winding" connections applies less than full voltage to the first-step coils, magnetic noise can never be less on part-winding than for a linestart—and is sometimes much greater because of flux harmonics.

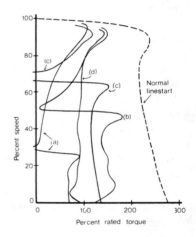

	Percentage full winding inrush	Noise, dB
Linestart	100	91
(a)	68	95
(b)	72	93
(c)	59	97
(d)	68	109

Figure 3-29: The simplest part-winding starter circuit. This "3-3" arrangement uses two contactors, each rated for half the motor current. It is usable for either star or delta windings. The first step contactor is "1M"; the "2M" contactor energizes the remainder of the winding.

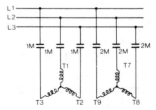

Any of these winding connections can use the same basic starter circuit, of which there are fewer variations. Most usual is the so-called "3-3" starter, so named because it uses two identical 3-pole contactors, as shown in the diagram of Figure 3-29.

Selecting connections

What determines which connection to use? Part-winding starting in general may be expected to meet any of these sometimes-conflicting criteria, depending upon the user's needs:

(1) Quiet starting, without undue vibration or "groaning." This is of special importance for such applications as office air conditioning systems. (Note the noise variation in Figure 3-28.)

(2) Minimum starting current.

(3) Enough torque so the drive reaches a high speed on the first step, minimizing both noise and power system disturbance when the second half of the winding is energized.

(4) Ready adaptability to standard stock 230/460 volt motors.

(5) Safe acceleration of load inertia reasonable for the motor size.

Which of those conditions is most important has varied from time to time. Inrush current got the most attention at first, particularly after the 1955 issuance of NEMA Publication No. 151, "Polyphase Motor Starting Current Rules," drafted by a joint committee of industry and public utilities. It called for automatically controlled motors supplied from "combined light and power secondaries" to have locked rotor current not exceeding 65% of the NEMA Design B value. The intent was to minimize randomly occurring light flicker in commercial buildings caused by automatic switching of sizeable air conditioning loads on power systems of limited capacity.

Unfortunately, the lowest inrush part-winding connections weren't necessarily the quietest. As power systems continued to become "stiffer," the emphasis shifted to noise reduction, leading to other part-winding schemes that achieved lower motor noise level at the expense of higher inrush. Like the earlier versions, such motors all tended to have "cusps" or "saddles" in the first step speed-torque curve (Figure 3-28a, b & c), preventing acceleration of any load to full speed. Arrangements providing acceleration to ⅔ speed—though sometimes noisier—became more popular than those allowing only ¼ or ½ speed on the first step.

Goal Number (5) was especially troublesome. One might suppose that because its starting current is much lower than the linestart value, a part-winding motor would not get as hot during a start as would a full-winding design. But it doesn't work out that way. The total heat generated in bringing a drive up to speed is constant (if we can neglect load torque and consider only inertia of the rotating parts). So the motor stator and rotor must absorb the same amount of heat whatever the starting method.

Winding temperature rise, however, is not just a function of the heat developed. It also depends on how much copper there is to absorb that heat. On the first step of a "half-winding" start, only half the total stator copper is in the circuit to absorb the initial heat. So it's reasonable to expect its temperature rise to be twice as great. Looking only at stator current: During a linestart, all the first step coils carry half the normal inrush. On a part-winding start, those same coils carry 65% to 80% of that same inrush. So the I^2R loss in those coils will be $(.65/.5)^2$ or 1.7 to 2.5 times as great.

If load torque is too great to neglect, total motor heating will depend not only on current squared, but also on starting time. If a high load torque lengthens acceleration time on the part-winding connection, total heat (and temperature rise) go up still more. This is especially significant at the lower speeds during starting, when the heat is building up at the highest rate. Hence, although the load many motors can safely start is limited only by heating in the rotor, part-winding motor accelerating capability is limited by stator winding temperature rise as much as four times greater than for a full-winding linestart.

Overcoming that drawback was one reason for the development of the "⅔ winding" or "4-2" starting method, the most popular of several variations in which the winding is not divided into equal halves. The 4-2 starter, using one 4-pole contactor plus one 2-pole contactor, energizes two-thirds of the winding on the first step, as shown in Figure 3-30.

The 3-3 starter works, with equal current through each contactor, for any of three motor designs:

(1) 6-lead single voltage star as in Figure 3-29;
(2) the same except delta;
(3) 9-lead dual voltage star, part-winding starting on the lower voltage only.

But for a standard 9-lead dual voltage delta-connected winding, the two con-

Figure 3-30: This is the "⅔ winding" use of a 4-2 starter. On the first step, two-thirds of all the motor phase groups are energized, rather than half. Harmonics are minimized if the coil groups are connected "alternate pole" or "top to bottom" in such a way that one of the two parallel paths in each phase contains only north poles, the other only south poles. Line currents on the first step will be unbalanced—nearly equal in two lines and half again as great in the third line, averaging about 65% of linestart inrush.

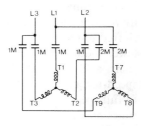

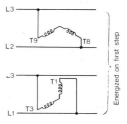

Coping with Starting Conditions

tactors do not carry equal currents. One of them must be oversized for the horsepower. This is explained by Figure 3-31.

The 4-2 starting circuit of Figure 3-32 is suited to either delta or star-connected motors, and can be used to energize half the winding on the first step instead of ⅔ if the motor is so arranged. It also reduces first-step noise and vibration, as well as allowing a motor to accelerate a heavier load safely. And both contactors carry the same amperes.

Eventually, the NEMA joint committee emphasis on inrush current disappeared. Today, NEMA Standard MG1-14.36 says only that "a commonly used connection results in ... approximately 60% of normal locked-rotor current. ... If actual values of torque and current are important, they should be obtained from the motor manufacturer."

Figure 3-31: Right: What happens when the 3-3 starter is applied to the lower voltage of a dual voltage delta-connected motor. Once the motor is up to speed with both contactors closed, each half of each phase in the winding carries the same current, A_2. But contactor 1M must carry a much larger total current than contactor 2M. The vector diagram below shows how the individual currents combine at different phase angles to produce that result.

A = vector sum of A_2 and A_2
A_1 = vector sum of A_2 and A
A_T = vector sum of A_2 and A_1 = 2A

Mathematically:
$A = \sqrt{3}(A_2)$
$A_1 = 2.64 A_2$
$A_T = 3.46 A_2$

Therefore, $A_2 = A_T/3.46 = .289 A_T$
and $A_1 = 2.64 (.289 A_T) = .764 A_T$

So instead of carrying half the line current of the running motor, contactor 1M must carry more than ¾ of it.

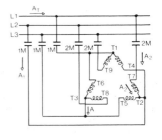

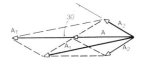

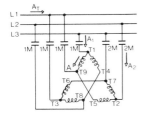

Figure 3-32: Left: How the 4-2 starter is used with a dual voltage delta winding, on the lower voltage connection, for part-winding starting with the same running current through all contactor poles. This avoids the need for one contactor to be oversized, as it must be for the arrangement of Figure 3-31 (above).

A_T = total line current $A_1 + A_2$ and $A_2 = A_1 = 1.73A$

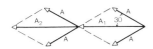

It is now commonly assumed in part-winding applications that the main objective is not to reduce motor current throughout the entire acceleration, but only to lower it long enough (on the first step) to permit operation of system voltage regulators. Within a few seconds after quick-acting "step regulators" sense the first-step inrush, they can restore much of the dip in system voltage caused by the current. When the rest of the winding is energized, the second increment of inrush current will have less effect.

However, to achieve reasonably quiet acceleration to full speed on the first step and minimize overall starting time, the so-called "double delta" connection was developed. (See Figure 3-33.) It was found that raising coil pitch to one slot under full pitch could reduce magnetic field harmonics enough to make the torque cusps disappear. The double delta is a true reduced voltage starting method, because all coils are energized on the first step, and each does see a lowered voltage. Winding heating is greatly reduced with all the copper in circuit.

Figure 3-33: Typical phase with 1M contactor closed is shown in the lower diagram. This is the "double delta" part winding circuit. All coil groups are energized on the first step, in contrast to all other part-winding connections. Each phase contains two groups in series, so that this is truly a "reduced voltage" method. Locked rotor torque on the first step for a standard motor connected this way may be only 30% to 40% of the full winding value, in the 25- to 50-hp range, which is why double cage rotors are often used.

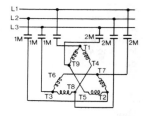

There are some tradeoffs. The higher coil pitch may "weaken" the motor so that accelerating torque is reduced throughout a start. In the 50- to 150-hp range, this has often required a special double cage rotor to restore the lost torque. The cost may be higher, and motor performance at full speed may suffer.

Three-step starting

Other part-winding starting circuits have been devised to bring a motor up to speed in three steps rather than two. Figure 3-34 is an example. The added complexity of such arrangements, involving three contactors with as many as seven poles all together, is seldom worth it, and they are seldom used. At least one 75-hp, 4-pole motor has been tested with four-step starting, energizing an additional ¼ of the winding on each successive step. Although inrush current was kept quite low, the magnetic noise was so great that the result could not be recommended for commercial use.

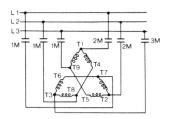

Figure 3-34: A "three-step" start using three separate contactors. First-step inrush current is about 40% of the full-winding value and about 80% on the second step. Wide variation in acceleration characteristics is possible by interspersing coils or splitting up phase groups. Those exotic arrangements are seldom used.

In 6-lead stators like that of Figure 3-29, can the two neutral points be joined together? Theoretically, such an interconnection improves magnetic balance, reducing the tendency of a somewhat distorted air gap to cause rotor pullover.

For 3-3 starting, the interconnection is permissible. If the user happens to install a 4-2 starter he might have on hand, however, the joining of the two neutral points will energize part of the second half of the winding on the first step. (See Figure 3-35.) So when rewinding or reconnecting such stators, it's best to keep the neutrals separated.

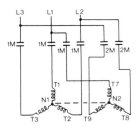

Figure 3-35: If a 6-lead star-connected motor is used with a 4-2 starter, interconnection of the two neutral points (dashed line) would cause part of the second half of the winding to be energized on the first step.

As for unbalanced magnetic pull, it certainly appears that the circuit of Figure 3-27c, having all the energized coils on the same side of the rotor, would generate much greater pull than would occur on a linestart. But that doesn't happen. Although there are distortions in the flux pattern, the magnetic field does extend all the way around the air gap. And it is a weaker field than would be present with the entire winding in circuit. That's why current and torque are reduced on the first step. So if noise and vibration are within acceptable limits for the application, there is no more danger of pullover than for any other motor, and no reason to avoid part-winding on that account.

2-pole motors

Can 2-pole motors be successfully connected for part-winding starting? Yes, although there is little need for it; overall cost is likely to favor some other starting method. The greater harmonic content of the 2-pole magnetic field, compared to that of slower speed motors, results in poor starting characteristics for the simple half-winding connection.

Examination of the 4-pole groupings in Figure 3-27 shows that there are always two separated "A" phase groups, some distance apart around the stator, energized on the first step. The same is true of the other two phases. The separation may be ¼ of the periphery, or ½, or something in between.

But in a 2-pole diagram, no matter how the groups are arranged, the entire "A" phase energized on the first step is a single group at one location. Noise, vibration, and torque cusps usually render a part-winding starting circuit unacceptable for such coil grouping.

One might conclude that the more coil groups per phase there are in a winding to begin with, the less harmonic distortion might result when a part-winding connection is used. That conclusion is correct. The torque cusps that cause many 4-pole part-winding motors to "hang up" or "crawl" during acceleration are greatly reduced or entirely absent for a 10- or 12-pole design.

Therefore, one way to solve the 2-pole problem is to split the groups. A couple of successful methods exist. One appears in Figure 3-36. Note that each coil group is split into halves, only one half being energized on the first step. (This has also been done with 4-pole windings, resulting in some improvement in first-step accelerating torque. But not every attempt has been successful.) Part-winding inrush current is typically 65% of the full-winding value, and locked-rotor torque about 80%. Starting is quiet. This connection was first used more than thirty years ago for a 150-hp air conditioning motor, about which the compressor manufacturer later said, "Everything seems to be running smoothly on this job and our interest in these part-winding motors is therefore accelerated."

Figure 3-36: A successful half-winding starting design for 2-pole motors. Each coil group is split in halves, one of which is energized on the first step (indicated by the heavy lines).

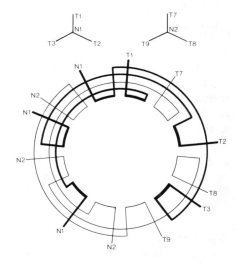

The best results are achieved with a high coil pitch—typically 60% to 70%—which may cause undesirably long coil extensions for some ratings. That has also helped with other speeds. One 25-hp, 4-pole rating was able to reach full speed on the first step, with inrush of 55% to 70% of the NEMA value, when wound with 100% coil pitch—an intolerable condition for a large machine. See Figure 3-37 for another effect of coil pitch variation, on the 75-hp design of Figure 3-28.

The 2-pole split group connection is not adaptable to an odd number of stator coils per group, because each group must be divisible into two equal parts. It will therefore work for 36, 48, or 60 stator slots, but not for 42 or 54.

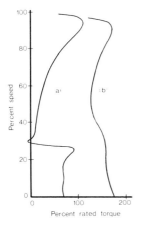

(a): coil pitch
1 & 12, 440 v.,
95 dB noise

(b): coil pitch
1 & 15, 480 v.,
91 dB noise level

Figure 3-37: An example of the effect of coil pitch and field strength on first step acceleration. The version at (b) has better torque and lower noise, but inrush kVA drawn from the line is 10% higher. The version at (a) is the same as shown as (a) in Figures 3-27 and 3-28 (page 126).

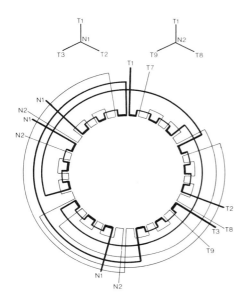

Figure 3-38: A parallel star, 36-slot, 2-pole stator winding diagram using the "fully interspersed" or "alternate coil" connection. Each coil is connected as though it were a separate phase group. The two halves of the winding are so arranged that currents are balanced when they are paralleled.

Both star and delta designs are usable. Acceleration to a high speed on the first step may or may not occur, and should not be counted on.

A second useful 2-pole part-winding connection appears in Figure 3-38. It's more expensive—particularly for random winding—because every coil becomes an individual group, with more numerous and complex interconnections. But except for that interspersing, the principle is much the same as for the other method: split each group in half. Again, the number of coils per group must be even. Test results reported for this method, in the 75- to 300-hp range, include:

(1) Locked rotor torque 75% of the full-winding values; current 80% to 85%.

(2) Noise level on part-winding start no higher than for full winding.

(3) No serious torque cusps. (See Figure 3-39.)

Very few 2-pole part-winding motors have been built in recent years.

Figure 3-39: An "alternate coil" connection makes it possible for a 2-pole part-winding motor to accelerate to full speed on the first step.

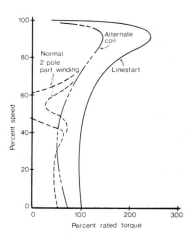

Regardless of motor speed, no part-winding arrangement is adapted to motors of relatively high voltage and low horsepower, such as 200 hp at 4000 volts. Such windings require a large number of turns per coil. Because part-winding needs at least two parallel circuits in the winding, the number of coil turns must be doubled compared to a series circuit, thus demanding impracticably small wire. Fortunately, the starting amperes of such a motor are so low that "reduced voltage starting" is seldom needed in any form. But when it is, autotransformer or reactor starters are the choice; neither part-winding nor star-delta will do the job.

Keep in mind that the objective in using reduced voltage starting to lower starting current is to reduce the strain on the power system. It is line current, not motor current, that is of concern. In most cases the two are the same, but in one common method of starting—the autotransformer—they are not the same.

Autotransformer starters are neither new nor exotic. Yet there is a persistent misunderstanding about them. Users tend to assume that the reduction in motor current on, say, the 65% tap is the same as the reduction in line current. They may be seeking a 35% decrease in amperes drawn from their system because their utility or some local system study told them that was required. They conclude that 65% voltage on the motor will do the trick, and call for 65% autotransformer starting.

But this unduly penalizes the motor designer in making allowances for starting a heavy load. No question about it—if he has to start his motor at 65% of rated voltage, he will lose 60% of its normal accelerating torque capability and have to produce a special, high-torque design to recover enough to do the job.

Redesign isn't really necessary. What will happen is that the current drawn from the line—which is the user's only true concern—with a 65% tap will actually be about $(.65)^2$, or only 42% of what it would have been without the autotransformer. This is overkill. It would be far more realistic to use the 80% tap instead, making the motor designer's job much easier, reducing motor cost, and still providing $(.80)^2$, or 64% of the original across-the-line motor inrush.

Look at Figure 3-40 to see why this is so. In any transformer, the input or primary current is related to the output current in inverse ratio to the respective voltages. If the primary-to-secondary turns ratio is 2 to 1, so that the secondary output voltage is half the input voltage, the output current will be twice the input current. In terms of the illustration:

$$I_M = 2I_L \text{ or } I_L = I_M/2$$
$$\text{and } V_L = 2V_M \text{ or } V_M = V_L/2$$

But since V_M is only half rated or line voltage, I_M is only half (approximately) the normal motor starting current. Thus, I_L is only ¼ of the I_M that would flow at full voltage.

Saturation in the motor, plus the small magnetizing current required to excite the autotransformer, has some effect on these ratios, but the important point is that applying half voltage to the motor through an autotransformer reduces motor starting current (and kVA) to ¼ normal, not ½ normal.

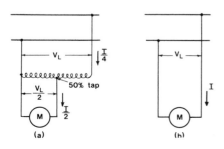

Figure 3-40: Motor voltage/current relationships during (a) autotransformer start on 50% tap, and (b) across-the-line full voltage start.

Solid-state starters

One of the more attractive developments of recent years, now available for motors 600 volts and below up through 400 hp or more, is the solid-state starter. (See Figure 3-41.) It uses SCRs or thyristors with controllable firing angles to produce a smoothly variable reduced voltage for starting any standard squirrel cage motor. Starter manufacturers claim to be able to bring loaded conveyor drives up to speed with motor inrush of 300% to 400% of rated current.

The first such units came on the market during 1965-67, although they were seldom economical at that time. There are now at least sixteen suppliers. In 1976, when the field was still considered to be in its infancy, there were less than half a dozen. Maximum voltage is 600. Although most manufacturers specialize in the 1- to 150-hp range, at least half of them offer starters from 450 hp to 1000, or even 1500.

What is a solid-state starter, and how does one work? The basic unit consists of three pairs of semiconductors (SCRs), one pair for each phase, connected in inverse parallel. Figure 3-42a shows one phase of a typical starter. Each SCR of the pair controls current flow alternately during the two halves of every cycle. No current can flow until a "turn-on" or "gate pulse" signal is supplied to the SCR.

Figure 3-43 shows what results from the proper timing of those gate pulses. If the pulse is received at the beginning of the half cycle, the voltage output of the SCR pair is a complete sine wave of essentially line magnitude. If the pulse appears at the end of each half cycle instead, there is no output voltage. And if SCR gating or firing occurs part way through the half cycle, a partial or reduced voltage appears at the output. Thus the motor terminal voltage is reduced below the line value, reducing both inrush current and developed torque as well. As shown in Figure 3-42b, some starters use an SCR-diode pair rather than two SCRs per phase. This gives half-wave, rather than full-wave, output voltage control.

A second essential element of the starter is the gate control that governs the SCR firing angle. The third element is the circuitry for adjusting the gate control in terms of user option or drive requirement. (See Figure 3-42c.)

Several makes consist of "controller" only. That is, a separate across-the-line type electromechanical starter is needed in addition to provide fusing, overloads, and contactor isolation between line and load. The electronic solid-state controller simply varies acceleration voltage as commanded. A complete solid-state starter, on the other hand, includes all the overload protection and other circuit options within a single unit.

Some of the possible operating modes are:

(1) "Ramping" of voltage from zero to the full line value linearly over a pre-set time range. (See Figure 3-44.) An alternate adjustment may permit higher locked rotor voltage for load breakaway.

Coping with Starting Conditions

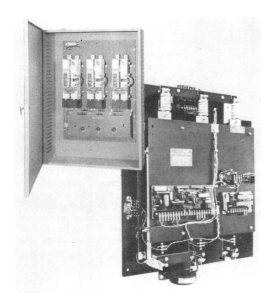

Figure 3-41: Typical solid-state reduced voltage or "soft" starters.

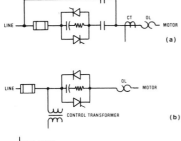

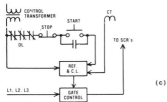

Figure 3-42: Left: Diagram (a) shows one phase of a solid-state starter, including SCR pair; bypassing contactor sometimes supplied to isolate motor from starter when desired and to bypass SCRs when the motor is up to speed; and current transformer (CT) to supply feedback signal to starter control circuits. At (b), an SCR-diode pair is used instead. Diagram (c) is one starter control scheme; C.L. is current limit control.

Figure 3-43: Right: One cycle of motor voltage. At (a), reactor or autotransformer starter lowers voltage (shaded wave) but basic waveform remains unchanged. At (b), solid-state starter SCRs "chop" the wave (shaded portion is supplied to motor) so its reduced magnitude is accompanied by a completely altered waveform rich in harmonics.

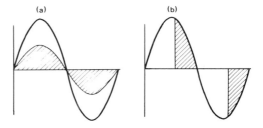

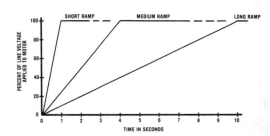

Figure 3-44: In this mode of solid-state starter operation, motor voltage is gradually applied or "ramped" up from zero. The motor may not develop enough torque to accelerate until the ramp time has largely elapsed, after which full voltage is applied. Acceleration time (normally 5 to 10 seconds) can be lengthened to 30 for some starters. There may be an "extended ramp" option for still more time before the full voltage comes on. This may help reduce shock but tends to overheat the motor—especially the rotor of a large machine. A linear or ramped acceleration mode produces the same sort of plot, with the variable being speed rather than voltage.

(2) Ramping of kVA from zero upwards, the time and current limits being predetermined after study of motor and load characteristics. The current limit may range from 100% to 500% of rated full load amps. This requires a current feedback signal to the starter from one or more current transformers as shown in Figure 3-42.

(3) Continuously varying voltage to maintain constant current to the motor for a manually adjustable time span.

(4) Adjustment of current and voltage to complete the acceleration within a controlled time. This uses a speed feedback signal from a tachometer in the drive.

(5) Options of controlled deceleration, jogging, or reversing.
There is a great deal of flexibility here. This is one reason for the great popularity of solid-state starters today. Other advantages are:

(1) No moving contacts, armatures, linkages, springs, etc., to wear out or burn out.

(2) Reduced size and weight.

(3) Longer life compared to electromechanical starters.

(4) Variety of self-contained protection available—phase failure or reversal, undervoltage, and overload.

(5) Suitability of the encapsulated components to dirty or corrosive surroundings.

Ski lifts, packaging machinery, coal conveyors, wire drawing lines, and air conditioning chiller drives are common applications for these soft starters. At least two large refrigerant compressor makers offer motor/starter packages using them.

Despite their very real advantages, however, solid-state starters have caused problems in some motor drives. Such problems can be the result of certain claims and assumptions made by some of the starter manufacturers through misunderstanding of basic motor behavior. Anyone involved in motor/drive servicing, the selection of drive components for a particular application, or the answering of questions about the operation of one of these starters should be aware of the possible difficulties.

No added torque

To begin with, "There is no free lunch." In other words, using a solid-state soft starter will in no way manufacture added motor torque. If a drive was hard to bring up to speed at full voltage, it will be much more difficult to bring up to speed if voltage is reduced at all during any part of the acceleration period. The motor doesn't care what exotic circuit is lowering its voltage; all it knows is that the reduction is there, and it will respond by producing less torque.

The relations between voltage, torque, and current in an accelerating squirrel cage motor are not widely understood in the control industry. For example, it's commonly assumed that torque varies as voltage squared. Depending on the size and type of motor, this can be greatly in error, as we saw earlier in this chapter. Applying 65% voltage to a motor's terminals, for example, will produce not 42% of its normal torque, but only 39% at best.

When that reduced voltage is sinusoidal in waveform as in Figure 3-43a, this decreased torque, along with lengthened acceleration time and increased motor heating, can be accurately predicted. The relationships are well known. The calculations are simple. From them, the motor designer can tell whether or not the load will be accelerated at all, and if so whether or not the motor will overheat in the process.

But the "chopped" wave output of an SCR system generates an entirely different and not readily predictable relationship between motor voltage and developed torque. Any odd waveform of the type in Figure 3-43b is equivalent to a whole set of sine waves each of a different frequency. We call these "harmonics." More of them are produced by the SCR-diode combination than by SCR pairs alone.

Unfortunately, the number of harmonics, their frequencies, and their relative magnitudes will be different for each SCR firing angle—for every value of effective voltage applied to the motor by a solid-state starter during an acceleration.

What do the harmonics do? For one thing, they produce extra heat in the motor windings. From tests on solid-state power supplies feeding loaded motors at rated speed and frequency, it's been established that winding temperature rise will be as much as 20% greater than for sine wave commercial power. During a severe start, this added harmonic heating coupled with that produced by the lowered accelerating torque may be enough to damage the motor or trip it off the line.

Nor is that all. Matters are worsened by another effect of the same harmonics. Some of them produce negative motor torque. Hence, the net shaft torque available from the motor may be greatly reduced even if the RMS or effective voltage magnitude isn't changed at all. One test showed that applying 85% of rated motor voltage with an SCR starter may cut torque 10% below what it would be with an 85% tap autotransformer starter.

The harmonics may also change the shape of the motor's speed-torque curve to generate more dips, or "saddles." The drive may then hang up unexpectedly at certain speeds until a change in SCR firing angle shifts the harmonic pattern.

This has two consequences for the drive operator. First, just because a similar drive has started safely in the past with some other type of reduced-voltage starter doesn't assure that it will work with a solid-state starter. Second, even if the user knows what voltage the solid-state starter will apply to the motor terminals, neither he nor anyone else can tell just how much motor torque will result—or if acceleration is even possible, let alone safe.

Experiment will prove the point, of course. Most of these starters have safety overrides so that if the high accelerating current persists for too long, meaning a failure to come up to speed, full line voltage will automatically be switched onto the output. The trouble is that the "soft" start then becomes "hard." All advantages of the flexible voltage control will be lost.

As an example of the misconceptions to be found in this field: Several years ago, one of the manufacturers published a paper explaining starter operation. Said the author, "All of us ... have been been used to thinking that torque is proportional to voltage squared.... However, when we deal with current instead of voltage, we find that torque is almost directly proportional to current."

But that is not the case. In his paper, the author showed a curve of locked rotor torque versus current for a 200-hp motor, most of which looked almost like a straight line. But such a plot on ordinary graph paper can be misleading. Furthermore, the basic nature of an induction motor simply will not support his conclusion. We know that:

(1) Locked (or accelerating) torque varies somewhat faster than voltage squared.

(2) Locked (or accelerating) current varies approximately as voltage.

(3) Therefore, it follows that torque must vary at least as current squared, at any given speed.

Figure 3-45 shows the actual tested relationship between locked rotor current and torque for a number of standard motors (mostly 4 pole) between 20 and 400 hp. Above about 300% current, the curves do look almost straight, as if current and torque were directly proportional to each other.

To see the true picture, however, these curves are replotted in Figure 3-46 on a logarithmic graph. Here the lines are perfectly straight over the entire range plotted. When any such logarithmic curve is a straight line of constant slope, the amount of the slope—the ratio between height and width marked off

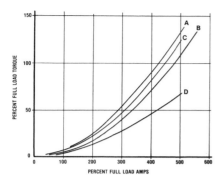

Figure 3-45: Left: Based on a large number of tests, these curves show how locked rotor torque is related to locked rotor current. Curves A and B apply for 20 to 100 hp; C and D cover the 200- to 400-hp range.

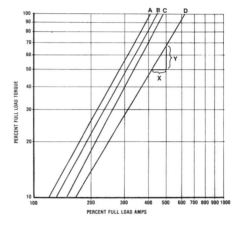

Figure 3-46: Right: Replotting most of Figure 3-45 on logarithmic coordinates reveals the unmistakable relationship: Torque varies as (current)$^{2.2 \text{ to } 2.8}$. The exponent in such a plot is equal to the slope of the straight lines, which is Y divided by X. If the exponent were 1.0, for torque to be directly proportional to current, Y/X would have to be 1.0.

by any selected length of the line—represents the exponent relating the two plotted variables. Here, the slope ranges from 2.2 to 2.8, proving that by actual test:

Locked rotor torque is proportional to:
$$(\text{locked current})^{2.2 \text{ to } 2.8}.$$

You may ask, "So what?" The importance of that example of misunderstanding is that it has caused some people to expect things of the starter which they will not get. Assuming that constant motor current during acceleration will generate constant motor torque leads to the belief that starting will be "softer" than it actually is.

Consider the typical motor of Figure 3-47a, for instance. This illustration shows what we are all familiar with—constant motor voltage, plus the accelerating current and torque curves that go with it.

Now suppose we want constant torque rather than constant voltage. This requires variation in the accelerating voltage. The curves of Figure 3-47b result. Note that the current curve still has the same basic shape, with a peak at locked rotor, although smaller in amount than when voltage did not vary.

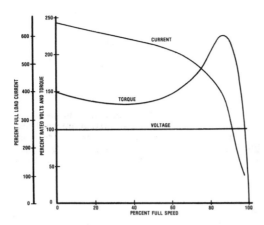

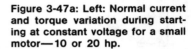

Figure 3-47a: Left: Normal current and torque variation during starting at constant voltage for a small motor—10 or 20 hp.

Figure 3-47b: Below, left: If motor voltage is adjusted throughout acceleration to produce constant motor torque equal to 100% of full load torque, we get this relationship. Note that locked current has been reduced from 650% to 520%, and the current versus speed curve shape has not changed much. For convenience in plotting, torque has been assumed to vary with voltage squared for these curves.

If anyone tells you that a starter will permit acceleration of a loaded conveyor—requiring 120% breakaway torque and at least 100% accelerating torque up to full speed—with only 300% current, using a NEMA Design B motor, don't believe it. It may be impossible even for a NEMA C design.

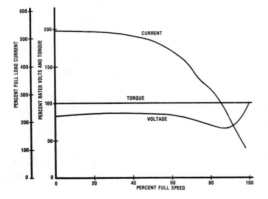

Figure 3-47c: Right: Now suppose we force the current to remain constant, at 400% of rated in this example. Do we get "constant torque"? Far from it. The torque peak at 85% to 90% speed will be the same as in Figure 3-47a, because current there automatically falls below the pre-set limit so that no voltage reduction is applied. The start is not nearly as "soft" as it might seem. Most solid-state starter literature, however, leads one to expect a fairly flat curve like the dashed line shown here. If we were able to accelerate at all with 200% to 250% current, which is unlikely, we might see such a torque curve—but not at 300% to 400% current.

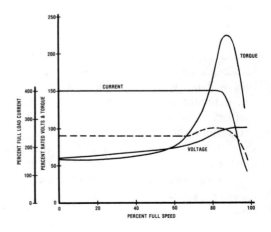

Next, suppose the current-limit control of the solid-state starter applies a constant current during acceleration. In Figure 3-47c, we can see that by the time we approach motor breakdown, the current has dropped below the pre-set limit of 400% even with full voltage on the motor. No voltage reduction at all is needed to maintain the current below its limit, so full voltage is applied by the starter. Result: the motor develops its full breakdown torque, producing the high torque peak on the graph—and producing conveyor belt stretch or other mechanical stress in the drive. From the misleading assumption explained earlier about how current and torque relate to each other, starter manufacturers have widely published the "typical" constant current torque curve illustrated by the dashed line in Figure 3-47c. This has no high peak. But it could be achieved only by a far greater drop in the current limit—which in some cases would push the locked torque so low that the drive would be unable to start.

All this may be unimportant for small drives. A standard 10-hp motor, for example, has so much accelerating torque compared to typical load demand that unsuspected harmonics may not prevent successful starts. Even today, most solid-state soft starters are being applied to fairly small machines. The vast majority of industrial drives, after all, are 125 hp or less.

But in the 450- to 1000- (or more) hp range, as Figure 3-48 points out, the actual motor torque—either NEMA standard or typical design—is far below what is normal for the lower ratings. Even for the latter, starter catalog information has tended to exaggerate what that normal is. One finds this kind of statement often repeated:

"For a typical NEMA Design B motor, the starting torque will be approximately 150% of full load."

This is misleading. The NEMA B standard locked rotor torque is 150% or more only up through 5 hp, 2 pole; 30 hp, 4 pole; and 10 hp, 6 pole. At 100 hp, the standard value drops to only 105%, 2 pole; 125%, 4 or 6 pole. At 500 hp, it becomes 70% to 80%—for NEMA Design B.

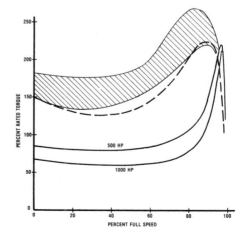

Figure 3-48: The shaded band is a composite of "standard" motor speed-torque curves assumed in their literature by several solid-state starter manufacturers. At or below the lower edge of that band, the dashed curve meets NEMA standards for 20- to 75-hp, 4-pole motors.

Far below any of these torques, the solid curves show what is standard for large motors. Obviously, such drives have nowhere near as much margin for voltage (and torque) reduction during starting.

It's true that motor manufacturers often provide more torque than these minimum figures. Still, no one should expect 150% or more throughout the entire range of standard motor designs with which solid-state starters can be used. Moreover, this value applies at zero speed. At 25% or 50% speed during a start, motor torque often dips well below the locked rotor figure.

Despite the added motor heating and the loss of accelerating torque, it's commonly claimed that solid-state starters are suitable for "any standard motor." Moreover, some suppliers have stated that because the soft start is so much easier on the motor, as well as on the starter itself (no moving parts to wear out), it is now possible to ignore the old restrictions on how often the drive can be started. "No limit on starts per hour," reads one sales bulletin.

This is dangerous advice. No one should blindly apply a standard motor (especially over 200 hp) for frequent plug-stopping or reversing, for example, in the belief that because the starter is solid-state there won't be any undue strain on the motor.

Starting across-the-line with 600% current subjects motor windings and the rotor cage to forces thirty-six times as great as during running, as well as high heating. If the starter lowers current to only 300%, forces will still be nine to ten times as high as during running. This is a great improvement but still can't be sustained an unlimited number of times without damage. The larger the motor, the more significant this is.

A 1972 nationwide industrial plant survey by the IEEE showed that in the 0- to 600-volt category the failure rate of electromechanical starters was 28% higher than the failure rate of the motors themselves. The solid-state starter should certainly improve that record. But nobody wants to see motor failures increase as a result.

It's often recommended that motor winding thermostats be used as protection against overheating on start. This will not work for any but the smallest motor ratings. Winding temperature sensors are too slow-acting for the rapid heat rise on acceleration. Furthermore, they cannot sense overheating in the rotor.

One solid-state starter manufacturer who has recognized the harmonic problem says this: " . . . Harmonics cause counter torques and winding heating that slightly decrease the efficiency of the motor." This is missing the point; drive "efficiency" during the brief starting period is of no importance. Starter efficiency while the drive is running, on the other hand, is something to consider carefully at a time when every needless loss of energy should be eliminated. The solid-state unit does use SCR power while the motor runs.

Typically, this loss is 1.5 watts per ampere per phase. For a 500-hp motor, that totals about 3 times 1.5 times 560 amperes, or 2520 watts. Total full load loss in the motor itself will be ten times that much. Therefore, starter losses—amounting to 2½ kilowatt-hours for each hour the drive runs at full load—are equivalent to at least a half percent drop in motor efficiency. This can be costly (typically $400 annually for a motor that runs half the time fully loaded). Putting it another way: some industrial users are now considering each kilowatt of drive loss to be worth from $1,000 to $3,000 in capital outlay. The SCR loss in some starters requires them to be force-cooled.

There will also be some small leakage current through the SCRs when the starter is off. This puts dangerous voltage at the motor terminals even when the drive has been shut down. To work safely on the motor circuit it is not enough to be sure that such a starter is locked out. If the starter is of the SCR-diode type, it will supply a constant d-c voltage to the motor terminals when off. Voltage magnitude will depend on the nature of grounding of the power system and the motor frame. Some solid-state starters include isolating magnetic contactors so that the circuit can be completely opened in the "stop" mode, and the SCRs can be completely bypassed in the "run" mode at full speed (eliminating power loss in the starter). But this is not a universal option. If separate contactors are added to obtain these advantages, their cost and reliability have to be figured into the job.

Other potentially harmful voltages are possible in these circuits. In Figure 3-42, note the capacitor/resistor combination connected across the SCRs. This is a transient surge suppressor, or so-called "dv/dt protector" normally furnished with any starter. It protects the SCRs from damaging voltage spikes that might be present. Unfortunately, the switching action of the SCRs themselves can produce such spikes at the motor.

Are these damaging to motor windings? Perhaps. It will depend on starter design. When the SCR fires to switch on each half-cycle of motor voltage, it tries to make that voltage rise from zero to some finite value instantly. Whenever such a change is attempted in an inductive circuit, high transient spikes result. Both the rate of rise of those spikes—the dv/dt ratio, typically 500 volts per microsecond—and their peak magnitude determine whether or not the motor windings will suffer. The motor designer can provide limiting values that will do no damage. The starter manufacturer should then be asked whether or not those limits can be met. If not, surge suppression may be needed at the motor. An alternative is for the motor manufacturer to use a higher insulation level, particularly between turns, but he seldom knows when his motor will be operated from this type of starter.

A final precaution about starter application: Many industrial motors today have power factor correction capacitors connected at the motor terminals. This is not recommended when a solid-state starter is used. Interaction between capacitors and SCRs during acceleration may damage the starter. One way around this is to connect a separate contactor in parallel with the solid-state starter, using it only to energize the capacitors. Any capacitor application should be checked out beforehand with the starter manufacturer.

Similarly, any friction brake wired to the motor will have to be reconnected. The initial reduced voltage supplied by a timed acceleration ramp mode of starter operation will not be enough to release the brake.

What about servicing solid-state starters? This can be more complex than for other types. It's easy to find an open or shorted contactor coil, broken springs, or burned or welded contacts. But if the SCR gating control isn't working correctly, finding the trouble may not be so simple. Various modules may have to be removed and replaced until the difficulty is located. Major trouble with the SCRs may not occur often but will be expensive when it does. The failure of one SCR in a pair may damage the other as well, and replacement can cost $1,000. Furthermore, some service men claim that the replacement SCR set may soon fail in its turn if the gate firing circuit module isn't replaced at the same time.

Why don't control and motor designers get together to coordinate their equipment? This has happened with a couple of manufacturers. Generally, though, the starter designer lacks the facilities to make design studies or tests of motor behavior with his control. The motor manufacturer seldom has access to a starter—especially a large one. Moreover, the entire technology is still new and changing. It may be a long while before adequate standards exist to cover major application questions.

Meanwhile, solid-state soft starters are valuable additions to the motor control field. Properly used, they can reduce operating and maintenance costs for a wide variety of drives. Making sure the application is proper, however, is quite different from what has been true in the past for other types of starting equipment.

4 Matching the Motor to the Power System

EVEN THE MOST carefully designed electric motor drive can be damaged or destroyed by a variety of electrical faults, overloads, or "power line pollution" such as transient surges or voltage unbalance. In this chapter, we shall see what some of these conditions can do, and how to guard against them.

As we have just learned in Chapter 3, one of the more important kinds of damaging motor "overload" can be overheating during acceleration. Industrial motor circuits are normally equipped with devices or systems to shut down the drive, or at least warn an operator to do so, before such damage occurs. The choice of protection is based on the manner in which heat is developed in the motor during both starting and running, as shown by the "thermal damage curve."

"Many industrial users don't seem to understand it. Yet it's the most useful motor protection tool we have." That's how one widely known electrical consultant has described the thermal damage curve of an induction motor. What is this curve? How is it derived, and how does an electrical designer use it for correctly matching other power system components to drive characteristics?

The nature of thermal limits

To understand this, let's review some basics. We know that any electrical device that carries current will get hot. If the current is too high, the device gets hot enough to "burn out." That's true for a fuse, a transformer, a cable, a switch—and of course for any motor. That behavior, normally destructive and to be avoided, serves a beneficial purpose in the fuse, thermostat, or circuit breaker. There, overcurrent heating triggers a corrective action. The "blowing" of a fuse is a desirable result because it protects other parts of the circuit that would be far more costly than a fuse to replace.

Although fuse action may make us think that circuit elements get hot only when current reaches a certain high level, some amount of temperature rise accompanies any value of current. The higher the amperage, the higher the conductor temperature. Some heat of course escapes to the surrounding air, insulation, or lamination steel. But once those surroundings get warm too, they can't absorb heat as readily, so the current-carrying conductor temperature itself must go up.

We can plot this in curves like those of Figure 4-1. When current is low, heat given off by the conductor will just balance the rate of heat production. Conductor temperature then "levels off" soon. If the current does not increase, the temperature will remain constant.

But if we now raise the amperes, step by step, we will produce heat faster than it can escape. Therefore conductor temperature will rise. Eventually we have so much heat that temperature cannot level off at all. Then, as time goes on with no interruption of the current, temperature climbs steadily until the conductor melts or burns open. Any insulation present will burn first. This is the familiar electrical fire situation. We've all seen the "roasted out" motor winding resulting from such a condition.

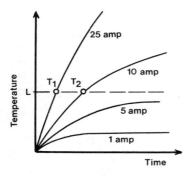

Figure 4-1: Basic time-temperature relationships for an electrical conductor. The "safe temperature limit," L, may depend on insulation life, conductor melting point, flammable atmosphere ignition point, etc.

If conductor current is low enough, the temperature limit is never reached. No damage occurs. As current is increased, eventually we find a value at which L is reached after a long while—time T_2. Increasing current further causes L to be reached in a short time, T_1. That is a basic rule for any circuit: the higher the current, the shorter the time during which that current can flow without damage.

As the curves show, overheating can follow either from extremely high current flowing for a short time, or from a lower current that flows for a longer period. This relationship exists in all electrical components. It is expressed in Figure 4-2—the "thermal damage curve." All current-limiting fuses have this characteristic, as do overload relays, cables, motors, etc.

In a simple form, this curve tells us the two basic truths of Figure 4-1. First, if current is low enough—the lefthand edge of the curve—the device can safely carry the amperes indefinitely. Its temperature will remain within the safe or "rated rise" limit for which normal insulation/conductor life is to be expected.

Figure 4-2: The relationships of Figure 4-1 are expressed in this "thermal damage" or "safe time vs. current" curve. Such a curve exists for any electrical device.

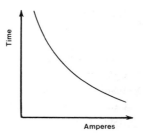

Second, the curve shows that as current is increased and the rate of temperature rise goes up, the allowable operating time must be reduced. Any circuit can stand a lower current for a longer time than a higher one.

Let's consider one more thing about Figure 4-2. Does it indicate that an amount of current even the slightest bit above the curve will cause instant failure? For a fuse, yes. If for any given period the amperes on the curve is exceeded at all, the fuse will blow. But for current-carrying insulated windings, such as transformer or motor coils, the situation is quite different. Copper or aluminum insulated conductors do not exhibit an abrupt, precise failure point. "Burnout" doesn't occur instantly. As the material is "overheated" by operating above the thermal damage curve, several things take place, separately or together:

(1) Conductor material loses strength. Mechanical stresses imposed by magnetic fields or centrifugal force can then decrease its life span below what is normally commercially acceptable. The "end of life" will be indicated by fracture, perhaps accompanied by some melting. (See Figure 4-3.)

(2) Insulation becomes brittle, carbonized, or cracked. That allows moisture or dirt to enter and "track" from conductor to ground. This takes time, but above a certain temperature it happens sooner than normally considered acceptable.

(3) Parts expand relative to one another so they come apart or distort. Example: rotor bars "lifting out" of their slots.

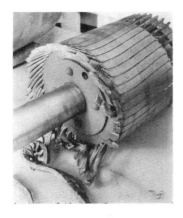

Figure 4-3: Typical result when a squirrel-cage rotor is operated long enough above its thermal damage limit.

Under extreme temperatures those troubles can develop in induction motors fairly quickly—within hours, even minutes. But more often they develop over a period of months. Some people in the industry speak of a motor as though it were a fuse, having a precise failure point at a particular time-current relationship. But that isn't true. Operating above a motor's thermal damage curve (sometimes called the "safe operating time versus current" curve) simply imposes enough thermal or mechanical strain on the motor so that the manufacturer is no longer willing to warrant normal life.

There are other important differences between the curve for a fuse and one for a motor. First, the fuse curve is "continuous." By that we mean that all points on the curve have the same physical significance. A small current through the fuse heats it a little. A large current produces much more heat. Continuously increasing the current, simply by lowering the load impedance of the circuit into which the fuse is series-connected, raises the fuse temperature until it melts. How much current flows is not governed by the fuse itself.

Motor heating and current flow, however, are quite different. The motor *does* control the current. Suppose it's running at full speed but supplying no horsepower to a connected load. The only current flowing in the windings is the idle or no-load amperes. That is a single value, fixed so long as voltage remains constant.

Now apply a shaft load. The current goes up, and as more torque is demanded, it goes up more. Winding temperature goes up too. This seems similar to the fuse situation, and it is. But as the load grows still larger, the motor suddenly becomes incapable of exerting the shaft torque being demanded, so it simply stalls. At once, current rises from the "overload" value to the locked rotor value. There it remains. It cannot go back down, and as long as rated voltage is applied it can't go up, either.

Why motor curves look different now

So, as Figures 4-4 and 4-5 show, the motor curve is "discontinuous." That part of it between the stalling point ("breakdown" or pullout) and locked rotor is not "real." There is no way the motor can operate stably in that region. Whether accelerating or stalling, it cannot be "held" at, say, 400% of rated current. It is meaningless, therefore, to think of that portion of the thermal damage curve as expressing any useful relationship between current and allowable operating time.

Nevertheless, for many years it was common practice to draw thermal damage curves for motors as in Figure 4-4. A motor is protected from damage by properly coordinating its curve with the similar curve for the fuse, relay, or

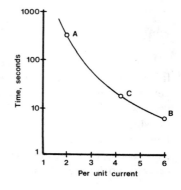

Figure 4-4: A motor thermal damage curve. Current at point A represents a running overload. If that overload is increased, somewhere around point C the motor stalls, current rising at once to point B, where it remains. Steady-state operation between C and B is impossible. (The current scale does not start at zero because we are not concerned with allowable operating time at any load below rated horsepower.)

Matching the Motor to the Power System

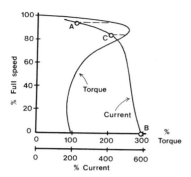

Figure 4-5: Speed-torque and speed-current curves for the motor of Figure 4-4. Note that point C is below the breakdown point.

breaker provided to take the motor off the line when it draws too much current for the time allowed. All such protective devices do have continuous curves. Hence, it seemed a logical compromise to use the same curve shape for motors.

How was that shape derived? For the lefthand part of the curve, down to about point C in Figure 4-4, a simple calculation balanced the rate of heat generation in the stator winding with the motor's cooling capability. Insulation can be allowed to reach 30°C to 50°C above its nominal temperature limit provided the overheating doesn't last long. So a "short-time overload temperature limit" is chosen, and the allowable time fixed, for any overload current on the basis of that limit.

Because running overloads in squirrel cage motors are always limited by stator winding heating, this portion of the curve is always what we term "stator limited"—that is, stator heating alone sets the limit on which the curve is based.

But what happens in Figure 4-4 when the machine stalls and the current rises at once to the locked rotor value? Depending on motor size and type, safe temperature in the stalled condition is more likely to depend on rotor heating than on stator heating. Bending of rotor bars, or expansion of end rings, can severely damage some large motors in ten or fifteen seconds. Even though stator winding current is five to seven times rated amperes when the motor is stalled, the rate of temperature rise in the rotor cage is often much greater than in the coils.

Thus, Point B on the curve is often "rotor limited." Furthermore, the lack of any cooling air flow at locked rotor produces a different relationship between heat generation and heat dissipation than is present during a running overload.

There is, therefore, no direct connection between Point B and the portion of the curve from A down to C. If we simply extended the latter on to the right, by calculating the allowable time for larger values of stator overload current, it would often not pass through Point B at all. For consistency, however, curves were drawn to do so. Recognizing the imprecise nature of all the calculations involved and the conservatism in the numbers used, an "exact" curve shape was not to be expected. Again, a motor is not a fuse.

In 1981, the IEEE recognized this by issuing Standard No. 620 to govern the construction and interpretation of motor thermal damage curves. It provides that the "discontinuity" of a motor curve—the separateness of its running and stalled portions—be recognized by plotting as in Figure 4-6. The single point "LR" denotes the one value of amperes that can flow at locked rotor, and the time for which such a current flow is safe for the motor.

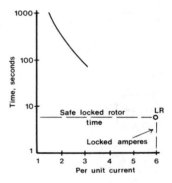

Figure 4-6: Here, a curve like Figure 4-4 has been replotted to omit the "unreal" or meaningless region between the breakdown and stalled conditions.

Of course, if terminal voltage can vary, as is common for large machines, the LR point will become a line, as in Figure 4-7. The reason is that when voltage is reduced, locked rotor amperes drops accordingly. The motor can sustain that lower current for a longer time. Because NEMA standards allow for motor voltage as low as 90% of the rated value (although that allowance is primarily intended to cover running conditions), it's common to show a locked rotor thermal damage line for the range between 90% and 100% voltage.

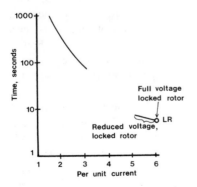

Figure 4-7: Otherwise the same as Figure 4-6, this thermal damage curve shows a range of safe locked rotor time corresponding to the locked rotor currents at several different voltages. (Safe time is assumed to vary inversely as voltage squared. Thus, at 80% voltage, the allowable time is: safe time at full voltage multiplied by $(100\%/80\%)^2$, or 1.56.)

The discontinuity or gap in the curve can be important in the protection of some drives during starting. Suppose we add the current vs. time graph for an accelerating motor to the kind of continuous thermal damage curve shown in

Figure 4-4. Figure 4-8 shows the typical result. There is nothing "safe" or "unsafe" about the accelerating current graph. It expresses no thermal limit. It simply shows how the actual motor amperes will vary with time as the drive comes up to speed.

If the load being accelerated has extremely high inertia or demands high torque, acceleration will take a long time. This can cause the two curves—accelerating current and thermal damage—to come dangerously close to each other, or even to cross. This means that during a normal start some part of the motor is overheating. A fuse, relay, or breaker chosen to match the motor thermal damage curve will therefore trip the machine off the line every time a start is attempted. This is called "nuisance tripping." Resetting the protective device to allow the start will risk motor damage because that device will then not respond until the motor gets too hot for safety. This kind of starting problem is shown in Figure 4-9.

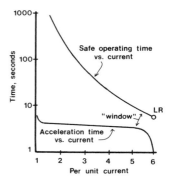

Figure 4-8: Adding to Figure 4-4 the variation of motor current with time during acceleration produces this graph. Since at any given value of current the actual time it flows is always less than the maximum allowable time it can flow, the motor will never overheat during a start. This safety is expressed by the "window," or minimum spacing between the two curves.

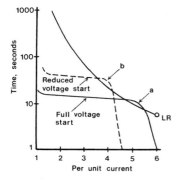

Figure 4-9: If load inertia is high, the motor takes much longer to start so that starting current flows too long for safety. At "a," the window between the two curves has disappeared. Reducing the motor voltage on start lowers the current, but lengthens the starting period still further, so that at "b" the drive is still in trouble.

Another potential protection problem arises when reduced voltage starting must be used, or when the power system voltage dips too low because of high motor inrush current. Either way, motor voltage is lowered during starting. That doesn't affect the thermal damage curve itself. But it does affect the accelerating time/current curve. Although it's commonly assumed that reducing motor voltage during starting will "make it easier" on the motor, that is seldom true. Yes, current is reduced. But starting time may be so lengthened that nuisance tripping is certain unless protective devices are bypassed or set too high for safety. (Again, see Figure 4-9.)

Both of these adverse conditions—severe starting load, or low voltage—often cease to be problems when we recognize that the trouble zone on the curves isn't "real." In that region where accelerating current most closely approaches the thermal damage line, the latter isn't really there. That's where we find the gap of Figure 4-6 or 4-7. When that gap is taken into account, Figure 4-9 becomes 4-10. The problem is gone.

Of course, the fuse or relay characteristics contain no gap. If they remain unchanged, the protection problem will also remain. That is avoided by using two separate protective devices having two different current-time tripping characteristics. One suits the stalled portion of the motor curve. The other is matched to the running overload portion. (See Figure 4-10.) In effect, they work in parallel, each one "taking over" in the region where its operating current is lower for the time span involved. The same approach is theoretically possible for a continuous motor curve like Figure 4-4, but in practice it's often impossible.

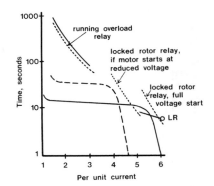

Figure 4-10: Changing the "continuous" thermal damage curve to that of Figure 4-7 causes problems of Figure 4-9 to disappear. The dotted lines represent typical relay or breaker curves that would provide starting and running protection at either rated or reduced motor voltage.

Options for severe starts

Sometimes the curves may be drawn with a refinement illustrated by Figure 4-11. During an actual acceleration, two effects tend to reduce motor heating compared to a stall. One is the cooling air circulation as speed increases. Both stator and rotor can then lose heat to the air two to five times as rapidly as when the motor is at standstill.

Matching the Motor to the Power System

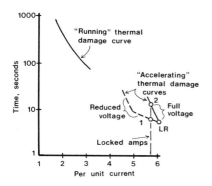

Figure 4-11: A way of accounting for the motor's ability to carry high current for longer periods during acceleration than at standstill. At any given value of current, point 1 on the "locked" curve is used as the text describes to arrive at point 2 on the "acceleration" thermal damage curve. Note how the shape of the latter suits the dotted protection device curves in Figure 4-10.

A second effect is the relative change in the rate at which heat is produced, expecially in the rotor, as the drive comes up to speed. Relatively more of that heat is generated during the low speeds at the beginning of the start. As speed builds up, the rate of heat generation decreases, because more of the energy drawn from the line is going into rotation of the parts. (See Figure 3-12, page 103.)

Here is a rule of thumb for using those effects to plot Figure 4-11. Assume the motor has reached one-fourth of full speed. From an acceleration speed-current curve of the motor (as in Figure 4-5), find the current at that speed. From the locked rotor thermal damage curve, as in Figure 4-7, find the safe operating time for that current. Then, divide that time by the value of slip (at 25% speed, slip is 1.0-.25, or .75; hence, divide the time by .75). Repeat for two or three values of speed (slip). Then plot the results as the "acceleration" line in figure 4-11. Note how acceleration allows the operating time to increase for any given current, largely because cooling is enhanced compared to locked rotor. The curve of Figure 4-11, sometimes even more than those of Figures 4-7 or 4-10, lends itself to selection of suitable protective devices.

Suppose that the load inertia is so great that no matter how the curves are drawn there is no way to fit a breaker or relay characteristic into the picture without nuisance tripping. Then the circuit must include some control feature to distinguish between a motor that is truly stalled (and must be tripped off the line quickly) and one that is on the way up to speed. One such option is a speed switch on the drive. That switch operates to bypass the short-time tripout function if the drive does reach a certain running speed within a predicted time.

There are other options. Certain types of relays can sense the change in motor power factor as it comes up to speed, versus the unchanged value if it remains at zero speed. Or a clutch of some sort may be used to let the motor reach full speed sooner, so that the acceleration and thermal damage curves become safely separated.

Do thermal damage limits differ for a motor that is "cold" compared to one that is already "hot" (up to full load operating temperature) when started? Usually not. Manufacturers differ on this, but the increasingly prevalent view is that only the added or "incremental" temperature rise produced by the start is important. That change in temperature causes the relative expansion of rotor parts, stressing the cage. Because that change or differential expansion will be the same commencing from either a hot or cold initial condition, the thermal damage curve will be the same for either.

It is, however, still recognized in NEMA standards (MG1-20.42) that a motor can safely start somewhat more often when it cools down between starts than when it remains hot. Stator insulation life is influenced by total temperature, not just the differential alone.

Can stator winding temperature sensing be used to "monitor" acceleration and counteract nuisance trips? No. Operation of a thermostat or thermistor circuit out on the end turns may lag as much as two minutes behind overheating within a stator coil. Embedded slot detectors react much faster, of course—but not instantly.

What's more important is the accuracy of measurement. Motors are designed for standard 40°C ambients. Suppose the stalled rotor cage begins to overheat while the stator winding has risen only to 70°C. To avoid thermal damage, we might set the temperature sensing system to trip at 80°C or 90°C. Our first problem is that at full speed and full horsepower load, the winding may operate at 110°C. Therefore, the temperature sensing system must be "locked out" or bypassed for normal running.

Our second problem is that most sensors or their electronic controls exhibit a tolerance of plus of minus 5°C on the reading. In this example, the temperature monitoring unit might trip at anywhere from 75 to 85 degrees. If the ambient temperature around the motor can vary 15 or 20 degrees from night to day or from season to season, we clearly have little chance of closely controlling what goes on in the rotor.

Finally, during a stall the rotor cage may be heating at the rate of 100 degrees per second. But stator coil temperature typically rises no more than 5 to 10 degrees per second. Even tiny variations in the heating time lag between coil and temperature sensor can result in large, potentially damaging "overshoots" in rotor temperature.

Therefore, although stator temperature sensing can serve to prevent too-frequent starting, that alone cannot be relied upon to safely guard against damage from a single start.

Some users and consultants are attempting to easily solve motor protection problems by prescribing some arbitrary relationship between "safe locked rotor time" and "acceleration time." For example, a specification might call for the safe time to be at least "5 seconds" longer than the starting time. A percentage margin might be dictated instead.

Such oversimplification is undesirable. As Figure 4-10 shows, relaying can be chosen to protect a drive even when acceleration time is substantially greater than the safe locked rotor time. Remember that, as Figures 4-5 and 4-9 show, motor current does not remain constant at the locked rotor value throughout acceleration. It drops off considerably long before full speed is reached. That always opens up more "window" or working space between the two curves.

Any arbitrary "safety margin" between locked rotor and acceleration times may sometimes be insufficient for proper protection. That depends on the shapes of the curves involved. Or, when the margin is adequate, it may be more than needed so that a special or oversized motor design is required at extra cost to provide a needless degree of safety.

From time to time, power system designers or motor users have pointed out that different manufacturers draw different thermal damage curves for "similar" motors, even for the same rating being quoted for the same application. Why aren't the different versions more nearly alike? Because not everyone builds the same horsepower rating in the same way. Some rotors are cast aluminum, others copper bar, with different safe temperature limits. Experience leads one manufacturer to set different short-time insulation heating limits than another. The ratio between stator and rotor resistance, which affects the relative rates of heating in different parts of the machine, can vary between designs.

Competitive motors do not exhibit the same performance in all other respects, such as full load speed, or power factor; why expect their thermal limits to be the same? There may be enough conservatism or safety margin to permit using one supplier's curve as a guide for proper protection of another supplier's motor—but it isn't recommended.

Analysis of motor circuit protection is up to the power system designer—or the plant designer, or consultant, whose job it is to select the proper motor, motor accessories, and protective equipment. But an understanding of what the thermal damage curve means, and how it may be drawn, is helpful to anyone involved in induction motor application.

Current transformers

Protective relays used in motor circuits, not only for starting conditions but also for running overloads or ground faults, commonly sense line current by means of current transformers (CTs). Energy-monitoring or machinery control circuits may also require motor CTs. These may be located in the motor terminal box itself. (See Figure 4-12.)

The current transformer's basic function is twofold. First, it transforms or reduces high motor line current (ranging from 30 or 40 amperes up to several thousand) to a five ampere level compatible with standard measurement or control circuits. Second, it isolates those "low voltage" circuits from the much higher voltage present on the motor leads.

Current transformers take several basic forms, some of which are shown in Figure 4-13. The wound and window types are seen most often, the latter being popular for several reasons. One is that because fully insulated motor lead cables passing through the CT window will act as a single primary turn, the transformer insulation need not be subjected to rated motor line voltage.

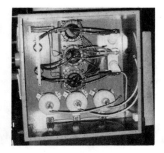

Figure 4-12: Typical installation of three window type CTs in a large motor terminal box (1500 hp, 4 pole, 4000 volts). These transformers, one in each phase, are connected according to Figure 4-18 (page 161) for differential relaying.

Figure 4-13: Several basic CT types. At upper left is the "window type." The round hole or window through the transformer is for a cable or bus bar carrying the primary current to be monitored; that forms the single primary turn for the transformer. The "wound type" CT (upper right) requires direct connection of the primary cable or bar to transformer terminals; the primary turn is built into the transformer. At lower left, the "bar type" unit combines features of wound and window designs. The high voltage window type (lower right) is intended for an uninsulated primary lead.

Some manufacturers do recommend against using standard 600 v window type CTs with unshielded primary cables above 2500 v. Capacitive coupling between the transformer (at or near ground potential) and the cable could produce local high voltage stresses in the cable insulation where it passes through the CT window. Nevertheless, it is widely accepted practice to use such CTs with unshielded motor cables up to at least 6.6 kV. Higher voltage CTs, constructed like those at the lower right in Figure 4-13, have an extra window insulation for use with bare bus primary conductors. Transformers of that kind, having a 15 kV primary rating, shouldn't be needed for a 4000-volt motor circuit. (Those in Figure 4-14 were installed in the motor terminal box by the user, in the field.)

Because no large terminal connections for the high-current primary circuit are needed, the window type CT is relatively small. Installation is relatively inexpensive. Removing it entirely from the circuit—for example, to disconnect motor leads for checking insulation resistance—is a simple matter.

Matching the Motor to the Power System

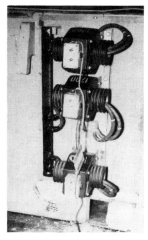

Figure 4-14: High voltage window type CTs, mounted on neutral leads in the field by the user of a large 4160 volt motor.

Finally, the window type CT ratio can easily be varied to suit different motor load currents. Passing a motor lead cable once through the window produces the single primary turn for which the CT design ratio applies. Looping that cable through the window twice provides two primary turns, cutting the ratio in half. Thus, a "400 to 5" ampere CT becomes a "200 to 5" unit.

However, primary cable size usually makes that impractical. It is much easier to change the turns ratio by looping the small secondary lead through the window instead. Depending upon the loop direction, secondary turns may effectively be either added or subtracted. (See Figure 4-15.) The effect on the turns ratio will depend upon the transformer design, but the figures in Table 4-I are typical (for a 300 to 5 ampere transformer).

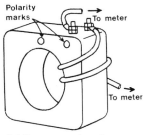

Figure 4-15: How to loop one secondary lead through a CT window to change the transformer ratio by adding or subtracting turns.

Adding secondary turns

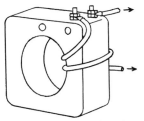

Subtracting secondary turns

Ratio	Number of secondary turns added	Ratio	Number of secondary turns subtracted
305:5	1	295:5	1
310:5	2	290:5	2
315:5	3	285:5	3
330:5	6	270:5	6
345:5	9	255:5	9

Table 4-I.

Polarity marks

This is one example of the importance of the "polarity marks," indicating relative direction of primary and secondary current flow, on all CTs. Once removed from the circuit for any reason, a transformer must be put back again with the same relative polarity.

These marks are most often small white disks or buttons on the outside of the CT winding or frame structure, one indicating one side of the secondary winding, the other associated with the corresponding side of the primary. A number of polarity marks are visible in Figure 4-13.

The transformer is normally connected so that both primary and secondary polarity marks are on the "upstream" or line side of the circuit—the side from which the current is flowing toward the transformer from the source (Figure 4-16).

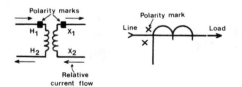

Figure 4-16: Polarity marks indicate consistent relative direction of current flow in CT primary and secondary circuits, as shown here. The right-hand view is a typical way of showing CT polarity on a diagram; one of the two polarity marks is for the primary, the other for the secondary.

For the "percentage differential" protection circuit of Figure 4-17, this rule is followed on the line side of the motor winding. However, the function of the relay here, which is connected to the secondaries of both sets of CTs, is to respond to any phase difference that develops between the line and neutral sides of the winding when a fault occurs somewhere between them. If line and neutral currents remain equal, as will always be true in the absence of any fault, they balance each other out in the relay, which consequently does not trip. The current balance results from reversal of the neutral-side CT polarity, as shown in the diagram. Unless that reversal is maintained, the relay will not operate properly.

For the "self-balancing differential" protection scheme (Figure 4-18), transformer polarity does not have that importance. Each transformer normally

Matching the Motor to the Power System 161

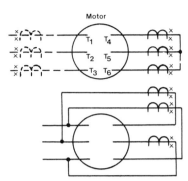

Figure 4-17: Some motors use this "percentage differential" CT hookup. One set of three transformers is located in the switchgear (left). A second set of identical CTs is mounted in the motor terminal box, in the motor neutral leads. All CT secondaries are connected into a common relay circuit. Any fault in the motor windings makes motor line and neutral currents unequal, "unbalancing" the CT secondaries and tripping the relay. The lower view shows how the same CTs are used with a delta-connected motor winding.

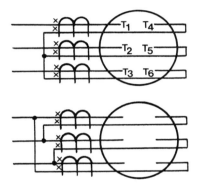

Figure 4-18: In this alternative differential relaying scheme, only three CTs are needed instead of six. Each one compares the line and neutral current within one phase. If there is no difference, the relay in the CT secondary circuit does not trip. The lower view shows a delta-connected motor using the same scheme.

sees in its primary the two equal and opposite-flowing line and neutral currents. If a fault renders those two currents unequal, net operating current appears in the CT secondary circuit, and the relay responds to the magnitude of that current. Relative polarity is not critical.

But other relaying circuits also require specific CT polarity, so be sure to observe and maintain carefully the original polarity for all CTs. Whenever motor leads are removed from window type CTs, they must be put back the same way—passing through the windows in the original direction.

Motor CTs are used not only for relaying—that is, to supply current to protective relay operating coils—but also for metering, to supply current to either instrument windings, for direct display of motor operating current, or to special electronic circuits (load monitors) which record or control motor load current. Because of the effect on long-term power cost determination, greater accuracy is expected of metering CTs than of those used only for relaying.

Recognizing that, ANSI has set up "accuracy classes" for standard CTs which reflect that difference. For metering service, ANSI specifies correction factors not to be exceeded—usually in the range of ⅓% to 1%—at two operating conditions: 10% and 100% of rated transformer amperes. For relaying, where currents far beyond the nominal transformer rating may be present during faults or overloads, an error up to 10% is allowable—but at currents as large as 20 times transformer rating.

How does a CT work?

Electrical theory teaches us to think in terms of voltage ratios when discussing transformers. A voltage V_1 on the primary or incoming power winding on a transformer distributes itself uniformly over a number of turns N_1 in that winding. Whether the secondary or output winding of the transformer is part of a closed circuit or not, a secondary voltage V_2 will appear across its N_2 turns, related to V_1 in the same proportion as N_2 bears to N_1. In other words, volts per turn in the transformer is a constant. Or, we can say: $V_1/N_1 = V_2/N_2$.

We have to change our thinking somewhat to understand a current transformer. Obviously, no current flows on the secondary side unless the secondary circuit is closed. There will be some secondary voltage regardless. The amount of current any transformer supplies to a load depends on the magnitude of the load circuit impedance. Since that impedance can be whatever we choose to make it, how can we speak of a standard secondary current rating for the transformer?

To answer that, bear in mind that any transformer tries to exhibit a (very nearly) constant value of volt-amperes which will be the same in both primary and secondary windings. That is, whatever the secondary circuit may be, if A_2 amperes flows through the secondary winding and its connected load circuit, while V_2 secondary volts appears at the secondary terminals, then the primary winding will see an input current A_1 and terminal voltage V_1 such that this equation is always satisfied (not exactly, but very nearly, because there must be some small current that exists only to magnetize the transformer core and does not appear as "load" current): $A_1 V_1 = A_2 V_2$.

Let's look at some actual numbers. Suppose we have a transformer with only two primary turns and 160 secondary turns. Connect those primary turns into some load circuit so that 400 amperes flows through them. From the equation given earlier:

$$V_1 = \frac{N_1 V_2}{N_2}$$

Substituting this in the volt-ampere formula, we get:

$$A_1 \left(\frac{N_1 V_1}{N_2} \right) = A_2 V_2$$

Since $N_1 = 2$ and $N_2 = 160$, we can solve for A_2 this way:

$$A_2 = A_1 \frac{N_1}{N_2} = 400 \left(\frac{2}{160} \right) = 5 \text{ amperes}$$

The design value of secondary volt-amperes (2.5 to 12.5 being typical) will be present when that secondary is connected to a "burden" or load impedance for which the CT is designed. This notion of "burden" is sometimes confusing, because the actual performance of the transformer—particularly the accuracy of the ratio between primary and secondary amperes—can only be predicted if the actual primary and secondary currents and voltages are known. There must be a match between the transformer characteristics and the impedance and power factor of the load on the secondary. Numerically, a CT burden is that impedance (including both resistance and reactance) of the connected secondary circuit, which may be a meter, a relay, several such devices in series or parallel, and the leads connecting them to the CT.

The higher the burden, the higher the volt-amperes supplied by the CT secondary when rated secondary current is flowing. This is easily seen from Ohm's Law. If current flowing is constant, the higher the circuit impedance (burden), the higher the transformer voltage must be to circulate that current. Constant amperes with higher voltage means higher volt-amperes.

Current transformers are designed with high magnetizing impedance. The core is magnetized on the lower portion of its saturation curve so that magnetizing current and flux vary as uniformly as possible with each other. Normally both are low. But they are relatively fixed in magnitude. This means that when the total current is low—when the transformer is working below its normal ampere rating—metering errors will be higher, because the relatively constant magnetizing amperes becomes a larger fraction of total transformer current.

Thus, it isn't wise to use a 300:5 CT, for example, for measuring primary currents of, say, 50 to 100 amperes, just as a 0-300 ampere scale on an ammeter will not yield high accuracy for currents of 50 to 100 amperes.

Short circuiting

One of the first things we learn in working with electric power is to avoid short-circuits. Fiery arcs, ruined equipment, blown fuses—these result from accidental shorts. So it may seem difficult to accept the necessity of purposely short-circuiting the secondary of any current transformer from which the normal burden is disconnected. Certainly we could never do that with any ordinary transformer. Why is it even possible, let alone necessary, with a CT?

Consider what takes place in a conventional power transformer. Whether it is a "step-up" or a "step-down" unit makes no difference. Primary voltage is relatively constant. When the secondary is open-circuited, that primary voltage acts only to circulate a small magnetizing current—and magnetic flux—in the core, inducing rated voltage at the secondary terminals. If we now close the secondary circuit through a load, some output current will flow from the transformer. That added amount of secondary ampere-turns in the unit will dictate a corresponding rise in primary ampere-turns, and the primary or input current increases accordingly. On short-circuit, the current will be dangerously large.

But in the CT, it is primary current, rather than voltage, which is relatively constant. That current is fixed by the rest of the primary circuit. Hence, primary ampere-turns cannot change significantly no matter what occurs in the secondary circuit. Short-circuiting that secondary will of course drive secondary voltage virtually to zero. Some secondary current circulates, but it cannot rise higher than the fairly low value balanced by the existing primary ampere-turns. Thus, no destructive overcurrent results on either side of the CT.

What happens if the secondary is not shorted but open-circuited, with full primary current flowing? Remember that with a properly chosen secondary circuit "burden," and rated secondary amperes flowing, only a very low secondary voltage is needed to circulate those amperes. So only a low value of magnetic flux is present in the CT's iron core.

That flux is produced by the net difference between the primary and secondary ampere-turns—normally quite small. The primary value is made up of a large number of amperes circulating through very few turns; the almost equally large secondary value includes many turns but few amperes.

Should the CT secondary circuit be opened with primary current flowing, the high value of secondary ampere-turns disappears entirely. The large primary ampere-turns, however, will remain (because primary amperes can't change). Without anything in the secondary to balance against that, those ampere-turns will abruptly increase the magnetic flux in the core to a dangerously high level—resulting in as much as several thousand volts across the open-circuited secondary. This can destroy transformer insulation or connected equipment, such as load-monitoring resistors. Moreover, anyone working on the circuit (normally considered a "low-voltage" circuit) is in danger of electrocution.

Secondary protector

For that reason, it is essential to short-circuit the secondary of any CT when its normal burden is disconnected. Several special precautions are available to deal with this situation. One is the so-called "secondary protector" (Figure 4-19), which is used as shown in Figure 4-20. Normally, the CT secondary current will flow through the low-impedance burden. Should anything happen to disconnect that while primary current is flowing, the transformer secondary voltage rises (as just described) enough to overcome the non-linear resistance of special heater disks in the protector, which then constitute the only load on the CT. When those disks get hot enough, the protector's internal thermostatic switch will trip to short-circuit the transformer. This cuts off the heating current, allowing the protector to cool and re-set. This cycle repeats itself as long as the normal burden remains disconnected. Because the protector impedance is higher than that burden, CT secondary voltage does rise when the protector is in operation, but only to the relatively safe value of 150 to 350 volts depending upon protector design.

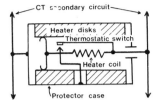

Figure 4-19: A secondary protector used to protect CT circuits against over-voltage from an accidental loss of the burden.

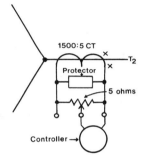

Figure 4-20: A compressor "load control" circuit commonly used with large motor drives. One CT, its ratio chosen close to rated motor line current, has across its secondary a protector of the type shown in Figure 4-19, plus an adjustable resistance burden. Secondary current flowing through that burden produces a voltage drop which is tapped off to a controller that adjusts the compressor to hold motor load within a predetermined range.

A second type of CT secondary protector is diagrammed in Figure 4-21. If secondary voltage rises above a low value (typically 10 volts), because the normal burden has become open-circuited, the zener diode triggers the thyristor which conducts through resistor R_1 to restore CT load current and drop the voltage.

Another protective feature, effective when the secondary circuit is likely to be disconnected only manually, is the "short circuiting" terminal block for CT secondary leads in the motor terminal box. These blocks carry sliding links which short-circuit the secondary whenever the leads to the external burden are removed from the block. Some transformers have such links as integral parts of the secondary terminals.

Whatever protection there may be, anyone servicing a motor having current transformers in the terminal box should assume, for safety, that the CT secondary is a "high voltage" circuit at all times that the motor itself is energized. Some motor specifications, written with that in view, call for complete

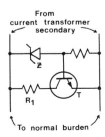

Figure 4-21: An electronic secondary protector used with CTs in some solid-state motor control circuits. Thyristor T conducts current through resistor R_1 when the Zener diode Z conducts as a result of secondary voltage above a safe limit.

separation of the CT secondaries from other control circuits such as temperature detector or space heater wiring.

That is not required, however, by the National Electrical Code. Article 300-3, Exception 3, states "control, relay, and ammeter conductors used in connection with any individual motor . . . shall be permitted to occupy the same enclosure as the motor circuit conductors." So it was only individual user preference, not a national standard, that resulted in the current transformer arrangement shown in Figure 4-22. This motor was a "high voltage" rating, and the CT secondary circuit was considered "low voltage." Specifications required the latter to be outside the main motor terminal box in a separate enclosure.

Figure 4-22: Inside the terminal box of a 1250-hp, 2-pole, 2300-volt compressor drive motor equipped with a load control circuit of the sort diagrammed in Figure 4-20. (Actually, only one phase passes through the CT window.) The resistor is in a separate terminal box attached to the main box (upper right); at the top of the picture is a CT secondary protector (see also Figure 4-19).

However, a proposed revision to NEMA Standard MG1-20.62 will require (for motors above 600 volts) that " . . . current and potential transformers located in the motor terminal housing shall be permitted to have their secondary connections terminated in the motor terminal housing [if] separated from the motor leads by a suitable physical barrier." The nature of that barrier has not been specified.

Some users have even called for the entire CT (not just its secondary wiring) to be in a separate box away from the high voltage main motor leads. This is normally impossible, because those motor leads must form the transformer primary itself. When both ends of each phase are brought out as for the circuit of Figure 4-17, the neutral connections and the CTs associated with them can be located in a separate box. But those neutral leads are at the same potential as the main leads, insulated to the same voltage class, so the separation is purely "cosmetic" unless the neutral is solidly grounded (which is rare).

Again, wiring within motor terminal boxes is exempt from Code provisions. When those provisions are written to prevent dangerous or damaging conditions, however, they will commonly be observed by motor manufacturers. One example concerning CT installation is Article 300-20. This calls for the grouping together of all phase conductors when the circuit passes through metal structures. If single conductors pass though a steel plate, for instance, the steel forms a closed path for the magnetic field surrounding those conductors. Induced current will therefore circulate through the metal around the conductor, causing local power loss and heating. That wastes energy, can cause serious current unbalance between phases, and can even pose a fire hazard.

Such a condition can arise when CTs are mounted in a motor enclosure. It's often convenient to mount the individual transformers on a single steel plate (as in Figure 4-12). Since each transformer must have the conductors of only one phase passing through its window, there must be an opening in the steel plate beneath each transformer through which only that one phase will pass. As the Code requires, therefore, the safe practice is to interrupt the magnetic path looping around that opening by cutting a gap in the steel. (See Figure 4-23.) In practice, it has been found that local heating becomes troublesome only when the phase current reaches at least 500 amperes; some switchgear manufacturers consider 700 amperes a usable limit. Also, for those relaying schemes in which both line and neutral sides of each phase pass through the same transformer, the oppositely flowing currents through each opening will cancel each other out electromagnetically, so the slots are not needed.

Figure 4-23: A steel mounting plate for three motor CTs. Conductors of one phase pass through each of the three holes over which a CT is mounted. The slot cut into each hole from the edge of the plate breaks the magnetic loop around each phase conductor which would otherwise form an induced current path. Slot dimensions are not critical; a simple sawcut will suffice.

A CT secondary circuit must always be grounded. Normally that is done at only one point. A second ground elsewhere on the circuit may permit stray currents to circulate through the "ground loop" and associated relay coils, leading to false tripping.

The reason for the ground is that electrostatic or capacitive coupling always exists between the primary and secondary sides of the transformer, as well as between the secondary and ground. This can "couple" high voltage onto an ungrounded secondary. Grounding the circuit prevents that.

When current transformers are mounted in a motor terminal box, the secondaries may or may not be grounded at that location. Look for such a connection. If it's there, make sure it is clean and tight. If it is removed for any reason during servicing, make sure it is restored.

Finally, don't try to test performance of a current transformer when there is doubt about its condition, accuracy, etc. Such testing is rarely possible in the field. It usually requires specific training in test procedures, plus special types of instrumentation and test equipment.

It's common knowledge that once a motor is up to speed, overload current (monitored via CT or otherwise) must be sensed in time to avert damage. However, it is less widely understood that how large an overload current can safely be depends not only on motor horsepower but on motor voltage. Unless voltage remains in phase balance and at the rated value, even "normal" motor amperes may not indicate a safe condition.

"We can't convince our customers; they tell us it can't happen." That comment came from a large eastern electrical service center, where an increasing number of motors were seen being damaged by single-phasing (the extreme case of voltage unbalance). Their operating practices, the users were claiming, prevented single-phasing on their system.

Saying there is no problem, however, will not make the problem vanish. How much voltage unbalance is dangerous for motors, and why? How is unbalance defined, anyway? To fully answer such questions, we have to look at theory developed many years ago by pioneers in polyphase circuit analysis.

We know that in a three-phase circuit there are three separate voltages, whether we measure between lines or from line to ground. Those three voltages are always 120 degrees "out of phase" with each other. That is, each passes through its maximum value, or its zero value, at a different time than the other two. It is this regular progression of "phase" voltages (and the associated magnetic fields produced in the winding connected to each phase) that generates the single rotating magnetic field of the induction motor.

We also know that those three voltages can be represented by three equally-spaced rotating vectors as in Figure 4-24a. If we were to stand off to the side of this pattern of vectors, viewing it "edge-on" as it rotates, we would see each vector in turn rise from zero to a maximum, decrease again, then drop below zero to a negative value. This rising and falling throughout one revolution represents one cycle of the voltage being observed.

Figure 4-24: Two equivalent ways to visualize the three voltage vectors in a balanced three-phase system.

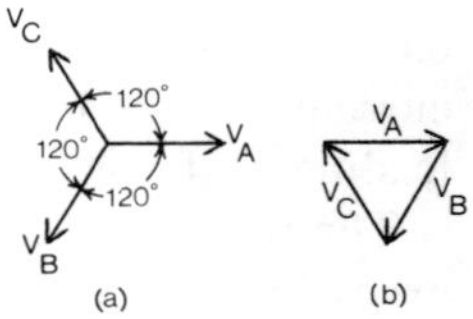

If the vectors are redrawn as in Figure 4-24b, there is no difference in the circuit behavior. But now we can observe that because all vectors are the same length and spaced 120 degrees apart, they will form a closed equal-sided triangle.

This is what we call a "balanced" three-phase system. Now suppose it becomes "unbalanced." What does that mean? In the unbalanced circuit, either the magnitudes of the vectors, or the angles between them, become unequal. The typical result appears in figure 4-25.

Note that the triangle is still closed. Without going too deeply into the theory involved, we should note that there are systems in which the triangle will not close—if the power system is either delta connected, or has a grounded neutral. When that happens, the circuit behavior becomes more complex.

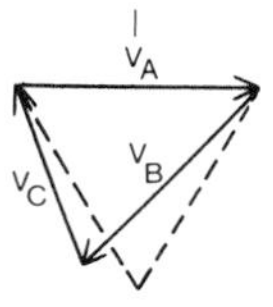

Figure 4-25: When the system becomes unbalanced, one or more of the three voltage vectors changes magnitude and angular position. In this particular version, Phase A happens to remain unaffected, while Phases B and C both change.

But for our purposes we can assume that the triangle will either close or that the "gap" in an open corner will be too small to worry about.

Skipping over the higher mathematics involved, we need only recognize that the vector system of Figure 4-25 can be exactly replaced by two separate systems of balanced vectors, as in Figure 4-26. One system rotates in the same direction as the actual vector system, and is called the "positive sequence" set. The second, much smaller in magnitude, rotates in the opposite direction and is therefore called the "negative sequence" set. Everything that happens in an actual motor or circuit can be calculated by normal methods based on the sum of the effects of both these sets of voltage vectors.

Figure 4-26: These balanced systems combined are equivalent to the single unbalanced system of Figure 4-25. At left are the "positive sequence" components; at right, the "negative sequence" components. The opposite phase sequences of these two systems can be represented either by an opposite rotation of the vectors (lower right) or the same rotation but opposite order of the vectors (upper right).

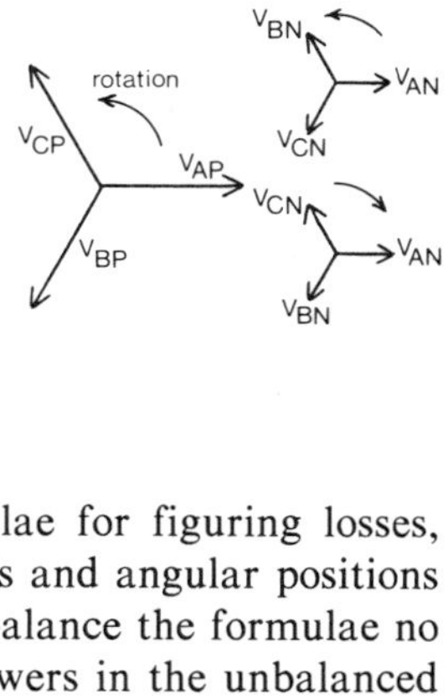

Symmetrical components

The advantage of this is that all our simple formulae for figuring losses, power factor, current, etc., depend on equal magnitudes and angular positions of the three vectors (as in Figure 4-24). Without that balance the formulae no longer work. The only way of getting performance answers in the unbalanced circuit is to use the two balanced sequence systems for which the formulae will still work. Those positive and negative sequence voltage systems are collectively known as "symmetrical components."

Let's take an example. For the actual unbalanced voltages of Figure 4-27, here's how we arrive at the balanced components:

(1) Positive sequence.

(a) Draw the "A" phase vector along a horizontal line, with a length to some convenient scale.

(b) From the arrow end of this vector, draw the "B" phase vector to the same scale, at an angle with "A" that is 120 degrees counterclockwise from its original position.

(c) From the arrow end of "B," draw the "C" vector to the same scale but rotated 240 degrees counterclockwise from its original position.

Figure 4-27: A typical set of unbalanced voltages in a 480-volt circuit.

(d) Connect the starting and ending points with a straight line, then measure its length. To the scale used, it will represent three times the positive sequence voltage. (See Figure 4-28.)

An examination of this result shows that for the typical unbalance situation we don't need to go through the exact process to find the actual positive sequence voltage. It will normally be just a bit less than the numerical average of the three actual voltages. This has only slight effect on motor loss and heating. $(V_1+V_2+V_3)/3 = V_p$

(2) Negative sequence.

(a) Repeat the first three steps above, except interchange the angles for the "B" and "C" vectors. In other words, rotate "B" 240°, "C" only 120°. This is because phase sequence for these vectors is opposite to that for the positive sequence components.

(b) The straight line connecting start and finish points will represent three times the negative sequence voltage E_n. (See Figure 4-29.)

Now, how do these component voltages affect a motor? First, and of least importance, the slight reduction in positive sequence voltage means a little higher current (and heating) in both stator and rotor, for a given load.

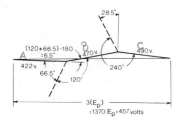

Figure 4-28: Using the angles and voltage magnitudes of Figure 4-27, this construction gives the positive sequence voltage E_p.

Figure 4-29: This graphical procedure determines the negative sequence voltage E_n for the same unbalanced system.

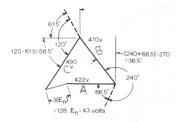

Second, and much more important, the negative sequence voltages produce a great deal more current and heating. One reason is that the motor impedance to the flow of negative sequence current is roughly the same as the locked rotor impedance. Thus, any given E_n generates 5 to 7 times as many amperes as the same amount of E_p.

Negative sequence current directly increases stator heating also. This is easy to visualize, even without thinking of symmetrical components. If voltages are unbalanced, line (or phase) winding currents will be unbalanced, too. Suppose rated current for a wye-connected machine is 7 amps per line. If under unbalanced voltage conditions the three line (and phase) currents are measured at 6, 7, and 10 amperes, the relative losses in the three phases will be:

Phase A: $(6/7)^2$ or 74% of normal
Phase B: $(7/7)^2$ or 100% of normal
Phase C: $(10/7)^2$ or 203% of normal

The total winding loss will be $(⅓)(74 + 100 + 203)$, or 126% of normal. In fact, it will be a little more than that, because as winding temperature goes up, so does winding resistance R in the I^2R loss formula. Phase C will probably fail first, and the motor will appear to have "single-phased"—which in effect it did.

Slight unbalance equals a big loss

The most recent study on this was performed in 1975-76 for the Water System Council, a trade association that was concerned with the problem of current unbalance on submersible well pump drives. It showed that the stator winding loss (not the temperature rise) in the hottest phase increased about 10 times as fast as the voltage unbalance; that is, a 4% voltage unbalance meant a 40% increase in loss. (In a small, random-wound stator, heat from the hottest phase tends to be "soaked up" by adjacent coils in the cooler phases. In a large, high voltage stator having formed coils, this won't happen as readily.)

Another problem is that the negative sequence voltage (and current) vectors rotate backwards. So do the associated magnetic field components. Therefore, the rotor cage bars are sweeping past them at practically twice normal speed. That generates bar currents of twice normal frequency. Because the "deep bar" resistance effect in the bars depends on frequency of bar current, negative sequence currents flow through a greatly increased rotor resistance. Bar heating depends on I^2R, so the negative sequence "I" coupled with a high value of "R" means high loss. The greater frequency in the rotor lamination teeth also means significant iron loss compared to the balanced voltage situation. Some of this extra rotor heat will be picked up by the stator winding. And in some small motors, excessive rotor heating may lead to early bearing failures from heat conducted along the shaft.

The first thing that will happen to the motor when voltage is unbalanced, then, is that it will start to overheat. The second is that it will lose torque. The negative sequence voltage produces a negative torque. Though small, that can be significant in some applications, particularly during starting (unbalance caused by unequal power system voltage drops may be much worse when the motor starts and inrush current is high).

Tests show that the torque a motor will develop over most of its speed-torque curve, compared to its torque with balanced rated voltage, will be about proportional to the square of the positive sequence voltage. Because that is nearly equal to the average of the three unbalanced line voltages, that average can be used to predict torque.

To avoid the trouble of figuring out actual negative sequence voltages or currents, NEMA many years ago proposed a simple formula which gives answers close enough for general use. It is this:

Percent voltage unbalance

$$= (100) \frac{\text{(maximum voltage deviation from the average)}}{\text{average voltage}}$$

As an example, for the situation in Figure 4-27, the average is (⅓)(422 + 470 + 490), or 461 volts. The maximum deviation from that average is 461 minus 422, or 39 volts. The percent unbalance is therefore (100)(39/461), or 8.4%. Thus, the "percent unbalance" from the NEMA formula is close enough for practical use to the "percent negative sequence voltage" in Figure 4-27, which is E_n/(average volts), or 43/461, which is 9.2%.

Another simple formula for voltage unbalance gives a still closer answer. It is:

$$\text{Percent unbalance} = \frac{82}{\text{Average voltage}} \sqrt{\Delta V_A^2 + \Delta V_B^2 + \Delta V_C^2}$$

in which

ΔV_A = difference between actual phase A voltage and the average
ΔV_B = difference between actual phase B voltage and the average
ΔV_C = difference between actual phase B voltage and the average

Looking at the same example:

$$\Delta V_A = 461 - 422 = 39$$
$$\Delta V_B = 470 - 461 = 9$$
$$\Delta V_C = 490 - 461 = 29$$

$$\text{Percent unbalance} = \frac{82}{461} \sqrt{39^2 + 9^2 + 29^2} = 8.8\%$$

NEMA also says (Standard MG1-14.34) that operating a motor at voltage unbalance above 5% (essentially the same as 5% negative sequence voltage) "is not recommended." Some users have mistakenly assumed that it should therefore be permissible to run any motor up to that 5% limit without difficulty. Published articles have stated "Up to 5% is OK."

But this is not correct. Any amount of unbalance makes a motor run hotter, so what results at 4% unbalance, or 2%, is potentially damaging. Usually, protective measures will work at lower values of unbalance. The NEMA standard simply means that once unbalance reaches 5%, the temperature begins to rise so fast that protection from damage becomes impractical.

The simplest protection, as proposed by the NEMA standard, is to derate the motor—to reduce its horsepower load so it can tolerate extra heating. Figure 14-1 of the NEMA standard gives the suggested derating factor. As an example of its use, a motor operating at 3% voltage unbalance (approximately 3% negative sequence voltage) should carry no more than 90% of its nameplate horsepower.

Because these unbalance limits are small, determining the amount of unbalance requires careful measurements. "Voltages preferably should be evenly balanced as closely as can be read on available commercial voltmeters," says NEMA. Granted, that kind of measurement may sometimes be difficult. For that reason it should always be coupled with other system checks that will reduce the likelihood of unbalance being present.

If you have a voltmeter of $\pm 1\%$ accuracy (by no means the best that is "commercially available"), an unbalance of about 1.6% could theoretically go undetected. If a clamp-on meter is used, even with considerable care, a 3% or 4% unbalance could exist without detection. That could be serious.

There are several ways to develop a derating curve. Most of them produce about the same result. One is based on many tests of a variety of motors over many years, establishing that for balanced voltages 15% increase in motor load results in a 25% to 30% increase in winding temperature rise. This means that:

$$1 + \frac{\text{percent rise increase}}{100} = \left(\frac{\text{percent load}}{100}\right)^{1.7}$$

But information developed by NEMA and various researchers indicates that when voltages are not balanced the percent increase in temperature rise equals about twice the square of the percent voltage unbalance. We can therefore say that:

$$\left(\frac{\text{percent load}}{100}\right)^{1.7} = \frac{(2)(\text{percent unbalance})^2}{100} + 1$$

For example, assume a 3% unbalance. This is an 18% increase in heating, so that:

$$\left(\frac{\text{percent load}}{100}\right)^{1.7} = \frac{18}{100} + 1$$

from which percent load = $\sqrt[1.7]{1.18}\,(100)$ = 110.2

To hold temperature rise down to the rated value requires derating to 100/110, or just under 91% of nameplate horsepower.

Similar calculations for other values of unbalance result in the curve of Figure 4-30, which is about the same as the NEMA recommendation. Here's

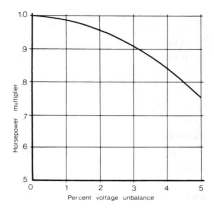

Figure 4-30: Developed from several sources, but in agreement with NEMA MG1-14.34, this derating curve suggests the reduction in motor load needed to compensate for overheating produced by unbalanced (negative sequence) voltage.

an easy way to remember the derating factor: Motor load should equal 100% minus the square of the percent voltage unbalance.

External devices

The alternative to derating, if voltage unbalance is only intermittently excessive, is to protect motors on the circuit with external devices. There is no easy way to do that with conventional overload heaters, voltage relays, or current relays. Remember that currents at full load will be unbalanced by as much as 6 to 10 times the voltage unbalance. Overload settings remain fixed until changed, whereas unbalance may vary from day to day. The result could be overprotection (nuisance tripping) under some conditions; loss of protection for others.

Do not assume that providing three overloads where there were only two, as now required for most Code installations, will automatically protect a motor against voltage unbalance damage. Local winding overheating can still occur even though currents do not exceed the setting of any one overload.

How difficult the problem may be is evident from this procedure that has been suggested for setting conventional overloads:

(1) Read the three line currents with the motor loaded.

(2) Find the sum of their deviations from average. For instance, in a motor rated 19 amperes full load, if the currents are 25, 16, and 19 amps, this sum equals (25-20) plus (20-16) plus (20-19), or 10.

(3) Raise or lower motor load until the average current equals the rated value minus one-third the sum of the deviations just found.

If the average current varies greatly, it may be necessary to repeat steps (2) and (3). Obviously, this kind of load adjustment will not be possible in some applications.

Continuing with the same example: Decrease load until measured currents average $19 - 10/3$, or 15.6. New readings are 21, 12, and 14 amps. The deviations from average are 5.4, 3.6, and 1.6, the average deviation being 3.5.

Subtracting 3.5 from the original average of 19 gives 15.5, which checks closely enough.

(4) Set protective devices to limit current to the values in each phase resulting from the final Step (3) adjustment.

NEMA recommends applying motors per the derating factor, then setting overloads at maximum current expected under the unbalanced conditions rather than the average current. While on the safe side, this may represent overprotection in some applications. But there is no simple solution to such a complex problem.

The recent development of solid-state electronic relays (Figure 4-31) has made it easier to protect motors against voltage unbalance. Inexpensive devices are available that combine such protection with phase failure or reversal as well as undervoltage or ground fault tripping. One make of "voltage monitor" will trip in 6½ seconds when voltage unbalance exceeds 6¼%.

Each type of relay has its own advantages and disadvantages. Those devices responsive to negative sequence *current* are exceptionally sensitive, because large currents result from fairly small voltages. But they require current transformers in the circuit. Other relays respond to actual negative sequence voltage, and these may not sense single-phasing under light load. "Phase balance" relays, which compare relative phase current magnitudes only, must be carefully applied to avoid nuisance tripping.

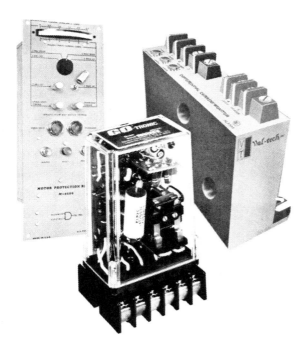

Figure 4-31: Typical solid-state unbalance or negative sequence sensing relays designed for motor circuit protection.

Whatever the type, relays may not trip for voltage unbalance below 4% or 5%. One typical characteristic appears in Figure 4-32. In that case, full motor protection obviously dictates either load reduction per Figure 4-30, or use of a motor having a 1.15 service factor (no longer standard above 200 hp) to avoid possible damage from conditions below the relay's lower limit of protection.

A 1.15 service factor motor will typically stand about 4.5% voltage unbalance with loss of only half its total insulation life, provided it does not operate above its nameplate horsepower.

"Voltage balance" relays may be of limited value, because they are pre-calibrated to trip whenever any one of the three line voltages drops to 83% of nominal. This can mean voltage unbalance of 12%, which is too high a limit for useful motor protection.

Winding temperature sensors are good general protection. However, they will not protect the average motor against loss of torque or possible acceleration overheating when unbalanced voltages are present during starting. More important, they cannot sense rotor overheating caused by negative sequence current. And the trip setting is usually fixed, except for large machines, allowing no adjustment for changing power system conditions.

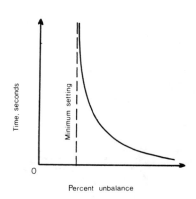

Figure 4-32: Tripping characteristic of a "negative sequence" relay. The minimum amount of unbalance will vary with the design, typically in the range of 5% to 30%.

Causes of unbalance

What produces voltage unbalance? These are the principal causes:

(1) A large single-phase load on the system. In offices or small shops, this is usually a lighting load. In a larger plant it may be welding equipment or an arc furnace.

(2) A blown fuse in one phase of a bank of power factor correction capacitors. There is growing use of such capacitors in industry today, and because an open circuit in the bank produces no load shutdown or other alarm, it may go unnoticed for a long time.

One serviceman reported to a plant engineering group, "My company was called in by one of the local plants of a large national manufacturer to do some testing. We found that, out of seven capacitor banks, only two were operating normally. The others had blown fuses. These blown fuses are normally caused by high surge voltages that come through from time to time. It would be well to have your maintenance personnel check the currents, not the voltages, in the lines feeding such capacitors at regular intervals."

(3) Unequal supply impedances. Low voltage bus duct with excessive impedance may cause this. An open delta transformer bank is another likely source, particularly during motor starting.

Here's a case history. Fortunately, as is now true for most drives, the motor was protected by three overloads. Called to check on a tripout of a 400-hp, 720-rpm, 440-volt pump motor, the serviceman made these measurements: voltages—460, 455, 443; currents—610, 480, 455 amperes (rated amps 490). According to the NEMA formula, with average voltage of 451, the percent voltage unbalance is only 2%. Note, however, that one phase is drawing more than 20% above rated current.

The unbalance was traced to an open delta transformer bank that wasn't supposed to be open. Checks by the utility on both primary and secondary had indicated nothing abnormal. But a service center investigation found one of the three units in the bank running cold while the other two were hot. An incorrectly made secondary connection inside the cold transformer had left it open-circuited after failure of a solder lug. After that was repaired, voltages returned to normal. Had there been only two overload relays on the motor circuit, an eventual burnout was probable, with no one suspecting its true cause.

Here's another example. A 25-hp, 230-volt submersible pump motor suffered rotor failure after restoration of service following a lengthy power outage. Two 25-kVA padmount transformers in open delta supplied the motor circuit. There was no failure of the stator winding.

Tests showed that even under "normal" conditions the voltage unbalance was about 1.2%, resulting in line currents of 56, 63, and 66 amperes (10.5% unbalance). When power came back on after the outage, abnormal single-phase air conditioning loads nearby unbalanced the voltages further. As is typical in most areas, the state public service commission imposed no voltage balance criteria as such, but other regulation limits dictated residential circuit voltages of plus or minus 5% of nominal; 7.5% for industrial circuits. If translated into voltage unbalance, these would be far beyond what any fully-loaded motor would tolerate.

The solution to this problem was a corrective autotransformer on the circuit, plus a solid-state voltage balance relay to protect the motor.

There has been some disagreement about how widespread such problems are. In 1955, a public utility spokesman made this claim: "There does not appear to be any need for protection against this [unbalanced voltage] condition. Systems should be designed to maintain a reasonable voltage balance.

Likewise, motor designs should be adequate for continuous operation with normal unbalance in voltage likely to be encountered on most systems."

That's a bit optimistic. In the first place, the words "normal," "reasonable," and "likely" aren't as specific as they might be. Second, there has been no repeal of "Murphy's Law." ("If something can go wrong, it will.") Third, motors themselves have certainly changed since 1955. The disappearance of standard service factors for many ratings, plus the "tighter" T-frame rerate designs, means that motors can no longer be abused as freely as once might have been possible. Finally, the unfortunate tendency towards maintenance cost-cutting means that troublesome operating practices may go unnoticed.

Can motors be designed for continuous service on unbalanced systems? Is there a "safety feature" or a winding change, or a difference in rotor construction, that can be "plugged in" to handle a specific amount of unbalance? The answer is no. The designer's only option is derating, making the motor oversize and cooler-running than usual. This has been successfully done for machines (up to 1500 hp 4160 v.) which were specified to operate within temperature rating continuously at 1½% voltage unbalance, with occasional periods of several days at 2½%. Again, the designer's only choice is to build in ample thermal margin based on Figure 4-30.

Transient voltages

One of the most troublesome and elusive operating problems for today's motors is the transient voltage surge, or "spike." At one time, these short-time severe over-voltages appeared only on lines subject to lightning strokes. There were few outdoor motors in those days, so winding failures from this cause were rare.

Today the problem is common, in both low and high voltage motor circuits. Transient surges originate from current-limiting fuse operation, from switching of capacitor banks used for power factor correction or voltage stabilization, from the operation of new types of high speed circuit breakers, and from various solid-state devices used to supply motors with variable frequency power.

Imagine such a transient as a "pressure wave" traveling down a pipe in a water system, as diagrammed in Figure 4-33. The two important properties of

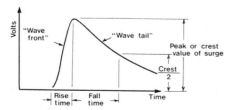

Figure 4-33: Transient surge terminology. Waves are typically identified by their rise and fall times; for example, a "1.5 by 40 microsecond" wave rises to its crest in 1.5 microseconds and tails off to half that crest value after another 40 microseconds. This shape is for medium or high voltage transients; in low voltage motor circuits (below 600 volts), surges are usually more complex and oscillatory in nature.

this wave or "impulse" are its rise time and its magnitude. Many motor cable circuits have a relatively low impedance to the passage of impulse waves, which therefore travel very fast down the line until they reach the motor winding. The "embedded" portion of the stator coils lying in the core slots presents a much higher "surge impedance" to the impulse, causing it to slow down quite suddenly. Just as the hydraulic pressure wave does when encountering a closed valve, that impulse produces a "hammer" effect in the line, piling up the entire magnitude of voltage across the first coil connected to the line lead along which the impulse has come. Recent tests show that within that coil the surge voltage will be unevenly distributed, some turns being stressed more than others.

Motor turn-to-turn insulation cannot stand nearly as much voltage per mil of thickness as can the ground wall (slot liner or layers of tape). An impulse which would not cause ground failure can therefore produce turn-to-turn failure. For example, an impulse with a peak or crest value of 4500 volts can be dangerous between turns in a 4000-volt motor, although a peak voltage of three times that much could easily be carried between winding and ground.

The problem is compounded by two factors:

(1) Such transient surges are extremely difficult to measure or monitor (though one method of identifying their presence will be described shortly).

(2) Even if able to withstand an occasional surge, turn insulation is stressed more and more by repetition of surges—and it's not unusual to find a winding undergoing severe impulses thousands of times during its life span.

Why is impulse rise time important? The shorter it is, the more vertical or "steep" the wave front, and the greater the hammer effect when the surge suddenly hits the winding. A steep-fronted wave will pile up entirely across the first coil at the incoming lead. But if that wave can be given a gentler slope, it will spread out over enough turns in series so breakdown becomes much less likely. As an idea of how short these rise times can get, 10 microseconds (10 millionths of a second) is considered fairly long and not troublesome. One microsecond is usually dangerous. Times of one-tenth to two-tenths of a microsecond are not uncommon.

Let's discuss several common misconceptions about voltage surges on motor circuits:

(1) Large, high voltage motors are more likely to see damaging surges, so they should get surge protection in preference to small machines.

This may be dangerously misleading. Studies in this country and in Europe have shown that "smaller and higher speed motors are more subject to the danger of overvoltages than those of larger size or lower speeds."

(2) Published insulation standards thoroughly cover all aspects of insulation system capability—thermal rating, voltage endurance, etc. Surely, then, the behavior of insulation exposed to transient surge voltages must also be standardized.

Unfortunately, this is untrue. The IEEE did propose such standardization more than half a century ago, but rotating machine windings are so different from those in transformers that as recently as 1960 the IEEE was forced to conclude that standard impulse voltage ratings were not possible for motors. Not until 1977 was there an IEEE standard for testing interturn insulation in motors. One problem has been a lack of agreement on the most useful test method. In a conventional overvoltage or "hipot" test, both frequency and rate of change of voltage with time are constants (the latter being zero). In some transient surges, both may be significant variables.

(3) Surges do cause trouble in medium or high voltage systems. But because of transformer reactance damping in power distribution networks, motors in low voltage systems shouldn't face a transient surge problem.

Transformer windings do hinder transmission of impulses between primary and secondary circuits. But damaging surges may originate on the low voltage side. Peaks up to 2500 volts, with wavefront rise time as low as 0.1 microsecond, have been reported in residential and commercial power systems. One cause is contactor bounce. Another is the output voltage "spike" in the waveform of some solid-state variable frequency power supplies.

(4) Even if surges are present on the feeder circuit, low voltage motors have much more insulation for their voltage rating than do high voltage machines, so that failure isn't likely.

Yes, small low voltage motors are relatively more heavily insulated in proportion to rating. But because the surge voltages they encounter may be relatively worse than in many higher voltage circuits, failures are far from uncommon.

In 1977, an engineer complained that his firm's entire plant air conditioning system was being knocked out by repeated 480-volt motor failures. After much detective work, including oscilloscope studies of the circuits, he concluded the culprit was transient surge impulse generation by fast-acting solenoids in an office copier, which had probably also damaged a 20-hp cooling tower pump motor more than a block away.

(5) If impulse voltages are such a problem, why do we see so few failures attributed to them? Very few winding breakdowns seem to be traced to surges.

The problem isn't that so many such failures occur, but that they are so difficult to guard against and so seldom predicted ahead of time. As we have seen, starting duty (or any of several other common sources of motor damage) is dealt with by well-established, conventional protection schemes. Surges are not so easily handled.

More important is the tendency of transient voltage damage to lead to other forms of winding breakdown. Eventual catastrophic failure completely masks the true origin.

After many years of wrestling with the problem of defining reasonable surge withstand capability for typical rotating machines, the IEEE has recently reported the conclusion that such capability is expressible by Figure 4-34. The problem for the user, of course, is ensuring that surges on his system fall within known areas on the chart. Both elaborate calculation procedures, and field tests, may help to do this. But more often, the known presence of surge-

Figure 4-34: In 1980, an IEEE working group concluded that the impulse withstand capability represented by this graph is "representative" of form-wound rotating machinery, proposing it as the "basis for development of an impulse capability standard" (yet to be written). Here's how to interpret the curve:

(1) If the machine will see surges of either magnitude or rise time extending into the area shaded with lines sloping downward left to right, wave-sloping capacitors should be supplied at the motor.

(2) If transient voltages in the area shaded with lines sloping downward right to left will be present, lightning arresters should be supplied. Some examples of the limits plotted; all values in kilovolts:

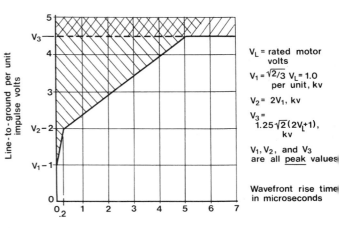

V_L = rated motor volts

$V_1 = \sqrt{2/3}\ V_L = 1.0$ per unit, kv

$V_2 = 2V_1$, kv

$V_3 = 1.25\sqrt{2}(2V_t+1)$, kv

V_1, V_2, and V_3 are all <u>peak values</u>

Wavefront rise time in microseconds

Motor rating V_L	Peak V_1	Peak V_2	Peak V_3
2.3	1.88	3.76	9.9
4.0	3.27	6.54	15.9
6.6	5.40	10.80	25.0

producing equipment on the system should lead the motor specifier to call for surge protection on the motor circuit (about which more later) as "cheap insurance."

Where surges come from

One of the often-unsuspected sources of high transient surges is the open-transition starter. Often, neither users nor service people realize the severity of voltage spikes generated in this way, or by duty cycles which involve any rapid opening and reclosing of motor circuits. To cite just one example: a 250/125-hp, 4/8-pole motor was applied to a mixer drive. The windings failed repeatedly, once after only 64 hours of operation. The operator couldn't keep his plant going; the mixer manufacturer was frantically begging the motor supplier for help in solving this mysterious plague of failures; the motor manufacturer in turn sent an application engineer halfway across the country to examine the installation. Up to this point, no one in the user's organization or among the service people involved had seen any clue to the cause of failure. Observation of the operating cycle revealed, however, that the motor was being switched in open transition from one speed to the other as much as 100 times daily.

When a motor circuit is opened, magnetic flux in the core cannot instantly disappear. It must gradually decay, sometimes taking as long as three seconds to die out. Until that happens, it continues to produce a generated voltage across the open-circuited leads of the motor. The frequency of that voltage gradually decreases as the drive slows down. If the circuit is reclosed during this period, the line voltage is placed directly across this remaining or "residual" voltage produced by the trapped flux. These two voltages are of different magnitude and slightly different frequency. As a result, combining them can generate a high resultant transient voltage sufficient to overstress winding insulation. Repeat this often enough, and the winding fails—often leaving no clue as to the cause.

A related side effect, which can be even more damaging, is the extreme transient currents caused by such open-circuit switching. These are inevitably accompanied by high transient torques. Shaft breakage, stators loosened in frames, rotors twisted on shafts, and end coil insulation damaged—all can result from those torques.

So beware of the extreme risk involved in switching of this kind, whether from one speed to another, from wye to delta, or simply in transferring a motor or bus quickly from one power source to another. The transition must either be extremely fast—within a very few cycles (preferably no more than five)—or else delayed long enough to allow trapped flux in the de-energized winding to decay. This *open circuit time constant* ranges from 0.1 to 3 seconds and can be obtained from the motor manufacturer. In general, the larger the motor, the longer its time constant.

Look for the possibility of capacitor switching on the system, too. If solid-state variable-frequency power is to be used, find out if the waveform will contain transient spikes.

What can be done to prove the presence of dangerous transients? Figures 4-35 and 4-36 illustrate a relatively simple and inexpensive surge detector which can be connected to a motor, or power system, to at least show what transient magnitudes are present. No rate of rise data, unfortunately, can be obtained this way.

Installed on 15 motors in one plant where high impulses were suspected, for a total cost far less than a single large motor rewind, such detectors showed that recurrent surges from 7 to 11 kV were reaching 4000-volt motors.

If vacuum breakers are to be used to control any motors on a bus, it is wise to ask for built-in surge suppressors in the control cabinets. This is a commonly available option. If it can't be done, there is a risk to the motors because vacuum contactors open and close their contacts so fast that current interruption is all but instantaneous. In inductive circuits, such a current interruption must be accompanied by a short-time instantaneous high rise in voltage. This "current chopping" condition is well known. Not so well known is the tendency of vacuum contacts, when closing, to generate high frequency impulse wave trains through making and breaking of arcs across the closing contacts.

These conditions have been investigated at length in several countries. Many users of vacuum contactors in motor circuits report no problems; others do. Differences of opinion continue to be brought out in IEEE conferences about

Matching the Motor to the Power System 183

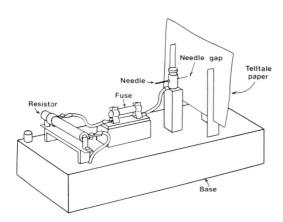

Figure 4-35: A homemade calibrated surge detector—overall length of base is about 12 inches. One detector is needed for each phase. Paper is any good grade of typing paper. Needle gaps are set as shown in Figure 4-36a, and paper is checked once or twice daily to see if it has been punctured by discharge between needles. If it has, a surge has occurred at least as high in magnitude as indicated by Figure 4-36a for the gap used. Increase the gap setting in periodic steps to find the highest surges being encountered. Several makes of transient surge detectors are commercially available. However, some are unsuited to the extremely short rise times that are most damaging to motors, and most work only on circuits rated 600 volts or less.

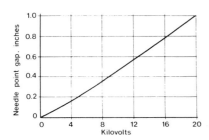

Figure 4-36a: Left: Relation between needle point gap in surge detector, and magnitude of surge voltage which will jump gap. Needles used are No. 00 double long sewing needles.

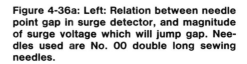

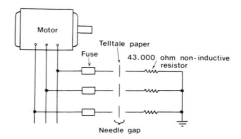

Figure 4-36b: Right: Circuit diagram for surge detectors wired to motor terminals. Needle holders are calibrated and easily adjustable so gap can be set precisely as in Figure 4-36a.

the vacuum contactor transient problem, but the weight of professional opinion is presently on the side of surge suppressors for such equipment on motor circuits.

Surge protection

At the motor itself, two types of devices can be installed to guard against surge effects. Most common on 4000-volt motors, they are available for 460- or 2300-volt machines as well. First is the *surge capacitor*. This is a 3-pole, 0.5 microfarad capacitor in a single can (above 5000 volts, each pole is separately packaged), housed in the motor terminal box, which usually must be greatly enlarged for that purpose.

This capacitor does *not* reduce the surge magnitude. Instead, it acts to lower the wavefront steepness by absorbing energy from the impulse so its rise time stretches out to 10 microseconds or more.

If high magnitude is the problem, *lightning arresters* are the solution. These discharge harmlessly to ground any surge above a certain voltage. One arrester is connected to each incoming line, also in the terminal box.

Together, capacitor and arresters give the motor a complete surge protection system. Because of their bulk at the higher voltages, they are sometimes specified to be located in a separate cubicle near the motor, rather than in the terminal box itself. But this is not a good idea. If the distance along the circuit from the protective devices to the winding exceeds about three feet, the added inductance of the cables will act to nullify the function of the capacitor, so that protection will be lost. Connections should be made as in Figure 4-37. The arresters could be located further away—even at the motor control. If they are at the motor, however, there can be no doubt that any line disturbance, of whatever origin, will be unable to reach the motor without reaching the arrester first.

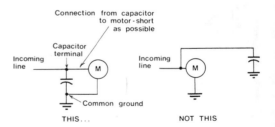

Figure 4-37: Surge protection capacitor must be connected so incoming voltage spike hits capacitor first, before reaching motor winding. If capacitor case is not solidly grounded to motor frame, surge may damage winding even if capacitor is close.

It may seem odd, after having described how capacitors can be used to prevent motor damage from transients, to discover that capacitors can also cause troublesome over-voltages on motor windings. But it is true. Capacitors provided on motor circuits to help raise system voltage or correct motor power factor can do exactly that if incorrectly applied.

Remember that trapped flux which remains in the motor after it has been disconnected from the line? If there is a power factor correction capacitor connected to the motor terminals, it will act to substantially boost the voltage produced at those terminals by that flux—just as capacitors can act to raise terminal voltage during motor starting. In some cases the boost can be so great that motor insulation—or even the capacitor insulation—can fail.

If the motor is connected to a high inertia load with little torque to help slow the drive down when the motor is de-energized, rotation at a fairly high speed may continue for some time, and the capacitor at the motor terminals may make the motor behave like a self-excited induction generator. A dangerously high voltage will be produced for some time.

Even if insulation failure does not occur right away, repeated operation of such a circuit may greatly reduce normal winding life. What is the remedy? If the total capacitor kVA connected to the motor terminals does not exceed the motor's magnetizing kVA, there will be no over-excitation and no dangerous over-voltage. Calculate the magnetizing kVA as simply 1.73 times rated motor kilovolts times idle amperes, and compare this with the capacitor rating involved.

Reclosing on a de-energized motor with connected capacitance is a problem because the presence of the capacitor may serve to maintain generated voltage—even if of a low enough magnitude so insulation is safe—higher or of much longer duration than without the capacitor. This makes the production of damaging torque transients on reclosure much more likely. Reclosing de-energized motor circuits of this type should always be on a time-delay basis, after checking with the motor designer.

If the capacitors are connected on the line side of the motor starter, none of these problems will occur. So if that option exists, it's the best choice.

Don't look for energy savings

What about the little black box "surge suppressors" widely advertised as both transient voltage protection and an energy saver in motor circuits? These are electronic devices with circuitry (Figure 4-38) that will "clamp" line voltage to remove dangerously high peaks. They're made for use in circuits operating below 600 volts.

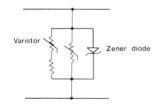

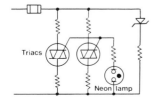

Figure 4-38: Typical surge suppressor circuits. The Zener diodes "clamp" voltage peaks. The patent on the upper circuit explains that the fixed and variable resistors are arranged so that "a minimal amount of energy is . . . utilized to maintain the protective circuit in a standby condition under normal conditions and in an efficient operating mode under abnormal transient conditions."

A nationwide controversy has arisen not from this protection function but from claims that the devices will save energy. "Reduce your electric bill by 9%-20% or more!" "We offer a system of voltage control which will cut kilowatt consumption by at least 10% (guaranteed)."

Those typical advertising claims "explain" the technology in terms like these: "These disturbances [surges] cause electrical motors (e.g. in air conditioners) to operate less efficiently than they are capable of operating, causing us to use more power than we need to."

A technical paper by one engineering consultant has correctly explained that "surges" of the sort shown in Figure 4-39 can increase iron loss in motor core laminations. By clipping those surges, the suppressor eliminates that increased loss.

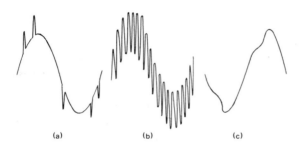

Figure 4-39: Voltage waveforms showing distortions troublesome to motors. The most common transient voltage surge is the occasional "spike" shown at (a). It can harm insulation but does not represent measurable energy loss. The "hash" at (b) can produce such loss—but this rare condition is far more likely to do damage in other ways. Waveform distortion at (c), caused by harmonics, causes substantial motor over-heating and energy waste, but is not a surge condition at all and does not threaten insulation.

But remember that a moderate saving in motor iron or core loss makes little change in efficiency. For 3-phase, 4-pole induction motors of 10 to 100 hp, only about one-fourth of the total loss is in the iron. For 2-pole machines the percentage is even lower.

Besides, how likely is it that any power system disturbance would produce continuous "hash" on the circuit? If it is there only for occasional brief periods, will the saving amount to anything? The consultant conceded that "while laboratory conditions may theoretically prove that transient voltages do act to increase energy consumption, in the real world of electrical networks, transients are usually the result of random switching phenomena, and this problem must be addressed on a statistical basis." That is why, in his 1972 paper on power system problems, a distinguished engineer contended that "Transients do not significantly affect motor losses."

Several IEEE studies have shown that high voltage transient surges are rare on most power systems. Yet one consultant claims to have measured 300,000 surges per hour on a small commercial service. In such a situation, surge suppressors would certainly be a wise investment—but to save the equipment, not to save energy.

One surge suppressor sales pitch contends the extra core loss due to surges heats motor windings to raise the copper loss. But stator I^2R loss in a typical machine would have to double before the motor's total power wastage rose even 10%. To cause that, through doubling winding resistance, would require a 250°C temperature rise. "If this explanation is valid," said one engineer, "it means that motors suffering from too much transient activity can be detected simply by turning out the lights and picking the motors that are glowing cherry red."

Here is another supplier's explanation of how suppressors can save energy on a motor feeder:

"The opening of contacts in relays, switches, and motor starters causes arcing which in turn results in dirty contacts and increased contact resistance. Surge suppressors can reduce arcing and thereby prevent the problem of increased contact resistance.... To show how much energy is wasted by just a little bit of contact resistance, consider the case of a 5 horsepower, 120 volt compressor motor and just ¼ ohm of resistance in each supply leg:

$$I_{motor} = 56 \text{ amps}$$
$$\text{Power loss} = I^2R \text{ total} = (56)^2 (0.25 + 0.25) = 1568 \text{ watts}$$

Assuming an 80% power factor, the motor is consuming 5376 watts. Therefore, the 1568 watts wasted in the dirty contacts represents a 29% increase in energy consumption...."

Commented a utility representative at a recent Southwest Electric Exposition: "The biggest flaw in this explanation lies in the seemingly reasonable assumption that contact resistance could actually increase to ¼ ohm. If this were the case, each of the 'dirty' contacts would be producing an amount of heat equivalent to more than 15 fifty watt soldering irons laid side by side.... A starter relay subjected to that much internal heating would probably burn itself up ... long before having much effect on an electric bill."

Recently a manufacturing plant made a test using the scheme of Figure 4-40. For several days, the timer cycled the motor on and off the normal shop power. A commercial surge suppressor was connected as shown. Then the identical test was performed without the suppressor. Result: identical core iron temperatures with identical kilowatt hour usage for both tests; no difference in motor noise or other behavior throughout. Conclusion: the suppressor had no effect whatever on motor heating or power consumption.

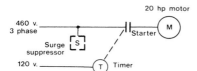

Figure 4-40: This cycling test (25 minutes run, 5 minutes off) of a standard motor on a commercial power system showed no difference in motor power usage or heating with or without the surge suppressor.

Suppressor prices have ranged from $75 to $500. A separate installation at each breaker panel in a commercial or industrial system is often recommended. In one large department store the total suppressor installed cost exceeded $12,000. They might well be justified if you expect those devices to prevent damage on the circuit. But don't expect the kind of utility bill reductions that could quickly pay off the investment.

Extravagant power-saving claims for suppressors have come under attack by business and utility groups, regulatory agencies, and the courts. In one Eastern state, one supplier must now include this note in all retail contracts and distributorship agreements:

"There is no scientific proof that the use of this product will result in appreciable reduction in electrical energy consumption."

While court cases and hearings continue on the merits of surge suppressors as energy savers, and utilities warn their customers against optimistic advertising, potential users should carefully investigate actual conditions on their power systems and the probable effect on motor drives. It appears that notable power savings are seldom likely. By all means, consider the suppressor as useful protection of a motor circuit from dangerous overvoltage. But don't count on it as an energy saver.

5 The Motor and Its Environment

THIS PHASE OF APPLICATION engineering involves both the protection of the motor from hostile surroundings and the protection of the environment from harmful effects produced by the drive. Essential to motor protection is the nature of its enclosure.

Motor enclosures

Basic motor enclosures include 20 different "standards," most of them seldom seen in normal industrial service. NEMA has established two broad enclosure classifications, universal throughout the motor industry: *open*, and *totally-enclosed*. Numerous different types of construction fall within these basic categories. Listing some of the more common versions:

(1) *Drip proof:* NEMA definition: "A drip proof machine is an open machine in which the ventilating openings are so constructed that successful operation is not interfered with when drops of liquid or solid particles strike or enter the enclosure at any angle from 0 to 15 degrees downward from the vertical."

Comment: "Downward from the vertical" needs some explanation. To many people, "vertical" may already mean "downward." Figure 5-1 illustrates the intent.

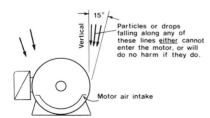

Figure 5-1: How the NEMA drip proof definition is interpreted.

Note that the definition does *not* say that particles or drops *cannot enter* the motor from the stipulated direction; only that if they do enter they will not interfere with the motor's successful operation. NEMA goes on to define a way of demonstrating that success, by subjecting the motor to a water spray at the specified angle for an hour, then performing a reduced-voltage overpotential test on the winding followed by a period of unloaded running at rated voltage.

Where used: Drip proof motors are adequate for most relatively clean indoor areas, where any dripping or falling material comes from directly overhead. Typical installations might be in well-kept machine shops, commercial buildings, elevator penthouses, separate motor rooms in steel mills, warehouses, etc.

Note: The advent of sealed or encapsulated insulations, plus NEMA emphasis on winding condition as the only criterion for successful protection, has led some users to apply drip proof motors with special windings in severe environments. They have been used outdoors, in chemical plants, and in places where floors are often washed down with high pressure hoses.

But winding protection is not the only concern. Severe corrosion or loss of lubricant can disable a motor as quickly as a roasted winding can. So a drip proof machine is not well suited to serious moisture or chemical contamination, no matter what insulation is used.

(2) *Splash proof:* NEMA definition: same as drip proof except the 0- to 15-degree angle is extended to 100 degrees. (See Figure 5-2.)

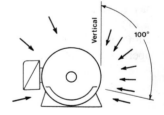

Figure 5-2: The splash proof motor must be able to exclude, or safely tolerate, particles approaching over a much wider angle, the 100° direction being typical of water droplets splashing up from the adjacent floor.

Comment: Again, material coming at the motor from the specified angle is not excluded from entering. The intention is that particles splashing or rebounding from the floor will do no harm to the machine, but they may still get inside.

In the smaller NEMA frame sizes, the splash proof motor is today almost extinct, following a period of declining usage which began 20 years ago. The degree of protection available in the more economical drip proof design makes that motor competitive environmentally with the splash proof. In the most severe service, a totally-enclosed machine would be required anyway. Even in larger sizes 500 hp and up, the splash proof enclosure is seldom seen today.

Where used: Food cleaning and processing plants, pump rooms, locks and dams, sewage treatment plants, etc.

(3) *Guarded motor:* NEMA definition: "A guarded machine is an open machine in which all openings giving direct access to live metal or rotating

parts (except smooth rotating surfaces) are limited in size by the structural parts or by screens, baffles, grilles, expanded metal or other means to prevent accidental contact with hazardous parts. Openings giving direct access to such live or rotating parts shall not permit the passage of a cylindrical rod 0.75 inch in diameter."

The definition adds the dimensions of certain probes which, when inserted through the guarded openings, must not be able to reach bare live parts, the wire of the winding, or any "hazardous rotating part."

Comment: A guarded motor, drip proof or splash proof, is intended to protect personnel coming into frequent contact with it—not to protect the motor itself. However, screens or grilles meeting the "guarded" definition are often employed to exclude leaves or rodents from the motor interior and thus protect the machine.

Where used: Marine installations; school laboratories; bench tool drives; municipal plants under special safety codes; may be required by OSHA in some commercial or industrial locations.

(4) *Weather-protected:* NEMA definition: "Type I—a weather-protected Type I machine is an open machine with its ventilating passages so constructed as to minimize the entrance of rain, snow, and air-borne particles to the electric parts and having its ventilating openings so constructed as to prevent the passage of a cylindrical rod 0.75 inch in diameter.

"Type II—a weather-protected Type II machine shall have, in addition to the enclosure defined for . . . a Type I machine, its ventilating passages at both intake and discharge so arranged that high-velocity air and air-borne particles blown into the machine by storms or high winds can be discharged without entering the internal ventilating passages leading directly to the electric parts of the machine itself. The normal path of the ventilating air which enters the electric parts of the machine shall be so arranged by baffling or separate housings as to provide at least three abrupt changes in direction, none of which shall be less than 90 degrees. In addition, an area of low velocity not exceeding 600 feet per minute shall be provided in the intake air path to minimize the possibility of moisture or dirt being carried into the electric parts of the machine."

Comment: The Type I motor is seldom more than a drip proof guarded machine with an extra outside bearing seal. The Type II, on the other hand, obviously requires complex housing construction. It is costly, and seldom available below 300 or 400 hp.

Where used: Outdoor installations almost exclusively — for pipeline pumping, power plants, refineries, chemical plants.

Totally-enclosed motors

The general NEMA definition for a totally-enclosed machine is "one so enclosed as to prevent the free exchange of air between the inside and the outside of the case but not sufficiently enclosed to be termed air-tight." There are many versions of this. Among the more common:

(5) *Totally-enclosed non-ventilated (TENV):* NEMA definition: "a totally-enclosed machine which is not equipped for cooling by means external to the enclosing parts."

Comment: Some of the smallest sizes of TENV motors have used smooth cast iron or steel housings. Up to 2 or 3 hp the motor may be in the same frame size as a drip proof unit. But as frame size increases, the enclosed heat-producing volume of copper and iron inside the housing goes up much faster than does housing surface area, which is the only heat-dissipating contact with the outside air.

So ribbed or finned housings take over, providing much more surface area for a given frame diameter. Even this is not enough beyond a certain size, so the TENV motor is almost never sold above the 400 frame diameter and is most common in the 256 size or smaller. As an example of how severely this enclosure restricts available horsepower, tests on one large 1200-rpm design showed that with external fan cooling, the parts could carry nearly 600 hp; without that cooling, operating as TENV, the motor could continuously carry only 300 hp.

Where used: Extremely dusty, dirty, or wet locations—foundries, steel mills, mines, paper mills—especially where space is tight, or the noise produced by an external cooling fan might be objectionable. In some places dirty air may foul ventilating passages of a fan-cooled motor, which gives the TENV another advantage.

(6) *Totally-enclosed fan-cooled (TEFC):* NEMA definition: "a totally-enclosed machine equipped for exterior cooling by means of a fan or fans integral with the machine but external to the enclosing parts."

Comment: Normally, these use a single fan located at the end opposite the shaft extension. Though it is not specified in the NEMA definition, for obvious safety reasons the fan shroud or guard around the fan (needed to channel the fan discharge air properly over the motor frame) includes a screened or grilled air inlet. Some chemical dusts, such as potash, even with rapid flow of air through this opening, will "pack" on screen wires or bars, gradually blocking the opening. Similar buildup may also occur inside the fairly narrow space around the edge of the fan guard through which the air discharges along the housing. This is also true for fiber flyings or lint in textile mills. Motors for such service may need special fan shrouds with passages designed to minimize clogging.

In the smaller sizes, particularly below 75 hp, the TEFC motor frame may be the same as for a drip proof motor of the same horsepower. Above that, the TEFC unit will be larger, because it can't dissipate heat nearly as well.

Where used: Much the same as TENV.

(7) *Water-proof:* NEMA definition: "A totally-enclosed machine so constructed that it will exclude water applied in the form of a stream from a hose, except that leakage may occur around the shaft provided it is prevented from entering the oil reservoir and provision is made for automatically draining the

machine." The drain may be a check valve or tapped hole in the lowest part of the frame.

Comment: NEMA suggests one way of testing such an enclosure, by directing a 65-gallon per minute stream of water from a 1-inch hose nozzle at the motor from ten feet away, for at least five minutes. However, there is no specified means of checking the degree of water penetration. One manufacturer has imposed insulation resistance tests following the hose spray as a way to decide whether the windings got wet. Note that completely excluding water—especially as vapor—is not to be expected; otherwise no drain would be needed.

These motors, like the fan-cooled or non-ventilated versions, need space heaters to eliminate internal condensation, unless the motor runs continuously, and sometimes even then depending on the surrounding climate.

Where used: Marine service, as in conveyor drives for self-unloading bulk cargo vessels; severe paper mill duty; dairies; canneries.

(8) *Explosion-proof:* NEMA definition: "A totally-enclosed machine whose enclosure is designed and constructed to withstand an explosion of a specified gas or vapor which may occur within it and to prevent the ignition of the specified gas or vapor surrounding the machine by sparks, flashes or explosions . . . which may occur within the machine casing."

Comment: This is the definition for motor enclosure to be used in a Class I hazardous location as defined by Article 500 of the National Electrical Code (NEC).

How is a motor enclosure design approved for such service? Based on certain design principles involving explosion pressures, bolt stresses, and so on, the motor manufacturer produces a structural package which he then submits to Underwriters Laboratories (UL) for approval. Unless that design is quite similar to one already tested, UL will require a sample motor for testing. This is placed in a sealed chamber containing the gas or vapor involved in the area class. Then, an internal explosion is touched off by an igniter tapped into the motor housing, to see if the blast spreads outside the housing.

If the test is passed, all the design details are incorporated into a "Label Service Procedure" book at the manufacturer's plant. He is authorized to build and sell such motors, each one of them bearing a UL label certifying its acceptability for the hazardous location class. Spot checks by UL inspectors ensure that production units continue to include the design features that passed the original test. This means two things to application engineers, or maintenance/repair people:

(a) Changes in materials or parts are generally not permitted. This may include winding temperature rating.

(b) Labelling of a motor for one type of hazardous atmosphere does not necessarily qualify it for another.

Where used: Petroleum and chemical plants or pipelines; gasoline pumps; natural gas compressors; hospitals or laboratories using ether or similar compounds.

(9) *Dust-ignition-proof:* NEMA definition: "A totally-enclosed machine whose enclosure is designed and constructed in a manner which will exclude ignitable amounts of dust or amounts which might affect performance or rating, and which will not permit arcs, sparks, or heat otherwise generated or liberated inside of the enclosure to cause ignition of exterior accumulations or atmospheric suspensions of a specific dust on or in the vicinity of the enclosure."

Comment: Notice the differences between this and the explosion-proof machine, though in the industry both types are commonly spoken of as "explosion proof." The dust-ignition-proof version must *exclude* the contaminant from the motor interior. Therefore, internal explosions cannot occur, though they are to be expected with the explosion-proof design because it is not vapor-tight.

Second, the exclusion applies not just to explosive mixtures but also to amounts of dust which "might affect performance." Hence, special dust seals are needed to keep dust out of the bearings. A motor that overheats or stalls through loss of lubrication can be a severe hazard in an environment of explosive dust.

Finally, dust *on the surface* of the housing can be ignited if the housing gets too hot. So a major part of the UL testing of a new design is the "dust blanket" test, in which a specified layer of the dust involved is built up on the motor surface to make sure it will not ignite when the motor is at rated temperature.

Labelling and repair considerations are the same as for the explosion-proof motor.

Where used: Grain elevators, coal handling equipment, wood processing plants, feed and cereal mills, sugar refineries, chemical plants. The NEC designates such locations "Class II."

Behind the UL label

Despite growing concern with safety stemming from the activities of OSHA, those working with motors are sometimes unsure of the nature and function of the explosion-proof motor enclosure. Why is it built as it is? What is the distinction between "Division 1" and "Division 2"? Who decides when a motor must be explosion-proof? What part does UL play in motor enclosure design? What is UL, anyway?

There has long been confusion in industry about the various insurance and code-enforcing agencies concerned with product safety. For example, specification writers to this day often call for motors to meet the requirements of the "National Board of Fire Underwriters" (NBFU). Others believe that the National Electrical Code covers motor construction, and is issued by "insurance interests," or by UL. (See Chapter 1, pages 2-5.)

To set those matters straight: The National Electrical Code grew out of an 1897 national conference on "Standard Electrical Rules." Since 1911, the document has been sponsored by the National Fire Protection Association

(NFPA), a non-profit, voluntary membership organization of individuals and groups interested in all phases of the protection of life and property from fire. Public utilities, design engineers, electricians, government agencies, contractors, and manufacturers—all are represented. It is not an insurance organization of any sort.

Aside from classifying the types of hazards, the NEC does not establish how motors should be built. Wiring inside motors or terminal boxes is not covered by the Code, which specifically exempts wiring inside apparatus.

Provisions of the Code are enforced by fire marshals, electrical inspectors, or other persons charged with that duty by local lawmaking bodies that have adopted the Code in statute form. In some places State or Federal laws may govern.

Nowhere does the NEC call for UL labelling on explosion-proof motors. Aside from certain stated features, the Code leaves it up to the enforcing authority's judgment what shall constitute an "approved" motor design for the application. But it is widely recognized that UL has the background of test experience to support such decisions. So the proper UL label is what the inspector usually wants to see.

Underwriters Laboratories is an organization set up to perform tests and calculations which local authorities will accept as proving the capability of motors to safely operate as expected. UL's approval requirements are broadly spelled out in their Standards 674A (Class I locations) and 674B (Class II).

From its tests on product samples, UL writes inspection or "label service" procedures the manufacturer is to follow in production. A staff of more than 500 UL inspectors makes spot checks to see that these procedures are being followed.

Some UL standards were tightened in several ways for motors built after December 1974. All newer motors labelled for Class I Group D service must either maintain surface temperature below 215°C at any load, as shown by type test, or contain temperature sensors to shut down an overheating motor. The 215°C limit had previously been 280°C.

The temperature sensors will usually be hermetically sealed thermostats having normally closed contacts. Other types, such as resistance detectors or thermistors, may be used only in addition to the thermostats—not in place of them (thermocouples, however, are not allowed at all).

These small thermostats can't interrupt line current directly except for the smallest integral-horsepower motors. Above 200 hp—normal limit for a Size 5 starter—the thermostat contacts (typically rated at 345 volt-amperes a-c only) cannot even break starter coil current. Hence, larger motors now carry a nameplate or instruction sheet giving the thermostat contact rating, plus a wiring diagram for "the use of an intermediate control circuit relay."

A second new requirement is that motor nameplates display a "temperature code" (first adopted by the 1971 National Electrical Code). This code symbol appears in Table 5-I taken from NEC Table 500-2b. If type testing shows that motor surface temperature exceeds 215°C, or if no tests have been performed, then the motor must carry a "caution" nameplate worded thus:

"Caution—To prevent ignition of hazardous atmospheres, this motor should not be installed in an area where vapors or gases having an ignition temperature less than (number)°C are present."

These changes placed an added burden not only on the motor manufacturer, but on the user as well. Thermostats must be properly wired into a control circuit to be effective. If the "intermediate control circuit relay" is required, its schematic must include fail-safety. And the caution nameplate must be heeded.

Maximum temperature		
°C	°F	Code
450	842	T1
300	572	T2
280	536	T2A
260	500	T2B
230	446	T2C
215	419	T2D
200	392	T3
180	356	T3A
165	329	T3B
160	320	T3C
135	275	T4
120	248	T4A
100	212	T5
85	185	T6

Table 5-I. Temperature code symbols now required on motor nameplates for UL-labelled machines.

Motor construction

As for motor construction, we recognize that strong housings, extra bolts, and long, closely machined seal fits are needed where explosive vapors are the hazard. Inevitably, such vapors will find their way inside the motor frame, where an electrical failure could ignite them. The purpose of the special construction features is to keep the resulting minor explosion within from touching off the flammable atmosphere outside. Any changes in these parts, substituting low-strength bolts, or drilling holes to mount identification plates, may reduce the strength of the assembly so the initial explosion could rupture the case, spreading flame to the outside. Tool gouges or burrs on ground faces or seal fits leave gaps through which flame can pass.

This is why UL so long declined to sanction rework on UL labelled explosion-proof motors in the field. Such repairs can now be made, but only after steps have been taken to satisfy UL that they will be properly done.

The need for limiting the frame surface temperature in labelled motors is also well understood. Too hot a surface would ignite the outside atmosphere, or the coating of flammable dust that builds up on a Class II machine.

But one point about the Class II application is not so well understood. Penetration of grain or coal dust into the motor interior wouldn't cause an explosion. Unlike the flammable vapor situation, special shaft/bearing seals aren't needed to confine flame developed inside the motor. So why are such seals needed on Class II motors? As UL standards make clear, it is necessary to avoid damage to bearing lubricant which can cause bearing overheating. Many dusts are highly absorbent. They harden up a grease quickly. As the bearing starts to run hot, the process of grease deterioration speeds up.

Dust explosions

In a grain handling facility, the results are often catastrophic. This problem was spotlighted late in 1977 by four grain elevator explosions which killed 55 persons. But it isn't that new a problem. Explosions in grain dust killed 75 people in Japan between 1962 and 1975; in Britain there were 20 "catastrophic losses" during 1974 alone. Overheated bearings, according to speakers at the first international symposium on grain dust explosions in 1977, are a "major threat" in elevators.

This means not only that proper dust seals on motor bearings in such atmospheres are of vital importance, but also that bearing and seal maintenance needs more attention on UL labelled Class II motors than on many other applications. Remember that surface cleaning plus proper thermal overload protection will take care of winding and frame overheating. But they won't help the bearings.

When to require an explosion-proof motor is up to the NEC enforcing authority. The machinery builder or plant operator can't make that decision. To quote one UL official, "Section 90-4 of the Code clearly indicates that interpretations as applied to the acceptance of specific installations are the responsibility of the local inspection authority having jurisdiction."

Thus, it is not up to service personnel to make such interpretations, either. Beware of anyone in plant operations who tells you that "it's OK" to put an unlabelled motor in place of an explosion-proof unit, even temporarily. Don't take part in attempts to switch to a motor that is "just as good, but doesn't happen to have a UL label."

Unfortunately, even plant designers don't always fully realize the nature of explosion hazards. Here's the dust problem again, for instance, in the complaint submitted by a reader last year to an electrical business magazine:

"Over the years ... we have always used TEFC motors ... in our fertilizer blending plant construction ... to guard against corrosion and fouling due to the dust in these plants that blend ... potash, ammonium nitrate and other inert materials. The dusts produced by these operations ... are not of an explosive or combustible nature. But a local electrical inspector recently stated that the equipment in such plants should be Class II, Group G rated.... We do not agree with that opinion...."

Ammonium nitrate "inert" and "not of an explosive or combustible nature"? On the contrary, it has been known for decades that ammonium nitrate is highly dangerous. It was the explosion of two shiploads of ammonium nitrate fertilizer that pulverized Texas City in 1947, killing 468 persons and causing $50 million damage. Points out an NFPA staff member, "It is especially dangerous to permit contamination of ammonium nitrate with oil," which can certainly happen if the material is used around motors lacking proper bearing seals.

In finely powdered form, a great many ordinary or "non-flammable" materials become deadly explosives, not only more dangerous than dynamite but much less predictable. In November 1976, a chewing gum factory in New York was blown apart by explosion of magnesium stearate dust used as a non-stick coating on the gum. Six workers were killed. (This material had never been considered dangerous, and was not listed as such in the handbooks.) The cause of the blast was a machinery failure resulting in high vibration that stirred up the dust, followed by a bearing/shaft failure which ignited it by friction sparks.

Knowing the environment

When planning drives or helping choose motors, then, the application engineer should work with the industrial customer to promote safety by making sure the nature of the environment is properly considered before the inspectors come around.

What does the UL label really mean? It signifies that the motor design and manufacturing practices have been proven by test to provide explosion protection. Of course, every single motor built is not inspected or explosion tested by UL. One might possibly slip through with a defect. However, that is probably less cause for concern than is the tendency to ignore the labelled motor in the field. Users sometimes assume a labelled motor doesn't need the maintenance other machines might, that it won't get in trouble as readily as motors less precisely designed and constructed.

But explosion-proof motors are specified to protect lives and property. Failure of other motors may cause inconvenience, loss of profit, or unexpected shutdowns, but they probably won't destroy the plant or the entire community. So explosion-proof machines need, if anything, more careful maintenance, not less.

All applications in a particular atmosphere do not present equal hazard, of course. Some locations are always dangerous, others only occasionally so. The distinction between Division 1 and Division 2 simply recognizes that the motor's surroundings may either be a hazard at any time under normal operation (Division 1), or only under some abnormal or intermittent condition (Division 2). In Europe, there is a growing use of statistical methods to calculate the chances of an explosive mixture being present in the air. When the chance is extremely small, special explosion-proof construction isn't required. Why spend the money to prevent all possibility of an explosion when the likelihood of explosive conditions is so slight?

Division 1 & 2 classifications in this country are a step in that direction. In gasoline pumping, for example, if vapors can only be present rarely, "by accident or unusual operating condition" such as when a pump seal fails, then a Division 2 classification can be assigned. An open motor can be used instead of a labelled explosion-proof machine, or a TEFC (totally enclosed fan-cooled) design. When dust is involved, the Division 2 location is one in which the explosive particles will not normally be suspended in the air so as to cause an explosion, but may settle on a motor so thickly that it overheats, or could ignite from an arc or spark on a surface.

Class I—hazardous gas or vapor	
Chemical group	Principal chemical involved
A	Acetylene
B	Hydrogen
C	Ether, ethylene
D	Gasoline, benzene, alcohol
Class II—flammable dust	
E	Metal dust, esp. aluminum and magnesium
F	Coal, coke, carbon black
G	Flour, starch, grain dust
Class III—flammable fibers or lint	

Note: Motors designed and approved for Class III service are seldom encountered except in textile mills. Because of the difficulty of building any motor to confine explosions in Groups A & B, these are even more rare. Some motors up through the 445 frame have been approved for Group C. Each of the three Classes is subdivided into Divisions 1 and 2 (see text for explanation). Strictly speaking in NEMA standard terms, only Class I motors are "explosion-proof," in that vapor explosions can safely occur inside. Class II motors are "dust-ignition-proof," built so they will neither ignite external dust nor permit dangerous amounts of dust to reach the interior. Group F has essentially been absorbed into Group G by the latest NEC.

Table 5-II. Hazardous area classifications defined by Article 500, National Electrical Code.

Division 2 motors cannot have high-temperature space heaters in them. Heaters will most often be energized when the motor is shut down, and at that time there is no ventilating air stream to disperse flammable vapors that might rarely be present. Nor can arcing devices like speed switches or unsealed thermostats be used. These might operate frequently, depending on the circuit, presenting a potential hazard even though the flammable substance might seldom be present.

An electrical failure in the winding might occur any time, it's true. But this is considered a slight enough risk in Division 2 to be justified.

Don't look for a UL label on motors designed for Division 2 service. Users sometimes request this, but the UL labelling procedure covers only Division 1 construction. Despite the absence of the label, Division 2 motors include some construction features not strictly spelled out by the NEC but known to be good practice and in some areas required by inspecting authorities. One such feature is the "non-sparking" fan, made of aluminum, bronze, or plastic. These fans must not be replaced by steel. The object of the special material is to prevent friction sparks should any small stones or metal objects get into the air stream and bounce off the fan blades. (It is also claimed that this prevents sparks if a fan becomes bent or broken and itself strikes against other metal parts inside the motor. However, the usual result of that is an early winding failure which makes plenty of sparks in itself.)

In many hazardous areas, there are alternatives to explosion proof motors. There are several reasons for this.

First, rapid growth in number and variety of hazardous chemicals is making motor application more difficult. Before 1970, UL conducted tests to add 15 basic chemicals (including vinyl chloride, butadiene, and propylene) to its listed atmospheres in Groups B, C and D. However, by 1975 the NFPA standard listing flammable chemicals included over 1300 substances. More are being developed almost daily. Explosive properties of most of these compounds within a motor enclosure remain unknown. In order to establish those properties, as an NFPA expert pointed out:

"1. We must know the ignition temperature. . . .

2. We must know the maximum pressure that can be generated by an explosion. . . .

3. We must know the maximum safe gap . . . the widest gap for a particular length of opening which will prevent ignition of an explosive mixture outside an enclosure by an explosion inside. . . ."

Although NFPA Standard 325M gives ignition temperatures, only exhaustive testing can produce the remaining two criteria. That's not feasible for hundreds of different materials. Meantime, UL labelling of a motor for any atmospheric group remains invalid in any atmosphere not yet tested for inclusion in that group. A user must then install a motor other than the labelled version.

Another reason for different motor design approaches is that above about 1500 hp the explosion-proof motor is seldom available. The general technology is known; the largest machine produced in this country, a 3000-hp unit, was built decades ago. But the cost is prohibitive; UL testing is expensive and time-consuming; and in the largest ratings, no practical design assures adequate safety.

If a source of uncontaminated air is available, a force-ventilated ("purged") machine may be an economical alternative. Clean air reaches the motor through ductwork. Figures 5-3 and 5-4 show the flow and pressure needed for typical motors. External blowers are required, rather than reliance on the motor's own internal fan action to draw air through the ducting, because complete purging of the motor enclosure is essential before the drive starts. Otherwise, there is a risk of ignition of explosive vapors that seeped into the

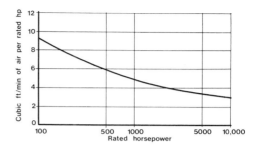

Figure 5-3: Right: Typical air volume needed to purge induction motor enclosure. See also Figure 5-8, page 208.

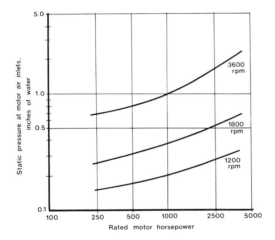

Figure 5-4: Left: Approximate minimum delivery pressure required for air supplying purged motors. Lower pressures could result in "starvation" at the inlet because of the fan action of the motor rotor itself; resultant negative pressure inside motor could suck in explosive gas.

motor enclosure while the drive was idle. These vapors must be swept away by clean air before the machine is energized. Also, using motor fans alone could result in pockets of negative pressure within the enclosure, permitting hazardous vapor to be drawn in.

The external air supply won't necessarily maintain positive internal pressure in the motor terminal box. Figure 5-5 shows a solution for that.

The applicable standard for "purged and pressurized" motors is ANSI C106.1, originally adopted by the NFPA in 1967 as NFPA Standard 496. It defines three basic types of equipment purging:

(1) Type Z, which reduces the hazardous area classification from Division 2 to nonhazardous.

(2) Type Y, which reduces Division 1 to Division 2.

(3) Type X, which reduces Division 1 to nonhazardous.

Here are a few of the more important features of a purged motor installation: Maximum "surface temperature" cannot exceed 80% of the ignition temperature of the surrounding atmosphere. Timing switches are required to prevent motor starting until ten enclosure volumes of purging gas have passed through the machine. (See Figure 5-6 for a typical motor control circuit.) The minimum time this takes must be marked on the motor. Interior pressure during operation must be at least 0.1 inch of water. For Type X purging, internal temperature detectors must be provided, connected to shut the motor down on overheating, while flow or pressure switches do the same job if the purging supply fails. (Again, see Figure 5-6.) Air must not be discharged from the motor into a Division 1 area.

The air needed to purge a motor room, to prevent buildup of an explosive atmosphere, will be much less than what's needed to carry away heat pro-

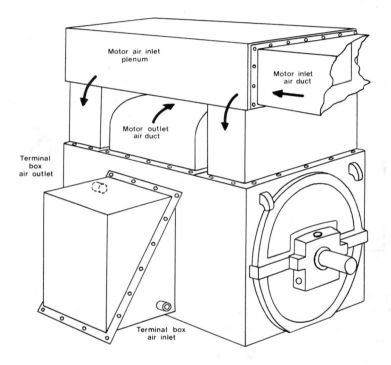

Figure 5-5: Air-purged motor for hazardous area use, with separately pressurized terminal box. Motors of this size must be double-end ventilated; some small machines could be purged by a single inlet at one end with an outlet at the other.

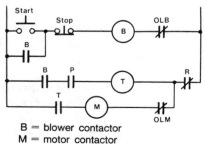

B = blower contactor
M = motor contactor
OLB = blower motor overload

Figure 5-6: Typical control circuit for purged motor. The motor starting sequence: Press start. Purging air blower starts, and assuming motor is cold and pressurized, both P and R contacts are closed so T coil is energized to commence time cycle. After time delay, T contact closes to start motor.

OLM = main motor overload
T = on-delay timing relay
R = main motor thermostat
P = main motor air pressure relay

duced by motors in the room. So the room will generally have its own pressurization system separate from any used with individual machines. How much room air is needed has been widely debated. The NFPA suggests a minimum of one cubic foot of air per minute per square foot of solid floor area. But the final selection must take account of probable leakage from the room.

An alternative to air purging, if there is no clean source or safe discharge area, is the Totally Enclosed Inert Gas Filled (TEIGF) design. This is the true "pressurized" machine, to which only enough cooling gas is supplied to keep the interior pressurized and thus exclude hazardous vapor. There is not enough flow to carry away motor heat loss. Hence, cooling must be provided by other means. Small TEFC motors will use their normal heat exchange systems; a small amount of inert gas (nitrogen or carbon dioxide) is piped into the motor interior under pressure, so that the internal atmosphere remains nonflammable.

Larger motors require an internal gas-to-water heat exchanger. Typically such a machine needs .03 gallon of water per minute per rated horsepower for inlet water temperature no higher than 90°F. Water flow controls are needed to shut the motor down if the coolant supply fails.

Large machines also require special shaft seals to limit gas leakage to 30 to 100 cubic feet daily. (See Figure 5-7.) Standard commercial bottles (Table 5-III), supplying gas through a pressure regulator and reducing valve, maintain at least two inches of positive pressure inside the motor enclosure. A low-pressure control shuts the drive down if the gas supply fails.

Another alternative to explosion-proof motor construction is the "increased safety" design. Such motors are common in Europe, under such standards as

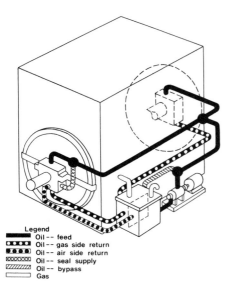

Figure 5-7: Oil/gas sealing system of typical large TEIGF motor. Oil sump has internal barrier separating oil containing entrapped gas from that open to air. Not shown: cooling system, gas controls.

Legend
Oil -- feed
Oil -- gas side return
Oil -- air side return
Oil -- seal supply
Oil -- bypass
Gas

Gas	Volume, cu. ft at 14.7 psi	Pressure, psi	Cu. ft. per lb. of gas
Carbon dioxide	430	900	8½
Nitrogen	215	2200	13.8

Table 5-III. Approximate inert gas data for commercial bottles.

the German VDE 0.70/2.61, "Specification for the construction and testing of electrical apparatus for use in explosive gas atmospheres." The concept is comparable to Division 2 classification in the United States. An increased safety design reduces to a minimum the risk of either sparking or dangerously high surface temperature. Tests must show that no part of the motor reaches a specified ignition temperature within a certain time at locked rotor (five seconds for large machines). All sparking accessories are safely enclosed.

As one European authority expressed it, "'Increased safety' is the [design] technique that hunts out the weak spots in a construction, applies extra safeguards and thus makes it virtually impossible for a spark to occur." This differs from the U.S. tendency to "hunt out" the weak (hazardous) spots in the atmosphere—thus stressing area classification rather than apparatus safety classification.

Such thinking has led some engineers to propose statistical analysis of equipment surveys, resulting in a MTBF (Mean Time Between Failures) for electric machinery. If that proved to be long enough, the equipment could be usable in any atmosphere.

However, increased safety is not yet incorporated in U.S. standards. Continued international cooperation may eventually bring that about.

Conduit boxes

Another common feature in an open or weather-protected motor is "explosion-proof" conduit boxes. Inspectors may consider the boxes part of the wiring system. Replacing or altering such boxes later could lead to trouble. Yet neither motor nor box will carry any kind of UL explosion-proof label. The reason is that the motor itself is not explosion-proof. The box itself may be, but neither motor manufacturer's nor UL's practices allow for labelling it as a single part. Any label on the motor must cover the entire machine, not just some parts.

Small Division 1 motors—445 frame and below—are usually built "dual label." That is, the standard construction suits both vapor and dust atmospheres, so the UL label covers both "Class I Group D, Class II Groups F & G." These are by far the most common explosive industrial environments.

As motors get larger, however, building the features for both Classes into one machine becomes difficult. Besides, bigger motors aren't usually sold off the shelf. It is more logical to design and build them just for the one type of atmosphere in which they are to be used. Class I Group D is most common.

A motor so labelled *cannot* be used in a dust environment. It might meet the necessary surface temperature limits, and be properly sealed, but without a specific Class II label, you can't be sure.

Even more important, a dust-ignition-proof machine labelled for Class II service doesn't belong in an atmosphere of gasoline vapors. Neither structural strength nor flame seal fits are necessarily adequate.

The larger the motor, then, the more important it is to be sure the labelling is right for the application. You may have an "explosion-proof motor," but not necessarily the correct one.

Another difficulty with large machines—above 1000 hp—is the size of the conduit box. Boxes acceptable for UL labelling may be too small for the connections when horsepower is high. If the voltage is high, there is seldom space inside the box for stress cone terminations. If voltage is low, the cables and lugs themselves get too big to fit within the box.

One solution is to use a larger box from a line of commercially available "explosion-proof" junction boxes, provided one can be found physically suited to mounting on the motor. There are two problems here. One is that if the box is shipped as part of the motor, it is subject to the UL motor label—which is supposed to cover all parts. But these commercial boxes have not been tested as part of the motor assembly, so the label cannot be used.

The second problem is that some such boxes, whether shipped on the motor or not, carry no UL explosion-proof label of their own. The box manufacturers have not had them UL-listed. So mounting one on a motor in the field is not necessarily legitimate either. Find out first if the resulting assembly will be approved by local NEC authorities.

What can be done if a large UL-approved box is mandatory? There are several approaches for the motor manufacturer and user. One is for the manufacturer to design a suitably-sized box, having extra gussets, heavy flanges, bolts, etc., to satisfy by calculation the available UL design guides for the atmosphere involved. This design can then be submitted to UL for approval. That can be a long, costly process. Another approach is to actually build the box, mount it on a similar motor, and have UL test it—a process even more expensive and time-consuming.

The third solution is to involve the Code authorities to begin with and get their agreement to accept the box arrangement, or even to waive UL labelling entirely based on their judgment. In reworking or servicing the equipment, maintenance people should look for solid evidence that one of these courses has been followed before becoming involved. The legal and financial problems of product liability are growing serious for industry.

Can better insulation substitute for some degree of motor enclosure? Not necessarily. Remember that insulation treatments can't necessarily eliminate failures from external causes—such as mechanical damage to a coil, the overheating caused by ventilation passages plugged with dirt, or the starter malfunction which leaves the motor on line during overload or stalling.

Wet, corrosive, or dirty surroundings cause some non-electrical problems, too, which no insulation system can solve. One is surface corrosion of motor parts. Accumulation of rust in the air gap, for example, can build up enough to "freeze" stator and rotor together; as much as 3/16 inch of rust can build up on unprotected steel within a year.

Dust can be harmful in three ways, besides blocking cooling air flow:

(1) By building up a blanket of thermal insulation on the windings which cause serious overheating.

(2) By filtering through seals into bearing chambers, fouling the lubricant, and ruining bearings.

(3) In larger, high speed machines, a heavy buildup of dirt on rotors or ventilating fans can throw the rotor out of balance. Magnetic or metal dust can also cut through the toughest insulation.

So the totally-enclosed machine is still needed for many applications despite insulation improvements. But there are some misconceptions about this enclosure. As already mentioned, such motors are not air or vapor-tight. All the fits between bearing brackets and housing in a totally-enclosed machine, no matter how tightly gasketed and bolted, will permit the surrounding air to "breathe" in and out with changes in temperature, especially during cyclic loading. As the air is breathed in, it brings its moisture/chemical vapor content with it.

A totally-enclosed motor can fill up with an inch or more of water inside in a short time as that vapor condenses. Users often ask for drain plugs to remove such condensation as it forms. But these cannot drain off all moisture. Furthermore, they have been known to work in reverse. One user installed several small TEFC motors with porous drain plugs in a sump, where water collected several inches deep on the floor. It ran *into* the motor enclosures through the drain plugs, until the windings failed. Other users have been greatly surprised to open up a "totally-enclosed" motor on the repair bench and have water pour out.

Therefore, any totally-enclosed machine to be used in a damp location should be kept dry with space heaters or winding heating. The more the motor may be idle, the more necessary such heating is.

Some users have sought to eliminate internal motor moisture problems by such specifications as this:

"All TEFC motors shall be equipped with shaft ... seals and gasketed terminal box in order to be classified watertight and to withstand 40 lb. hose test."

There are two basic difficulties with such an approach. First, there is no NEMA or other industry classification of "watertight." As we have seen, NEMA MG1-1.26 does define "water-proof," and outlines a typical test that may be used to establish the capability. But neither in the test nor in the definition is there implied any ability of such a motor to entirely exclude moisture in either liquid or vapor form.

Second, what is a "40 lb. hose test"? This description is unrelated to the NEMA test proposal. Furthermore, no criteria are identified for passage of

The Motor and Its Environment 207

the test. Will a successful motor show zero internal water after the test? How long should the test last?

When considering how much enclosure protection a motor may need, indoors or out, review these points carefully:

(1) Piping systems nearby: what do they contain? Steam? Oil? Acid? Are they located so leakage will fall directly into the motor?

At a 1976 workshop on motor repair problems, the speaker cited the case of a motor mounted beneath a leaking acid pipe. Asked why he didn't get the pipe fixed, the plant operator replied "It would cost too much." Meanwhile, the motor failed every few weeks, being removed and rewound in a different repair shop each time.

(2) Dirt and chemicals in the air: what are they? Do conditions change with the season, the time of day, or the process?

In auto salvage yards, dirt deposited on motor windings is a peculiar mixture of steel particles and fine upholstery fibers, forming a mass which locks itself tightly onto the winding surface.

In one plant, where motors were seldom cleaned, abrasive particles of alumina dust down to five millionths of an inch in size piled up so heavily inside the enclosure that, whenever a drive was started, dust fogged out in such clouds the motor appeared to have caught fire.

Whatever the enclosure options, a choice should be made only after careful study of the location.

Cooling air requirements

Like people, motors need air. Without plenty of clean, cool air to carry away its heat losses, no motor can long survive. Both the rerate program of yesterday, squeezing more power into a smaller frame, and the "energy efficient" design of today, depend on constant improvement in motor ventilation systems.

How much cooling air do motors need? For 80°C rise, 80 to 100 cubic feet per minute (cfm) of air per kilowatt of motor loss. Figure 5-8 translates that into air flow per rated horsepower for typical open motors (totally-enclosed machines won't be much different). For special motors (multispeed, high slip, variable frequency), which won't fit this pattern, consult the motor manufacturer.

This air flow requirement is based on several assumptions. Let's look at them carefully.

The first is that air entering the motor is not preheated by warmed air leaving the motor. There should be no "recirculation." It can have several causes. One is placement of special terminal boxes or machine components too close to air openings in motor enclosures. (See Figure 5-9.) Although most operators understand that blocking an air intake is risky, they do not always realize how easily obstacles near air outlets can deflect discharge air streams back into the cool incoming air.

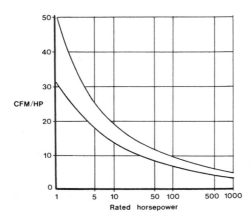

Figure 5-8: Standard open motors rated 80° C rise require about this much cooling air. High side of the band is for 2-pole ratings, low side for six poles.

Other common causes of recirculation appear in Figure 5-10. There is no rigid rule governing closeness of motors to walls; the safe distance depends on relative size and position of the air openings plus the motor size and speed. Some authorities recommend a minimum clearance of five feet. That may be impossible. But it would certainly be dangerous to let the clearance become no greater than the width of the air outlet opening facing the wall.

A second motor placed too close to the first will be just as harmful as a wall in the same position—perhaps more so, because the second motor, unlike the wall, is a source of its own heated air, being directed right back to the first machine. Thus, the distance between two machines having neighboring air discharge openings ought to be more than twice as great as the allowable distance from either motor to a solid wall.

Whatever their location, motors can also suffer from poor air circulation if the design is altered in the field. One such alteration is reversal of the intended direction of rotation. A frequently asked question about many motors is:

Figure 5-9: Beware of installations like this. The compressor builder here has mounted a control console directly in front of the motor enclosure air opening, only a few inches away. This is certain to restrict motor ventilation.

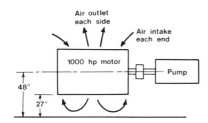

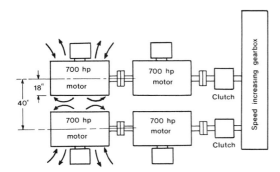

Figure 5-10: The hazard of air recirculation. Lower view shows a four-motor test stand. Motors were so close together that warmed air was forced back into motor intakes (see arrows), causing them to overheat. The problem was solved by installing ducts for forced ventilation. Motor in upper view is too close to wall, causing the same difficulty.

"Can we run it backward?" This is seldom a problem for motors 200 hp or smaller. But with increasing emphasis on high efficiency, motors are being designed for the most efficient movement of cooling air. For large ratings, that means any internal fans mounted on the rotor assembly are *directional*—they will move the proper amount of air when rotating in one direction only. In a 2-pole open motor under full load, reversing the rotation can overheat the winding within a half hour.

Even if the fans themselves are interchanged or replaced to suit the opposite rotation, in a large high-speed machine the rotor cage itself may be directional. The bars are often curved, "slanted," or "bent" to circulate air through axial core vents, and may perform poorly if rotation is changed. Depending on the design, the difference in winding temperature can reach 5°C to 10°C.

Another troublesome alteration is interference with the intended manner of internal air circulation. Many motors below the 440 frame diameter, and almost all larger ones, are double-end ventilated. That is, air enters around the shaft inside each end of the enclosure, passes through the core and winding, then vents out the center of the frame. Any attempt to convert such a motor to single-end ventilation, by causing all the air to enter at one end and leave at the other, is bound to fail, even if the amount of air is increased by an external blower. By the time the air reaches the outlet end of the stator, it's picked up so much heat that it does little good. That area will overheat.

In one example, a group of vertical machines was mounted on a rooftop where winter snow accumulated. Snow was drawn into the lower air intakes, causing winding failure. So the user blocked those intakes. That kept the snow out, but the winding failures continued because the lower end of the motor overheated. Temperature was 41°C above normal. Finally, the lower intakes were re-opened, and the internal air baffle removed at that end to convert those openings to discharge. Even then, winding rise remained 30° too high. (See figure 5-11.)

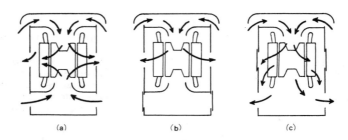

Figure 5-11: Vertical motor with double-end ventilation (a) is designed to draw in air at both ends (arrows indicate direction of flow). Snow being drawn in the lower intakes, off the surrounding roof, caused operator to block off those openings (b). Lower portion of winding then overheated. Removing lower air baffle as at (c) and using lower openings as air discharge did not help much, because air then reaching lower end of winding was already warmed by passage through rest of motor.

For small motors, simply converting the construction of Figure 5-11a to Figure 5-11c can increase winding temperature rise by 50%. Extra air from a separate blower will bring that down, but the outlet end of the motor will remain 10° to 15° hotter than the winding average.

Motor 'suffocation'

A second assumption about the air flow per Figure 5-8 is that it's free to carry motor heat entirely away from the surroundings. Even if a single motor sits alone in the center of a large room, so that recirculation of the kind shown in figure 5-10 is impossible, the discharged heat can still raise the entire room's air temperature gradually until the motor literally suffocates.

Hence, there must be enough "room air changes" to keep the overall air temperature below a design limit. That limit is normally 40°C. If a higher value is predicted, the motor designer can compensate for it by lowering the operating temperature rise.

Room air conditioning—or simply an intake/exhaust fan system—is helpful, but often costly. On the other hand, normal window or door "leakage"

The Motor and Its Environment

alone from a closed room is not likely to suffice, unless the space is quite large (such as an entire factory floor).

Figure 5-12 illustrates a drive for which this problem was overlooked. Repeated thermal tripouts of the large motor forced expensive rework to both room and drive foundation, to duct the heated air outside.

Figure 5-12: Although in a 20,000 cubic foot room, this 1250-hp synchronous motor overheated because there was insufficient air circulation to transmit motor heat outside the room (air equal to the entire room volume was drawn through the motor every five minutes). Concrete foundation and wall had to be broken through to duct the heated air away from the machine.

It was not necessary, however, to provide an intake duct. That may seem strange. If normal room leakage isn't enough to discharge the warm air, why is it enough to supply the needed inflow? The answer lies in pressure/flow characteristics of the entire system of room plus motor. Consider a typical room. Suppose it has a floor area 25 by 40 feet with an 8-foot ceiling. Internal room volume totals 8000 cubic feet. Suppose the walls are concrete block, with one weather-stripped door but no windows.

According to various design rules of a national heating and ventilating engineering group, if there is a 20 mph wind against two walls, an internal air pressure source supplying 200 to 250 cfm would be needed to keep any air from entering the room from outside. A conservative rule of thumb calls for providing a supply equal to half the floor area in cfm, or $1000/2 = 500$ cfm. That's how much air would otherwise leak into this "tight" room. Remember, there are no windows.

So the far larger room containing the drive of Figure 5-12, with more doors (including a large truck entrance with a gap below the door) and many windows, plus a high ceiling, could easily "leak" in the 4000 cfm needed by the motor. The machine's strong fan action easily pushed the discharge air through the low-resistance duct (only about 10 feet long) to the outside.

Without that duct, the discharge air flowing into the room would eventually cause outward flow through the same wall openings. But air would simultaneously be trying to enter. There wouldn't be room for both flows. Furthermore, by the time the warm air gets well away from the motor, it has lost most of its heat to the incoming air, so that eventually the entire room contents become too hot for the winding rating. Unless the atmosphere is explosive or extremely dirty, it isn't necessary to duct air both to and from a motor in a closed space—but at least one direction of flow should be ducted.

Piping the intake air

In most dirty or hazardous locations, it's the intake air that is piped. That can be done two ways. For a short flow path (20 feet has been a recommended limit), the motor's internal fan action can pull the air in through the duct, provided the duct is large enough. To limit the pressure drop in the duct so the motor fans won't "starve" themselves, velocity of air through the duct must be kept low, as Figure 5-13 shows. Higher flow speeds may produce objectionable noise. A duct area of ¾ to one square foot per 1000 cfm is ideal.

In rectangular unsealed metal ductwork, at least 10% leakage can be expected. Round ducts or pipes, feasible for small motors as in Figure 5-14, can use a type of jointing which cuts that leakage rate to 2% or less. Such ducts are also only half as expensive to install and cause less pressure drop at higher air flow velocities. However, there may not be room for circular pipes large enough in diameter to handle the cfm needed.

If the path must be longer, involves several bends, or is so small that air velocity exceeds the Figure 5-13 limits, then a separate air pressure source should be added upstream from the motor. Selecting this "external blower" is a job for an air handling specialist. A small diameter, high speed fan wheel takes the least space but demands relatively high driving horsepower. Its high speed may cause noise. A lower speed, large diameter fan for the same cfm will be cheaper to operate, and quieter, but also much bigger.

If discharge air is to be ducted away from the motor, it's necessary to make certain the motor plus upstream blower can overcome the additional pressure drop through the outlet ductwork. The system designer can calculate that outlet pressure drop, then add it to the nominal positive pressure at the intake, leading to selection of a larger blower capacity. Adding an outlet duct for a force-ventilated motor in the field, without checking that blower capability, can result in air flow reduction and consequent overheating.

Can a motor safely be overloaded if its supply of cooling air is increased? Theoretically, yes. But it may not be practical. If an external blower supplies forced air, which then circulates through the inside of the motor enclosure in the normal way, then Figure 5-15 shows how boosting the air flow can permit more horsepower output. If the air is cooler than normal (Figure 5-16), the load could be increased further.

There are several limitations to this. One is that the extra air may not reach local hot spots caused by the overload. Many sizes and shapes of air flow passages exist within a motor. Those where the greatest amount of heat removal takes place may be so restricted that simply raising the incoming pressure may not move much more air through them. Instead, the extra flow may just bypass the hottest parts of the motor to exit ineffectively through wider channels elsewhere—taking the "easy way out."

The Motor and Its Environment 213

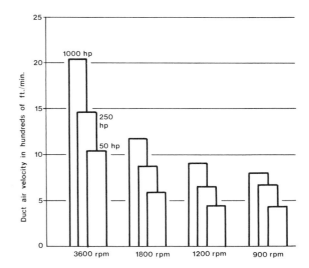

Figure 5-13: If motors are "self-ventilated" through intake ducts, with no separate blower, maximum duct air velocity should be approximately as shown here.

Figure 5-14: A 250-hp rubber mill motor ventilated through circular intake ducts.

Another problem is that the temperature drop or gradient across insulation or through laminations will increase as winding and core temperatures rise. The result can be severe local overheating that may not appear from readings of overall winding rise by resistance. Thermal expansion and contraction will get worse, as the load is cycled on and off or as air temperatures vary throughout the day or season. Finally, lubricant/bearing heating may be excessive because not enough of the added cooling air can be channeled over the bearing housings.

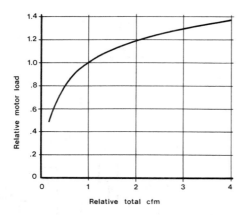

Figure 5-15: Based on many tests of both open and enclosed motors reported over a 40-year period, this curve shows how the amount of cooling air influences motor loading. Note that even moderate overload requires a great increase in the air needed.

What about motor enclosures not adapted to forced ventilation? At one time or another most of us have been in some motor room where the plant electrician or operator has set up a fan to blow air over a "hot" machine so it could keep going during a hot day or a temporary overload. (See Figure 5-17.) That seldom works for any length of time. If the motor is an open type, air blowing over its exterior from one direction is unlikely to reach surfaces at which efficient heat removal can occur. If it's a TEFC type, the extra "wind" can actually make things worse, by interfering with the natural flow over the motor frame produced by the shaft-mounted fan.

In one extreme situation, a motor was kept cool under severe overload for hours by spraying water on it—but that's a different situation in many ways. For one thing, the water was carried inside the motor to reach windings directly. Furthermore, water is a far more effective coolant than air. A liquid remains in contact with surfaces more closely than any passing air stream can.

Figure 5-16: Combining information from six publications including NEMA MG1, this curve shows how motor load must be reduced when ambient temperature exceeds the standard 40° C. It also shows the overload possible when ambient is lowered, although that can be risky (see text).

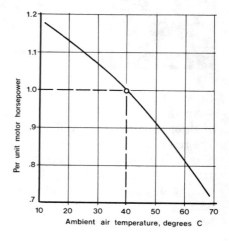

So be cautious in using increased air flow as a means of holding down winding temperature under overload. For moderate overloads, with a motor enclosure adapted to forced ventilation, it can be effective, but actual winding temperature should be watched closely.

Figure 5-17: Operators here have a large electric fan blowing on a 700-hp drive to keep its temperature down. Extra air supplied this way seldom does much good.

When the air supply is decreased, motor load must be reduced sharply. Remember that not all motor heating losses drop off rapidly when load is reduced. Iron (core) loss remains high.

A final condition for proper motor ventilation is that the air be clean. Visible clouds of sand, coal dust, or other contaminant in the air are clear danger signs. But seemingly harmless "ordinary building air" is surprisingly dirty. Users sometimes overlook the amount of such dirt that can be deposited within a motor via the tremendous air volume passing through the machine.

Factory air may contain from .02 to 0.2 pound of dust per million cubic feet. Suppose .03 is assumed. If only 10% of that dust is left inside a motor by its cooling air stream, at 100 cfm per kilowatt of motor loss, a 500-hp motor running 5000 hours annually would accumulate nearly three pounds of dirt per year. Wet or greasy windings, combined with particularly clinging dirt such as textile fibers, gypsum, potash, or alumina, would greatly increase that total. That's why many motor rooms use filtered air.

At least four different types of air filter have been used in motor ventilation. The choice depends upon the amount of air needed, the nature of the environment, and the size and shape of the motor housing or frame openings and passages to which filters can be fitted. (See Figure 5-18.)

Motors need relatively more air than other heat-producing electrical equipment. Motor efficiency is lower than that of a transformer, for example, and there is less heat-producing coil surface directly exposed to air. Hence the air flow through the motor must be greater. Restriction in the flow path must be

Figure 5-18: A large outdoor motor equipped with air filters inside the housing. They can be removed and replaced, without stopping the motor, through the vertical access covers near the top of the machine.

minimized. Filters, therefore, must allow free flow of large amounts of air without too much pressure drop.

The usual motor air filter is one of two types (see Figure 5-19 for a cost comparison):

(1) The "permanent" viscous impingement filter, made of crimped layers of fine wire mesh coated with adhesive to trap dust in the passing air stream. It is easily cleanable for re-use indefinitely. (See Figure 5-20.)

(2) The glass fiber mat "throwaway" type, designed to be used once and replaced when dirt-loaded. This is the least expensive filter to buy but the most costly to maintain, because it cannot be cleaned and re-used but must be periodically replaced. It contains layers of glass fiber mesh as does a household furnace filter.

Most throwaway filters are designed for fairly slow-moving air—300 to 350 feet per minute. Filtration depends on the air moving slowly enough for dirt particles to lodge against filter fibers and be retained there. If a motor needs 2000 cfm of air (typical for a 500-hp rating), required surface area of such filters would be at least 2000/350, or six square feet. It would be difficult to

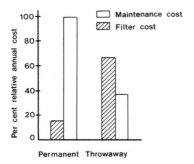

Figure 5-19: Though cleaning permanent filters may be expensive, the investment is relatively low. This cost relationship is typical for a four-filter bank in a large motor. It can change considerably in extremely dusty locations, requiring filter servicing every few weeks (as in one newspaper pressroom heavily contaminated with ink mist and paper dust).

Figure 5-20: Typical high velocity filter panels of the "permanent, washable" type most often used in motor ventilation.

fit that large a filter area into the motor enclosure. Some filters of this type are designed for higher air flow velocity so that total filter face area can be kept relatively small.

The important things to remember about the throwaway filter are:

(1) Do not replace a high velocity filter with one of the same size designed for lower air speed. The result will be a restricted air flow and probable winding overheating, loss of filter efficiency in dirt removal, or both. For example, a typical 300-hp motor may need 1800 cubic feet of cooling air per minute. With high velocity filters, two 16- by 20-inch panels will be ample. But if low velocity filters are substituted, three such panels would be needed—and the motor enclosure will make no provision for that many.

(2) Be careful to observe any air flow direction arrows on the filter casing. These filters, like some metal mesh types, work best only when air passes through in the marked direction.

Filter Maintenance

Most motor filters are of the "permanent" or "cleanable" design. They are made of corrosion-resistant wire mesh, usually galvanized steel (although stainless steel, Monel, bronze, or aluminum are often used for special atmospheres). Special sizes are available, but the universal standard is the two-inch thick filter in multiples of 16, 20, or 25 inches on a side. The wire mesh is held within a formed metal casing, which in turn fits in a supporting frame built into the motor enclosure. One motor may contain from two to a dozen filters.

Clean air filters themselves will add about 0.1 inch pressure drop so that some air restriction results. Added winding temperature will seldom exceed 5°C. However, if the motor was not originally designed for filters, adding them later (as in Figure 5-21, for example) may be risky. Check with the motor manufacturer before doing so.

What happens when filters "clog up"? Actually, only a throwaway type filter clogs. The spaces between fibers become progressively plugged with dirt so air can no longer pass through. The filter must then be replaced.

In most atmospheres, filters do not clog in the sense of becoming blocked so

Figure 5-21: Filters added "after the fact" to a splashproof motor. The size of filters required exceeded the available space, so special intake boxes were built to house them.

that air can no longer pass through. The layers of mesh are coated with an adhesive to which dust particles stick as the air rushes past. When all the exposed adhesive surface has been covered with clinging dirt, no more can adhere, so that dirt in the air from then on will pass on through.

Air flow does become restricted (Figure 5-22), which is why a sensitive pressure gauge (Figure 5-23) will sometimes be supplied to show the operator at a glance when filters need cleaning. But the main effect of filter "loading" is to allow dirt to freely enter the motor.

In some surroundings, a filter can actually become plugged. Certain kinds of dirt will produce heavy deposits of interlocking granules on the wire, even in a fast-moving air stream. One such material is coal dust. In a Western power plant, a 3000-hp motor, equipped with filters, was not properly maintained. Coal dust plugged the filters, winding temperature detector circuits were bypassed, and the motor endured more than a year of severe overheating before finally burning out.

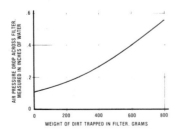

Figure 5-22: How air flow becomes restricted as a metal mesh filter "loads up" with collected dirt (20×20×2 inch panel).

Never brush the dirt off. This packs it into the mesh surface, making thorough cleaning more difficult.

Some throwaway filter designs can be vacuum-cleaned for re-use. Filters that look alike may not work the same way or require the same maintenance, so find out what is appropriate for the particular make involved.

Figure 5-23: Manometer or "draft gauge" installed on a large motor to sense increasing pressure drop across air filter as it becomes dirt-loaded. Such devices can be provided with alarm contacts.

One make of filter can be cleaned by simply flushing with a hose, using hot water when possible. But in most cases, because the adhesive itself is necessarily a sticky material, filter cleaning requires a solvent. Never use caustic soda or other strong alkali. Agitate the filter in a hot solution of trisodium phosphate (1½ pounds to five gallons water), household detergent, or a cleaner such as Oakite No. 20 (three pounds to five gallons water). It is especially effective to boil the filter for five minutes.

Then, remove the filter and look through it toward a bright light. If the cleaning has been thorough, you will see no cloudy areas. Don't worry about discoloration of the wire mesh surface.

Dry the filter thoroughly. Then, recoat it with the necessary adhesive. Most filter suppliers offer a suitable material. If this is not convenient, apply a heavy (SAE 30-50 weight) motor oil by dipping, spraying, or heavily brushing the oil onto both sides of the filter. Total immersion in the adhesive best ensures penetration to the interior. Pre-heating the adhesive to about 150°F will lower its viscosity to speed up draining. After the filter is completely drained, it is ready for re-use.

How can you tell when filters need cleaning? Watching the winding temperature is one way. This may be difficult, though, if either motor output or ambient temperature varies considerably. (Temperature detection/alarm devices are always desirable in any filter-equipped motor as protection should filter servicing be too long delayed.) A temperature sensor in the motor's outlet air stream is subject to the same limitations.

The most reliable option is a pressure gauge or switch to sense the pressure drop across the filters. When that reaches ¼ to ½ inch of water, it's time for filter cleaning. If such devices aren't furnished, check filter condition frequently until a reasonable cleaning schedule can be established through experience.

Some motors, especially those built without provision for filters, have had various kinds of filters installed in the field. One is the foam type. Because plastic foam can easily be cut into many shapes and sizes, it seems best suited to this "retrofit" application. (See Figure 5-24.) However, no such filter should ever be installed without making certain that it will not restrict air flow too much. Some foams cannot filter effectively at high air velocities. Others may impose too great a pressure drop for the motor fans to overcome.

Figure 5-24: This splashproof motor used air filter foam (indicated by arrows) cut to fit inside the air intake openings around the top of each bearing bracket, thus saving the space and cost of a large fabricated filter enclosure.

Filter foam breaks the air stream up into many small jets which change direction rapidly. Adjacent closed cells or pores in the material trap the heavier dirt that is unable to follow the air around all the corners. This material is normally used dry, without adhesive. After the foam has been removed from its supports, it can simply be washed like a piece of towelling, wrung out, then allowed to dry briefly. If mounted in fixed frames, dry foam filters can be reverse flushed with water, or vacuum cleaned.

There are also "dry type" metal mesh filters of the permanent, reusable kind which use no adhesive. They are a low velocity design (typically 300 feet per minute; one 16- by 20-inch panel is rated at 640 cubic feet per minute of air flow). These are cleaned by rapping off excess dirt, flushing with warm water, and drying.

Some motors, particularly d-c machines or variable speed drives even below the 440 frame diameter, include separately motor-driven fans to supply cooling air. This is necessary because when the machine runs well below top speed, its own internal fan action cannot circulate enough air. Figure 5-25 shows one such machine.

Maintaining these "air conditioning systems" requires not only cleaning, and filter servicing, but making sure the auxiliary fan drive is working right. Check for missing fan blades, vibration because of dirt buildup, etc.

Figure 5-25: A 200-hp, 8-pole motor equipped with a separately powered external fan that has its own filtered air intake.

Heat exchangers

When motors must be totally enclosed but are too large for aluminum or cast iron finned frames, two basic types of heat exchanger can do the cooling. One is the air-to-air or "tube type." (See Figure 5-26.) Used in both horizontal and vertical motors, explosion-proof and otherwise, this has a large number of smooth-walled tubes which are heated as the internal air from the hot windings and core flows over them. That heat is removed from within the tubes by an outside air stream from an external shaft-mounted fan.

Figure 5-26: A totally-enclosed "tube type" motor. The entire tube cluster or bundle is removable as part of the lift-off upper housing. In most such machines, the tubes are thin-walled aluminum or copper-nickel, five-eighths of an inch to one inch in diameter.

These heat exchangers are easy to keep clean. The tubes have no inner or outer fins. The smooth surfaces do not catch dirt. And the outside air moves at least 30 miles an hour through the tubes so that nothing settles out. Only occasional tube inspection is needed, by shining a light in one end, plus cleaning of the screen or grill that admits air to the external fan.

The other type of heat exchanger requires more attention. This is the air-to-water or "cooler" used on TEWAC (Totally Enclosed Water to Air cooled) motors.

No tubes are used here. Instead, internal air is circulated by rotor or shaft fans through (usually) a top-mounted air-to-water heat exchanger from which the heat is removed by a steady flow of cooling water. (See Figures 5-27, 5-28, and 5-29.) Unless internally pressurized or purged, this type of motor cannot be used in hazardous atmospheres. The amount of water needed can be estimated from Figure 5-30.

Thermodynamic principles dictate that the cooled air circulated from the heat exchanger back to the rotor can never be any cooler than about 12°F *above* the temperature of the incoming water. Therefore, inlet water at 32°C (90°F) is about the highest that will allow motor rating to be the same as for standard drip proof ambient air of 40°C. Water at 105°F (41°C) is not uncommon, and it will evidently require "derating" the motor just as a drip proof machine in 45°C to 50°C ambient air would have to be derated. The unavoidable restriction in internal air flow through the heat exchanger also results in some derating compared to an open motor.

Figure 5-27: Cross-section through TEWAC motor (side view) showing internal air circulation through cooler, mounted at top.

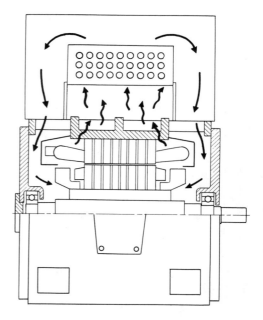

The Motor and Its Environment

Figure 5-28: A 1250-hp TEWAC motor with top-mounted cooler. The heat exchanger, partially withdrawn in this view, slides into the motor housing like a desk drawer.

Figure 5-29: Coolers may also be mounted on the sides of the motor frame, or in the bottom as with this 600-hp machine.

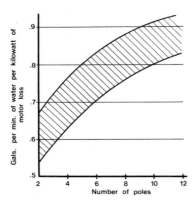

Figure 5-30: Approximate amount of cooling water needed by TEWAC motors depends upon motor internal heat loss, as shown here. A flow of .03 gallons per minute per rated motor horsepower is typical.

This represents an important difference in temperature rating method between the TEWAC machine and any other type of "enclosed" motor. Whether open or TEFC, other motors use the surrounding or ambient air to cool themselves. Therefore, any such machine carries a temperature rise rating based on the temperature of that ambient air.

The TEWAC motor, on the other hand, is *not* cooled by ambient air. True, some small amount of motor heat probably does escape by radiation through the walls of the enclosure. That will vary somewhat with ambient air conditions. But the effect on motor winding temperature will be negligible.

So TEWAC motor temperature rise is given with reference to its internal air temperature, governed by the cooling water, rather than with reference to external air temperature which has no influence. A TEWAC design nameplated 80°C rise will remain the same for any reasonable ambient air temperature. This contrasts with open or TEFC motor rating practice in which any change in ambient above or below the 40°C standard requires a corresponding revision downward or upward of the nameplate temperature rise, and consequently a change in motor electrical design. Figure 5-31 compares the thermal rating structure of open, tube-type, and TEWAC designs.

Figure 5-31: Comparison of temperature rating method for open (top), tube-cooled TEFC (middle), and TEWAC (bottom) motors.

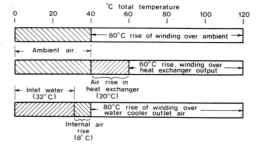

The Motor and Its Environment

Though most common at 1000 hp or more, this enclosure has been supplied down to 300 hp. Several cooler designs have been used. Some are of the removable tube type, in which each individual finned water tube can be removed from the cooler and replaced; others use plate fins locking all the tubes together. (See Figures 5-32 and 5-33.)

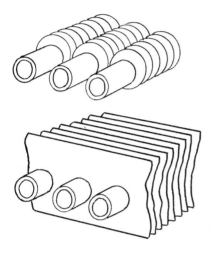

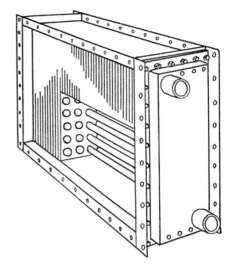

Figure 5-32: Left: Individual cooler tubes (top) compared to plate-fin design (bottom).

Figure 5-33: Right: Typical plate-fin cooler for TEWAC motor, partially sectioned to show the tubes. In foreground is removable supply header containing inlet and outlet pipe connections.

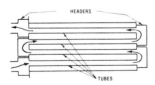

Figure 5-34: Left: How water flows through cooler tubes (arrows) is determined by header baffling.

While cooling water flows through the tubes in a sometimes elaborate pattern designed for most efficient heat transfer (Figure 5-34), the motor internal air is blown over the fins by normal fan action. Unlike the tube type exchanger, these coolers are subject to two operating problems that do require regular attention:

(1) Clogging of tubes. This is caused by corrosion—the water is often laden with chemicals which damage tubes—or by settlement of sediment in the water. The typical water flow velocity is only about three miles per hour.

(2) Leakage. Most often a result of corrosion, this allows water to drip or run inside the motor enclosure. A cooler-equipped machine may contain a leakage alarm system, with baffles or troughs to collect such leakage and channel it to a pressure switch. This closes an alarm circuit when a certain water level is reached. Electronic humidistats have been used to monitor water vapor or droplets in the motor's internal air. Some coolers are built with double tubes. When the inner tube, which contains the water, develops a leak, water can escape only into a space between inner and outer tubes, from which it flows to a sight glass or alarm device.

Motor coolers normally have removable "headers"—the bolted-on covers at each end. (See Figures 5-28 and 5-33.) With each header removed, it is easy to look through each tube to be sure the flow path is clear. Sediment or corrosion products can then be cleaned out of the headers themselves.

The importance of regular cleaning is emphasized by one cooler failure caused by ammonia—a substance damaging to certain copper alloys. Numerous locusts were picked up in the cooling water supply and lodged in the tubes where they died. The decomposition of their bodies created enough ammonia to destroy the cooler tubes.

For normal tube cleaning, push an air lance through each tube. Wash the tube sheets and clean out the headers. A pressurized water lance works best for more severe tube fouling. These tools can be improvised, being simply smooth pipes fitting fairly closely inside the tube diameter, with water or air pressure supply connections at one end, and a series of small holes around the plugged other end which is passed through the cooler tubes.

For extreme fouling, pass a rotating non-metallic bristle brush through each tube. Flush away loosened deposits with a garden hose. Or soft rubber plugs can be driven through the tubes by air or water pressure. Do not use wire brushes or other abrasive tools that can scratch or nick the tube interior. Such scratches form focal points for the start of corrosion, so that leaks may soon develop at those points.

The outsides of the tubes can be cleaned with a steam jet, followed by compressed air drying. Besides dirt, look for bent fins which can restrict air flow past the tubes.

For highly corrosive water, the cooler may be fitted with replaceable zinc anodes designed to "sacrifice" themselves through electrochemical corrosion while protecting the cooler components. Proper maintenance includes removal of the hard crusts of corrosion deposits from anode surfaces, usually by wire brushing, to expose fresh zinc surface to the water. Similarly, all the joints between the anode and the supporting cooler structure must be kept clean. These parts complete an electrochemical circuit, as in a battery, and without good electrical contact throughout the anode cannot function. Any anode more than half eaten away by corrosion should be replaced.

When a leaking tube is found, the normal procedure is to drive plugs into each end to block it off. When the replaceable-tube cooler was common in electrical apparatus, leaking tubes could be replaced in the field. However, this expensive and specialized procedure was seldom justified, because if one tube becomes corroded it is likely that others will soon be leaking too. A replacement of the entire unit is indicated. So simply plugging bad tubes in the meantime is far more economical. About 5% of the tubes in a cooler may normally be plugged without too much loss of performance. The cooler manufacturer can provide repair kits containing ready-made plugs, or advise on how to make them.

The noise problem

We have considered many ways of protecting a motor from its often hostile surroundings. Sometimes safeguarding those surroundings from the motor is of equal importance. The potentially damaging influence is *noise*.

See if this doesn't sound familiar:

"In the early 70s, a comprehensive environmental protection program was launched by the government.... The program comprises guidelines ... aimed at effective control of the increasingly urgent problem of environmental deterioration. Of outstanding importance is the law for the protection of the Environment from Air Pollution, Noise, Vibration and Similar Effects.... This law stipulates the noise emission of existing electrical machines must be reduced and that low-noise designs must be adopted for new machines."

No, that wasn't from OSHA. It's from a 1980 West German publication. The noise problem is worldwide.

In Europe, as in the U.S., electric motors drive almost all machines throughout all industry. It's natural, therefore, that the noise they make has to be considered in any attempt to reduce noise in the workplace.

There are three ways in which application and maintenance engineers may help solve noise problems:

(1) By realizing how the sound output and the acoustics of the surroundings each contribute to overall noise, so all noise problems aren't blamed on the motor alone.

(2) By carrying out repairs or rewinds so as not to create a "noise producer," and in knowing how to diagnose motor ailments that could result in needless noise.

(3) By knowing what can and cannot reasonably be done to change the nature or the magnitude of the sound emitted by various types of motors, so that such undesirable noise control side effects as overheating do not occur.

Nobody expects a plant operator to be an acoustical expert. However, to avoid the pitfalls of noise problems, he or she must have basic knowledge of what makes a motor "noisy" and what to do about it.

First, we have to know what "noise" is. Dirt is sometimes called "matter out of place." Similarly, noise may be described as "sound out of place"—sound that is unwanted. We have only to overhear teenagers and their grandparents discussing music to realize that one person's "sound" is another's "noise," and that everyone finds certain sounds more irritating than others.

The sound from an electric motor, as from any other machine, possesses two measurable properties: magnitude or intensity, and frequency or pitch. Most sound is a complex mixture of many individual tones, each of its own magnitude and frequency, which combine in ways that are annoying or otherwise depending to a great extent upon the listener. More than any other item of motor performance, then, sound level is subjective—that is, it may or may not bother you.

But because we can set standards or compare designs only through some means of measurement, there are standardized methods and instruments for measuring both pitch and intensity of the sounds emitted by electrical machinery. What we hear as "loudness" of a sound is not really its measured intensity, unfortunately.

The human ear responds differently to sounds of different frequency, even though they have the same intensity. A rumble sounds different from a scream. So sound measuring instruments are "weighted" or biased electronically to respond to sound intensity ("sound pressure") to a different degree at different frequencies. These network weightings are identified by letter, the one in common use today being the "A weighting" (because it so nearly imitates the ear's response).

What the instrument indicates, then, is the number of units of sound pressure or "decibels" determined in accordance with the A weighting, abbreviated "dBA." The higher the measured dBA at a given frequency, the "louder" that frequency should seem to a listener.

Pitch is dealt with by filters within the instrument that "pass" or respond to only certain narrow ranges of frequency known as "octave bands." Depending upon which standard is used, there are from six to eight such bands within the range of human hearing capability (about 20 to 20,000 cycles per second or Hertz, abbreviated Hz; see Figure 5-35).

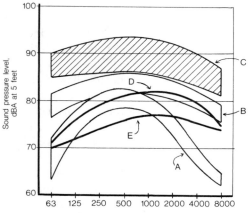

Figure 5-35: Ten to fifteen years ago, 2-pole open motor sound levels generally fell within the bands shown here: "A" for 5 to 150 hp, "B" for 300 to 700 hp, "C" for 800 to 5000 hp. Acoustical treatment to the enclosure today can generally achieve the level of Curve D up to about 2500 hp; of Curve E, up to 300 to 1250 hp. Still further improvement may be possible in special cases.

Octave band test

This kind of "octave band test," in which the dBA are measured for each band, is useful because it identifies the frequency range within which the most intense sound is being produced. That allows the best selection of corrective measures. Obviously there's no use making changes in a motor that will reduce high frequency emissions when the most intense sound is being produced at a low frequency. The concept of octave bands is explained in the box below.

What is an "octave band"?

An octave band is a range of sound frequencies spanning one musical octave. To cover the important range or audibility for electrical machinery, IEEE Standard No.85 defines seven such bands, denoting each by its "center frequency," which is separated by half an octave from the "edge frequencies" marking the boundaries of the adjacent bands above and below. These are the seven bands:

Center frequency, Hz	Edge frequencies, Hz
125	90,180
250	180,335
500	335,710
1000	710,1400
2000	1400,2800
4000	2800,5600
8000	5600,11200

It's common industry practice, however, and suits the available test instrumentation, to add an eighth band with center frequency of 63 and edge frequencies of 45 and 90. (NEMA Standard MG3 omits that band as well as the 8000 Hz band, using only six.)

An "overall" sound level is also useful in assessing the total effect on a listener, or the total noise contribution of a motor to an area containing other machinery. The whole is greater than the sum of its parts. That is, the overall dBA reading will always be higher than the highest single octave band reading. The total effect isn't an "averaging" of the highest and lowest bands, but an accumulation of all their effects. The lowest single band reading may not add much, but it always adds something to the highest band. (See Figure 5-36.) Although the addition is a complex process, a simplified way of doing it is shown in Figure 5-37.

There's much more to the science of acoustics, of course, but those are most of the basic principles governing electric motor noise. Four remaining facts about the way sound is transmitted are helpful when working with motors.

One of those facts is familiar to all of us. A listener's perception of sound diminishes as the source gets further away. Hence, no dBA reading is of any use unless taken at a measured distance from the surface of the noise source. Figure 5-38 shows how the decibel value decreases as that distance increases.

For a long time, the standard sound test distance in this country has been five feet. Metric standards elsewhere use a one meter distance, however, so

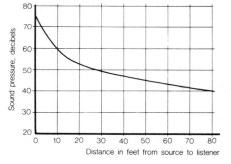

Figure 5-36: Octave band and "overall" noise figures look like this for typical 50-hp, 4-pole open and TEFC motors. Note that the overall figure always exceeds the highest individual octave band reading. Motors of other speeds, and other sizes, will show the same result, although the curves will have different shapes.

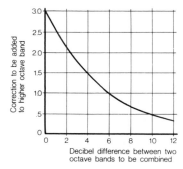

Figure 5-37: The correction factor to use when adding up octave band levels to get an overall sound pressure level. As a simple example, suppose three bands of 85, 82, and 94 dB are to be combined. Put the lower two together first: the difference between 85 and 82 is 3. From the curve, the "correction" at a difference of 3 is about 1.8. Add that to the higher band to get 86.8. Now combine this with the third band—the difference between 86.8 and 94 is 7.2. From the curve, add 0.8 to 94 and get 94.8 as the overall.

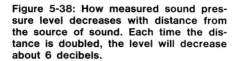

Figure 5-38: How measured sound pressure level decreases with distance from the source of sound. Each time the distance is doubled, the level will decrease about 6 decibels.

that three feet is becoming common. As a rough rule of thumb, add 2 or 3 decibels to a 5-foot reading to get its 3-foot equivalent. Unless the distance is known, there's no way to compare tests on different machines or under different conditions.

A second fact about sound transmission is also familiar. What we hear as sound is the result of a pressure wave travelling through the air from some object to our ears. That wave emanates from the source in all directions, not just towards us. And wherever it strikes a solid surface, such as a wall, ceiling, column, pipe, or another piece of machinery, it will bounce off. Depending on the sound frequency involved, that reflected wave may reinforce the incoming wave, increasing its intensity. Reflection may also direct sounds toward listeners in unexpected directions. Or it may bring sounds from an unexpected source to an otherwise quiet area.

Directionality

A third complication in sound transmission is "directionality." A motor is not a "point source" of sound, but a complex assembly of surfaces and separate noise generators. The external fan of a TEFC machine is located outside one end of the motor. While drawing air in through a screened opening at that end, it is discharging air towards the other end. The rim of the fan wheel is shrouded or hooded. Therefore, most fan noise will be blown along with the air stream, seeming loudest directly opposite either end of the motor.

Other sounds the machine produces may seem louder from the side. With an open, high speed motor, the greatest noise is usually perceived opposite the ventilating air openings in the frame, whereas at low speed the magnetic noise may seem loudest at the center of the frame. That's why standard motor noise test procedures involve taking data with the test microphone at five different locations—opposite both ends, both sides, and above the motor.

The fourth, and last, complication of sound transmission—the one that often causes the most difficulty in deciding what motor sound level to specify for a new installation—is that within any room, plant area, or building, the sound output of all noise sources within that space will add together in a complex manner to give the overall sound level perceived by a person within that space. Both reflection and directionality will have their influence.

Just remember this: When two or more motors are installed in a room, the overall sound level will always be higher than for the noisiest single motor alone. No matter how quiet a second motor may be, it always adds something to the first one. There's no averaging—only an accumulation.

If a local ordinance or standard requires a room sound level not above 85 dBA, it will do no good to specify motors having a maximum sound pressure level of 85 dBA. Even 80 dBA for each motor could be too high. And the noise contribution of the driven machinery itself hasn't been considered yet.

Users or consultants, finding a certain dBA level in a design code, may simply write that into a motor spec without realizing the need to account for the cumulative effect of multiple sources of sound. For example: the designers of a chemical plant were bound by code limits of 85 dBA throughout a motor room intended to house two 800-hp, 3600-rpm motors driving ammonia compressors. The designers stipulated a motor noise limit of 85 dBA and received motors tested at only 80 dBA. But in the finished room, after plant startup, the sound pressure level reached an unacceptable 90 to 95 decibels—which the operators assumed was the fault of the motor design.

NEMA standard

There is a NEMA standard, MG3-1974, which tackles this complex problem, using 14 pages of text and illustrations to describe the estimation of overall sound pressure level in a room resulting from the behavior of a motor installed there. The details are too extensive to go into here. Even the construction, and the nature of the driven machinery, will have an influence on the result.

Because sound pressure waves cannot exist in isolation, but are always influenced by the surroundings of the source, a sound pressure level reading measures sound intensity only in the specific test environment. Testing a motor in the field won't give the same readings as a test in the manufacturer's plant. That doesn't mean that either test is "wrong."

For that reason, NEMA standards, and some user specifications, give motor noise in *sound power*. This is not the same as sound pressure, although it uses the same decibel units. Sound power measures the intensity of sound in a nonreflective environment. It is an absolute measure of the acoustical energy radiated from an object without regard to the surroundings. But because no physical test is possible except in an actual physical environment, no instrument can take sound power readings directly. These must instead be calculated from sound pressure data, using correction factors based on the nature of the environment. Typically, says NEMA Standard MG1-20.49 Part C, the sound power will be higher than the dBA sound pressure by these amounts:

If the measurement distance of microphone from major machine surface is 1 meter, the decibel increase in sound power compared to sound pressure is 10 to 13; if the distance is 2 meters, the increase is 15 to 18; and if the distance is 5 feet, the old standard, the decibel increase is 12 to 15.

Remember that typical sound level meters read only sound *pressure*, not *power*. Motor nameplates will sometimes be marked with sound power values, especially in the smaller horsepower ratings, so make sure you know the units of measurement before drawing any conclusions about the numbers themselves.

What does OSHA say about electric motor noise? Few sections of the Federal regulations are as misunderstood as this one. OSHA's requirements concern only the overall sound pressure level within a workplace—within the working area where humans are exposed to the sound as part of their jobs. No limits are imposed on the sound output of any piece of machinery, electrical or otherwise. Therefore, it is not possible to satisfy OSHA by asking that any electric motor have a sound level "in accordance with OSHA standards."

These two paragraphs from the January 1981 *Federal Register* sum up OSHA's position:

"Noise is one of the most pervasive occupational health problems. It is a byproduct of many industrial processes. Exposure to high levels of noise causes temporary or permanent hearing loss and may cause other harmful effects as well. The extent of damage depends primarily on the intensity of the noise and the duration of the exposure.

"There is an abundance of ... evidence that protracted noise exposure above 90 decibels causes hearing loss in a substantial portion of the exposed population...."

That is the reason for the OSHA exposure rule, first issued as part of the Walsh-Healey Public Contracts Act in 1969, and since included in Paragraph 1910.95 of OSHA standards, and shown in Table 5-IV.

Permissible noise exposure

Duration per day hours	Maximum sound pressure level, dBA
8	90
6	92
4	95
2	100
1½	102
1	105
½	110
¼ or less	115

Table 5-IV.

The decibel figures in Table 5-IV are "time-weighted averages" (TWA) rather than absolutes. That is, a worker could safely be exposed within a 12-hour period to 4 hours at 90 dB plus 8 hours at 85. Or he might safely work 6 hours, 4 at 90 dB plus 2 at 95. (See box at the bottom of this page.)

Beginning with a 1974 proposal, OSHA has been considering revising the 8-hour exposure limit of 90 dBA down to 85. Because of insufficient information about the need for such a change, or the benefits from it, and faced with a cost estimate of $31 billion for industry to meet the lower TWA level, OSHA

How to derive a time weighted value by combining periods of exposure to noise.

The formula is:

$$\frac{H_1}{L_1} + \frac{H_2}{L_2} + \ldots \frac{H_n}{L_n} \leq 1$$

in which H_1 = the number of hours' exposure to sound level No. 1;
L_1 = the maximum allowable number of hours' exposure to sound level No. 1;
H_2 and L_2 = the same values applicable to sound level No. 2;
and so on. Example:

Suppose sound level No. 1 = 90 dB and sound level No. 2 = 85 dB. Then, L_1 = 8 hours and L_2 = 16 hours. (The figure doubles for each 5 dB decrease in sound level.) If H_2 = 8, then:

$\frac{H_1}{8} + \frac{8}{16}$ must equal 1 (or less), from which:

H_1 must equal 4 hours.

decided to postpone that revision. Industry did, however, have to expand its programs of hearing tests, emphasis on personal protective equipment, and monitoring of workplace noise.

To emphasize that OSHA's interest is only in worker exposure, the agency has stated: "If the employee is not present while high sound levels are being generated, OSHA is not concerned." Does that mean any level can be tolerated in an unattended location? No, because the general public may also be exposed to the sound—and that is the province of the Environmental Protection Agency, as well as local authorities. (Noise control laws now exist below the Federal level in most localities.)

Test conditions

All standard sound level tests on motors are taken at no load. Why shouldn't motor sound level be determined under load? What is the effect of full horsepower output on the sound produced?

It's well known that electromagnetic noise produced by rotating machinery is generally "voltage dependent." That is, it originates with the magnetic field, which is largely determined only by the motor's applied voltage. The aerodynamic noise—fans, or siren effects—depends only upon speed. Neither voltage nor speed changes significantly with induction motor load. Therefore, motor sound output should not change significantly as horsepower increases.

There are a couple of exceptions. One is that the "internal voltage" or "counter e.m.f." in the winding will actually decrease somewhat with rising load current. This tends to somewhat reduce voltage-dependent noise at full load. At the same time, though, the ampere-turns producing leakage flux in the machine will increase with load current, which may have the opposite effect. In some designs, then, there can be a small increase with load of the magnetic noise in certain octave bands.

The other exception is the variable frequency drive. Here both voltage and speed may change with load. Furthermore, the solid-state power supply itself tends to introduce harmonics that influence magnetic noise production. Some such drives may be either magnetically or aerodynamically noisy, or both, at certain speeds and loads, but not at others. Users need to be prepared for this, because it may not be possible for any manufacturer to "design around" all the possible noise-producing conditions.

But for the general run of motors at constant voltage and frequency, tests and experience show little difference in sound level from no-load to full-load operation. This is fortunate, because noise testing under load can be extremely difficult—and not necessarily useful.

It is difficult because the three problems of sound reflectivity, directionality, and accumulation will enter the picture. To apply load requires coupling a dynamometer to the test motor shaft. That will bounce sound back towards that end of the motor to distort the measurements. Readings above, or at the sides, may be less affected. Furthermore, the dynamometer generates its own sound—probably of different frequency distribution. All readings in the vicinity will include both motor and dynamometer components. How can they be separated?

All that can be avoided, in theory, by extending a long shaft through the wall to a dynamometer outside the room containing the motor. Such a wall is reflective, too, but let's assume it is far enough away to be ignored. Unfortunately, few such test facilities exist. Remember, too, that whatever the loading arrangement, it will be quite different from the user's installation, with connected machinery of a different sort, differently located.

Some years ago, a machine tool builder made tests using a long shaft through a sound "barrier" separating a pump from its driving motor, in a NEMA frame size but of unspecified horsepower and speed. He then took sound level readings at various places near the motor, on one side of the barrier, as well as near the pump on the other side. He reported that the motor noise did increase signficantly with load—as much as 8 decibels overall. However, the degree of isolation provided by the barrier was questionable. The microphone positions during the test were not stated. So this experiment could be no more than a useful beginning for an investigation.

A series of octave band readings by a motor manufacturer, on machines 300 hp and larger, using dynamometer loading with all its limitations, showed variation both upward and downward in different octave bands at full load compared to no load. Many overall decibel readings increased 0 to 1½ at full load, but one machine actually showed an overall decrease. Twenty years ago, another manufacturer found that a typical 5-hp motor might show 6 to 8 dB increase at full load, but that a 40-hp machine showed no change. The larger motor was apparently unchanged because of its lower leakage path saturation and more uniformly distributed air gap flux.

Should an "energy efficient" motor be quieter? Possibly. Such motors tend to have lower magnetic flux densities because there is usually more lamination steel in the core stack. That tends to lower electromagnetic sound output. Some designs may boost efficiency by using a different ventilation scheme or fan design that lowers aerodynamic noise. In general, however, there is no relationship between efficiency (or power factor) and overall sound level.

Who decides if a motor is too noisy? The listener does. But whether or not a particular code or standard is met can be determined only by careful measurements with calibrated test equipment. Few operators undertake such work in the field. If you want to try, be prepared to spend considerable time (and money) learning the essential principles that can only be briefly summarized here.

Noise's origin

With that background, we can now consider where motor noise originates, and how to avoid creating a noise problem where none existed previously. There are three types of motor noise:

(1) Aerodynamic, caused by the flow of air through or around the machine;

(2) Electromagnetic, caused by the vibration of stator or rotor under the influence of magnetic forces; and

(3) Structural, caused by bearing defects or the movement of loose parts.

Most air noise comes from one or another of these sources:

(1) Rotor bar tips sweeping past stator end turns nearby.

(2) The "chopping" of fast moving air streams through rotor vent spaces as the rotor sweeps past similar vent spaces (and coil sides) in the stator.

(3) Fan blades.

Sources of electromagnetic noise include:

(1) Slot harmonics. These often produce a characteristic group of three frequencies, typically in the range of 900 to 1800 Hz or multiples thereof. They tend to be "pure tones"—that is, a single frequency much higher in intensity than any nearby frequencies. This is the "spike" or "screech" that is most annoying to the ear. (See Figure 5-39.)

In most motors, either rotor skew or special winding pitch will remove these frequencies —but not always, especially in some multispeed designs for which no slotting or winding arrangements will completely suit all polarities.

(2) Excessive variation in the air gap.

Mechanical noise may sometimes appear to be electrical in origin. For example, a bad air gap, with a so-called "moo" or "wow" low frequency fluctuating sound, rising and falling in intensity at rotor slip frequency, can result from misalignment between a 3600-rpm motor and its load. This is sometimes called a "slip beat" noise.

Other mechanical noises include the hissing or clicking of defective or improperly mounted ball bearings; the rattling or humming of air baffles or enclosure parts which are either loose or resonate at rotational frequency; and the "drumhead" effect of flat steel surfaces, amplifying a variety of small noises that might otherwise go unnoticed.

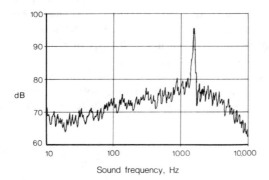

Figure 5-39: A "discrete frequency analysis" like this is produced by test equipment that scans the entire frequency range instead of measuring individual octave band levels. The purpose is to disclose any objectionable sharp spikes or "pure tones" as shown here and which are usually of electromagnetic origin.

Rebuilding pitfalls

When rebuilding a motor, how might you avoid intensifying or adding to those noises? Here are some things to watch out for:

(1) Redesigning with so high a flux density that core "exciting forces" become greater than the core stiffness can control. Movement produced by these forces, varying as the square of the flux density, is what produces magnetic noise.

This can happen when old machines are uprated. Such a horsepower boost should be expected to increase the overall motor noise level, so it must be approached with great caution if there are noise limits for the location.

(2) Introducing noise-producing harmonics into the stator field. This can be an unexpectedly severe source of annoying pure tones. Some of the causes are:

(a) Changing coil pitch. At least nine times out of ten this does no harm. But, on occasion, the design is sensitive to a certain harmonic. A chording change—even a difference of only one slot—can increase that harmonic such that noise goes way up within a certain octave band. Watch out for 2-pole coils below half pitch.

(b) Changing poles. A slot combination that is "good" for one polarity in one frame won't automatically suit that same polarity with a different stator slot, or on a different bore diameter. The reason is because although the magnetic noise frequencies generated will be the same, the core stiffness—its ability to resist the forces—may differ greatly.

And if you don't have evidence that the slot combination has ever been good for the polarity, rewinding for that polarity is asking for noise trouble.

(c) Using unbalanced fractional slot winding grouping. The degree of unbalance may be small. Its effect on overall performance may be largely compensated for by cutting out a coil, for example, but the dissymmetry in the magnetic field may still generate noise.

(3) Distorted air gap. This can originate from any of the following:

(a) Out-of-round or eccentric stator because of improper lathe cleanup, too much heat during winding burnout, or shifting in the frame (especially if a failure has occurred through rotor rub or a bad bearing).

(b) Out-of-round rotor caused by bent shaft.

(c) Eccentric rotor caused by shift in bearing position.

That can happen through improper shimming of large machine bearing pedestals, or through mismatching bearing bracket position on smaller motors.

(d) Excessive magnetic pull caused by too high a flux level.

(4) Unbalance. Causes: rebuilt rotors, fan changes, use of static balancing alone, or working with the wrong balance planes. This tends to generate sound at rotational frequency.

If there is a motor noise problem, how can you help correct it? A new installation that gives trouble right away will probably be settled between the user and the manufacturer. But assuming some on-the-spot cure has to be worked out, here are the options:

(1) Internal changes. Normally these will involve only aerodynamic noise—

the whistle, scream, or roar produced by the air stream. Be sure that's what the problem is before you start modifying parts, though. If the noise continues after power to the motor is turned off, with the load uncoupled, it is non-electrical in nature.

Banding or blocking rotor vent passages will help. Epoxy glass tape can be used to build up a band around the rotor outer diameter at each vent, which can then be heat-cured and (if necessary) machined to the rotor diameter. Depending on the vent spacer design, you may have to undercut the vent area slightly (and carefully) first, to make enough room for a band at least 1/16 inch thick. And don't use steel banding wire. It can overheat magnetically, and if it ever breaks loose the stator winding could be be destroyed quickly.

Steel discs or plates can be welded over air vent holes or spider passages at the ends of the rotor core.

Before taking any such steps, however, you *must* be sure that the motor has ample thermal capacity to carry its load without overheating when ventilation has been reduced.

It may also be possible to change fans to either blow less air or direct it more quietly. Perhaps fan outer diameter can be reduced. A change from radial or "paddle" blade construction to backward-sloping or curved blades will usually help. In some high speed motors, the spacing of blades around the fan circle can be changed from regular to irregular, with due regard for mechanical balance. This will minimize some noise frequencies.

Most fan changes will affect the total air flow, so watch out for possible overheating of the winding.

(2) Frame or structure changes. Generally these do nothing to reduce the sound produced by the machine. Rather, they absorb it before it can reach the listener. The usual procedures include:

(a) Adding external silencers or mufflers to air exhaust openings in the enclosure. (See Figures 5-40 and 5-41). Because of the directionality of sound,

Figure 5-40: Close-up of an air exit opening for a 500-hp outdoor motor, showing acoustical foam in place.

The Motor and Its Environment   239

Figure 5-41: Left: A 700-hp motor equipped with sound absorbing "silencers," or acoustic mufflers, in both inlet and outlet air openings.

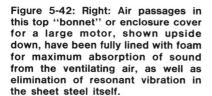

Figure 5-42: Right: Air passages in this top "bonnet" or enclosure cover for a large motor, shown upside down, have been fully lined with foam for maximum absorption of sound from the ventilating air, as well as elimination of resonant vibration in the sheet steel itself.

Figure 5-43: Inlet mute on a 200-hp, 2300 volt TEFC motor. The cylindrical chamber is lined with foam and contains extra bends in the air stream entering the fan. Similar units are on the market for motors down through the NEMA 250 frame diameter.

and because the air being blown out of the motor enclosure carries with it much of the sound generated within, these can be quite effective. Some further benefit often results from adding them to air intake openings also.

(b) Lining airflow passages in the enclosure with sound-absorbing material. (See Figure 5-42.) This is usually a plastic foam, which absorbs a wide range of frequencies; it may contain lead sheet which is an effective barrier to sound transmission through the enclosure structure itself.

(c) Adding a "mute" to the air inlet of the external fan of a TEFC machine. (See Figures 5-43 and 5-44.) These are readily available for motors down into the NEMA frame sizes.

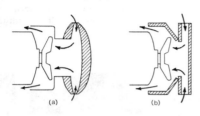
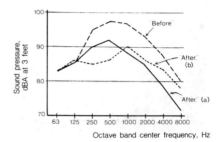

Figure 5-44: Users have reported noise reductions like this for 445 frame 2-pole motors using commercial mutes similar to Figure 5-43 (previous page). Arrows show direction of air flow.

More pitfalls

Two precautions need stressing. First, as with internal change to a motor's ventilation system, you must be certain that the motor has enough temperature margin to accommodate some reduction in cooling air. A 1- or 2-inch thickness of acoustical foam may greatly reduce the area of internal passages. The silencer, muffler, or mute will cut down on the amount of air reaching the windings. Allow for an added temperature rise of at least 5°C when making such changes. For some small motors, a 20° higher rise is possible. If the motor was designed to be unusually quiet, the manufacturer will have taken account of that. But a "retrofit" in the field can cause trouble.

Makers of sound absorbing material will not guarantee specific noise reductions from its use in or around a particular motor. Results in the field may be less than expected or necessary. There may have to be some "cut and try."

Foam or other sheet material must be selected to suit the environment. Unprotected glass fiber insulation, for example, may do well in absorbing certain sound frequencies from a passing air stream, but fast-moving air will soon begin stripping away the surface. Wet or contaminated surroundings may not only disturb the acoustical properties of some materials but may also destroy them chemically.

(3) Changes in the installation outside the motor itself. Building a sound-absorbing house around the drive (Figure 5-45) is the most common version, especially for small motors. Such a structure may not even need to include any sound-absorbing materials. If it does, they can often be added more effectively

Figure 5-45: Field experience shows that it may be possible to reduce motor sound level greatly by building acoustically treated "houses" like these. The type at (a) is for a TEFC machine, typically 250 hp. The construction at (b) provides sound-absorbing baffles at inlet and outlet air openings for a machine of similar size.

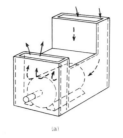
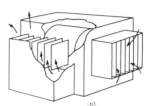
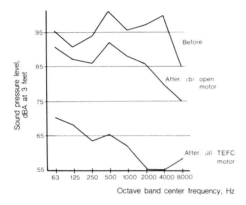

with less influence on motor cooling than inside the motor enclosure itself. But air entrance and exit openings must permit adequate ventilation for the motor rating.

In summary: "Noise" is of growing concern to industry. But much of the sound output of electric drives doesn't come from the motor, but from the driven machinery. Avoid blaming the motor to start with. If motor noise must be reduced in the field, be careful—a quiet motor that soon burns out is not necessarily better than a noisier one that keeps running. And whenever a motor is in for repair or redesign, don't create a new noise problem. Try to avoid those changes likely to increase the sound output.

6 Insulation and Windings

THIS CHAPTER is not a detailed survey of the specialized field of rotating machine insulation (or of winding design). Materials, systems, chemistry, and temperature rating have been dealt with at length in other books devoted entirely to electrical insulation. Instead, we will concentrate here on a few questions of concern to many users and application engineers.

Basic coil types

One such question is: Which is better, random-wound or form-wound coils, and what dictates the choice?

The so-called random or mush wound coil, using round wire laid in a random pattern of no particular strand orientation, is so economical that its supremacy in small motor windings is unchallenged. But in very large machines, above about 600 hp, the number and size of wires needed for heavy copper cross-sections rule out such windings in favor of the formed coil using rectangular ribbon or strap. In between these size extremes, both designer and repairman sometimes have a choice between the two types of coils.

The decision is based on motor voltage, motor performance, and the type of service. Let's look first at voltage. Class B (130°C), Class F (155°C), and Class H (180°C) mush wound insulation systems have been successfully used up to 1000 volts. A slightly longer and thicker slot cell, thicker phase insulation, and a fibrous over-wrap on the film-coated wire are adequate additions to 600-volt systems. The lead cable covering, such as Hypalon® or silicone rubber, can usually be operated up to 1000 volts between phases.

However, there's seldom any use for 1000-volt insulation. Some oil field pumping units are rated 460/796 volts, but the power system is normally "grounded neutral" so that the basic 460-volt insulation is enough. European mining machinery is using some 910-volt motors, powered from a 550-volt supply through a 1:1 delta-wye transformer bank. Adoption in the 1960s of revised U.S. Bureau of Mines equipment standards for this country permitted such usage in the U.S. for the first time.

It's at 2300 volts and above that high voltage mush windings could have widespread use. Lower winding cost would benefit rewind shops as well as manufacturers; the latter would also have the advantage of using smaller frame size for a given horsepower. It would be possible to build 2300-volt motors of lower horsepower ratings (and therefore smaller frame sizes) than is otherwise possible, because mush windings use round wire, available in small sizes, overcoming the problem of unmanageably small rectangular wire needed for some formed coils.

(Coils may be "formed" of round wire, of course. A typical example: a 150-hp motor rated 5250 volts, requiring a "crossover" coil of #13 square wire. This wire is hard to handle, because it tends to twist and distort the coil shape. The loss of cross-sectional area to the corner radii (Figure 6-1) makes the wire virtually round anyway. So round wire is called for. But such coils require great care in handling and taping to keep the strands aligned and the structure intact.)

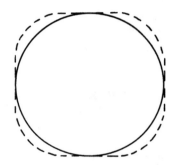

Figure 6-1: Enlarged view of cross-section of .072 (No. 13 AWG) "square" wire, shown dotted, and .072 diameter round wire, shown solid. Actual difference in area between the two is only 12 percent.

At least one major European manufacturer has used mush windings for years, up to 5000 volts. Others reportedly are trying it, or intend to. But some American experience has been poor. What are the problems? A major one is the formation of "crossovers" within the slot. That is, the individual strands cannot be made to align themselves in neat parallel rows to eliminate any place within the coil where wires cross each other. Such crosses may produce wire-to-wire voltage far above what the wire insulation will stand. Although this occurs in a 460-volt mush winding just as readily as at a higher voltage, the potential between wires when the turn near the beginning of a coil lies adjacent to one near the end of the coil is inherently five times greater for a 2300-volt winding.

New winding techniques could perhaps remedy this problem. One manufacturer has reportedly gained greater slot fill by pulsing high currents through the coils as they are placed in the slots. Strong magnetic fields produced between wires then force them to nest parallel in the slot, so more wires can be packed into the same space. Whether or not this could make higher voltage windings practical is uncertain.

Surface creepage is another difficulty with high voltage mush coils. It has been necessary to lengthen slot cells, and to tape the entire end coil area, to prevent arcing between the end of the core and "pinholes" or other slight defects in the wire covering. The condition is greatly worsened by dirt or moisture in service.

Insulation and Windings 245

Unlike many formed coil insulation systems, involving epoxy resins, the mush coil has no overall sealing coating to exclude contaminants. Encapsulation of the entire winding is a remedy for small NEMA frame size motors, but not for large machines where the problems of differential expansion and of long end coil flexing have not been solved by available encapsulants.

Another limitation to coating or sealing systems is this: Because the formed coil already has layers of insulation over the conductors, it's less sensitive to an additional thermal barrier and will get only a few degrees hotter if another layer is added over the part of the coil that overhangs the stator. The mush wound motor, on the other hand, with no such overall outer layer to start with, may run 10 to 20 degrees hotter with an added sealing over-coat, unless the materials' nature and thickness have been carefully chosen after consulting the motor manufacturer.

Some small motors, such as 125 hp in the 445 frame, have seen long service with 2300-volt mush windings. So despite the problems such windings are possible, provided: (1) the motor is small, with short, compact end coils; (2) insulation system design—particularly wire covering and creepage allowance—is conservative; and (3) the greatest care is taken to provide the highest class of workmanship and materials. This all adds up, of course, to a high-cost job—and one not without risk even if it passes routine tests.

The conclusion is that the present state of the art precludes high voltage mush windings under any but the most exceptional circumstances. Such windings have always been attractive to designers because, despite added insulation, taping, and more expensive wire, there is still a saving compared to the formed coil. (See Figures 6-2b & 6-2c.)

This brings us to the question of motor performance, where we have to consider which type of winding will have the lowest power losses and thus produce the maximum horsepower in a given frame. Mush windings run cooler, with lower copper loss, because the slot shape allows more copper to be used. (See Figure 6-2a.) Formed coil slots must have parallel sides to match the rectangular shape of the coil. Mush coil slots, not restricted by a fixed coil cross-section, are trapezoidal so that the parallel-sided teeth carry fairly uniform magnetic flux density. The difference is clear from Figure 6-3.

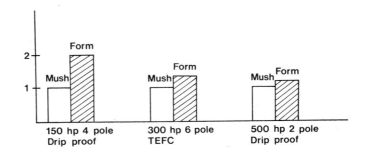

Figure 6-2a: This graph and the two that follow give a glimpse of the economics of the two types of windings. Here we see the relative copper weights for typical low voltage motors, mush wound vs. form wound in same frame size.

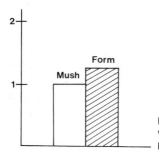

Figure 6-2b: Relative cost of complete motor, form wound vs. mush wound in same frame (300 hp 6 pole TEFC).

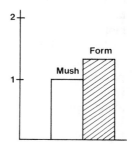

Figure 6-2c: Relative cost of rewind coils and insulation only for mush wound vs. form wound stator of same frame size.

Figure 6-3: Gross available slot area of semi-closed slot for mush winding is approximately 1¼ times that of the open slot at the left before making any allowance for insulation space.

Even if the mush winding enjoyed no advantage in copper, however, its thinner insulation would enhance cooling. The structural makeup of the formed coil depends on overall wrappings of tape or sheet material. These layers of tape on the exposed end turns form a "block" to heat transfer from copper to cooling air.

Similarly, slot insulation itself lets the mush winding run cooler. The minimum number of layers of tape giving adequate overlapping and dielectric strength, even at low voltage, makes a thicker ground wall than the cell used with mush windings. The tape must be relatively thick to stand the tension of tightly wrapping it around the coil. Typical slot insulation allowance for a low voltage mush winding is .023 on a side, or .045 total width; for the low voltage formed coil the figure is .078 total width. Thus the formed coil has almost twice the thermal barrier to heat flow between copper and core as does the mush coil.

Structural support

Under *application* we have to look at the physical stresses the winding will see because of frequent or severe starts, reversing, etc., and the ability of the two types of winding to resist those stresses. Many users, and some designers, feel that in a motor which is subject to severe duty involving unusual winding movement, formed coils should be used regardless of voltage or motor size.

But just what is "severe"?

Mush windings in general tend to have less structural rigidity than do formed coils. But mush coils can be compacted—an extreme example is the exaggerated flattening in the so-called "pancake" motor—to greatly increase their strength. Furthermore, when the end turns are laced together properly, they can be made to form a solid ring with greater bracing between coils, or "arch binding," than is sometimes possible with formed coils.

Even without special shaping, the mush coil is inherently shorter in overhang at the ends, and therefore has slightly more rigidity in a given frame size. Two reasons are: the large knuckle pin used to shape each end of a formed coil, necessary because of the rectangular wire resistance to sharp bends, and the impossibility of any "nesting" of strands at the end of the formed coil. For example, a typical 500 frame, 2-pole motor has end coil projection of 7.3 inches formed, and only 6.2 inches mush.

Vibration of unsupported cross-connections between coils is no problem with mush windings, because such connections are formed by loops of wire integral with the coil structure rather than by splicing protruding leads.

A final point in favor of mush windings, even in some applications involving repetitive starts or cyclic loading, is the likelihood of the mush wound motor being on a smaller frame—subject to lower inertia forces, able to use smaller rotor and stator diameters.

So it's not a blanket solution to motor operating problems to substitute a formed coil stator for a mush coil stator. It could even make things worse. Of course, when it is apparent that "fatigue" type failures are occurring because coil ends are too flexible, corrective measures are needed. Thorough compaction and lacing of the windings is one; another is impregnation with high-strength varnish, such as epoxy.

Surge rings of various types have been used to support mush coils. But they will not do much good unless the individual coils are compacted.

All these design and economic considerations have resulted in successful mush windings of large size. Certainly close to the "world's record" is a 400-hp, 10-pole totally enclosed machine having 120 coils in a stack length of 25 inches, and using nearly half a ton of copper in the stator winding.

On the other hand, where formed coil windings have been deemed necessary for rigidity, it is a mistake to replace them with mush coils without full knowledge of the application.

Also, of course, when the number of turns per coil is small and the slot extremely large, the number of wires "in hand" needed for mush winding becomes impossibly high. Individual wire gauge above #12 is not practical, and a well-equipped shop can seldom handle more than two dozen such wires in parallel.

One low voltage 2-pole machine, which could theoretically be mush wound, illustrates the limitations of such a winding. End coil length was too great for adequate rigidity. Electrically, the design required alternate 1- and 2-turn coils, which would have meant an impossible number of wires in parallel for mush winding, even using the largest wire and the smallest possible slot.

Insulation grading

Another question evidently in the minds of many who write motor specifications is this: Are motors being built today with "graded" insulation systems? Is this a problem?

We are not speaking here of the "voltage grading" or "stress grading" practice. Instead, we refer to the use of insulation thickness based not on motor nameplate voltage, but on the lower winding-to-ground voltage which may be achieved by special motor and power system neutral connections. At one time, the practice was not uncommon in Europe, but this now appears to be discouraged by the latest international standards.

In transformer manufacture, graded insulation has been normal practice for many years. The standards go back to 1899. Voltage stress within the windings, particularly that caused by the transient surges or impulse voltages common in high voltage transmission and distribution circuits, is not uniform. Close to the neutral point in a wye-connected winding, for example, the transient voltage to ground may be much lower than at the incoming line end of the winding. If the neutral is grounded solidly, or through a low impedance, it becomes possible to use thinner insulation in that area, reducing transformer winding size and cost without sacrificing reliability.

In motors, it's a little different, but the end result can be the same. Here is how one motor manufacturer put it some years ago:

"The use of grounded neutral on a wye or star connected motor reduces the voltage stress ... on motor insulation systems. This electrical equipment may, therefore, be built with an insulation system having a lower voltage rating with a subsequent cost saving."

This was derived, not for transients, but for steady-state operating voltage. Figure 6-4 illustrates the theory. It is voltage between winding and grounded core or frame that stresses the motor insulation. If the winding and power system neutrals are solidly grounded, no part of the winding in the wye-connected machine ever sees more than the line-to-neutral voltage. Because that is lower than line-to-line or nameplate voltage by the ratio of $\sqrt{3}$ to 1, the insulation system in theory need only be designed for $460/\sqrt{3}$, or 266 volts.

In the lower voltages, major application of this basic idea has been in oil field pump drives. Large groups of motors in the 10- to 75-hp range, scattered over a wide area within a working oil field, can benefit from a rated voltage

Insulation and Windings

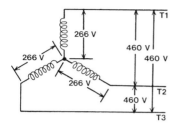

Figure 6-4: Diagram of wye-connected three phase motor winding showing neutral point N. Voltage between line and neutral never exceeds $1/\sqrt{3} = 57.7\%$ of the nameplate or line-to-line voltage. If the neutral is solidly grounded to the stator core, voltage between winding and core likewise never exceeds that 57.7% value.

above 600 (to keep down voltage dips when motors are started and reduce the cost of power transmission throughout the field).

This has been done for more than 25 years, as shown in Figure 6-5. Instead of having to build these machines with special 1000-volt insulation systems, or go all the way to 2500-volt formed coil taping, standard 600-volt motor insulation is usable.

A somewhat different and less controversial use of graded insulation in motors has been common with some manufacturers for many years. That is the use of thicker end-turn insulation for those coils at the ends of each phase group.

Recently, some large motor manufacturers have talked of extending this principle. It has been proposed, for example, that 6600-volt motors be wound with basic 4000-volt standard insulation, when intended for use on solidly grounded neutral systems. The maximum winding-to-ground voltage would be $6600/\sqrt{3}$, or 3820 volts, for which such insulation seems on the surface to be quite adequate.

But motors so designed, according to NEMA standards, would still get final over-potential tests of 2(6600) + 1000 or 14,200 volts for 60 seconds, coil to slot. Is it proper to use 4000-volt insulation in such a motor? Even the annual maintenance test of 9000 to 10,000 volts represents over-stress for 4000-volt insulation. The oil well pump motor practice of Figure 6-5 represents no such problem, because final and field tests are based on the 460-volt nameplate level rather than the 796-volt portion of the rating.

Afraid of the possible consequences of such insulation "grading" at the

Figure 6-5: Connection of 460/796-volt oil well pumping motor, on solidly grounded neutral power system. No part of the winding ever sees more than 460 volts between winding and stator core or ground. Hence, insulation can be the same as used in any standard 460-volt motor.

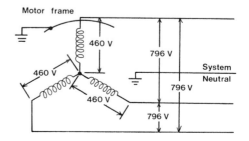

higher voltages, some users now issue motor specifications requiring that "graded insulation not be used," or that "the entire winding... be insulated for full nameplate voltage."

What do changes of this kind mean to the repair process? The first problem is making sure the repair or rewind insulation system is correctly chosen. How will that system hold up under the proper final test level following rewind? If you are in any doubt about what insulation was used in the original winding, use the values of Table 6-I as a rough check, or consult the manufacturer. Design practices will vary, so no absolute figures can be given here.

Rated voltage range	Insulation wall thickness, inches on a side	Layers of mica tape (5 mil tape thickness is typical)
2500-3000	.025-.030	3
4000-5000	.030-.050	4 or 5
6000-7000	.060-.080	6 or 8

Table 6-I. Typical insulation wall thickness for form-wound motor coils above 600 volts. Slot space available for copper must be calculated not only allowing for this thickness but also for manufacturing tolerance, amount of overlap between successive turns of tape, lamination stagger, etc.

If it becomes necessary to change insulation from a "graded" thinner wall to a system fully qualified for operation and test at the nameplate or line-to-line voltage, what happens to copper content and motor temperature rise? For a typical large machine going from 4000- to 6600-volt insulation, expect a 10% to 20% increase in temperature rise. If the design has that much margin, fine. If not, at least a 10% reduction in horsepower output must be made to avoid possible overheating.

Regardless of what may happen under maintenance test or steady-state operating conditions, this use of graded insulation also ignores the problem of impulse voltages caused by lightning, capacitor switching, vacuum breakers, etc. Depending upon the waveform of such surges—how fast the voltage rises with time at the motor's connection to the incoming line—these transient voltages will cause more or less severe stress on the insulation between turns of the motor coils.

Wave-sloping or "surge protection" capacitors at the motor, or the effect of line impedance, may lower that inter-turn stress to a safe level. Nevertheless, the entire surge magnitude will still appear between winding and core. On a 6600-volt system, surges having peak or crest values of 19,000 volts aren't unusual. If the ground wall insulation on the coils is designed for only 4000 volts, such surges repeated perhaps thousands of times annually represent a real threat to normal winding life.

The second problem is political. If you find that "graded" insulation has been used, how sure are you that the rewound motor will continue to be applied on the same grounded neutral power system? Is the motor being reconditioned for another application? Has it been sold, and does the new owner realize the basis for its design?

In the design of new power systems, according to a 1977 IEEE paper on the subject, there is a trend away from solidly grounded neutrals. Instead, there is greater use of resistance grounding—sometimes with high resistance—to limit ground fault currents.

Without a solid system ground, or if motor frame ground is not at exactly the same potential as system ground, the voltage between motor winding and slot or frame cannot be exactly known or controlled. Should one phase conductor anywhere in the system become accidentally grounded, for example, full line-to-line voltage will at once appear between the other two phases and ground, throughout the system—including within the motor. New, clean, dry insulation may stand that. Aged, wet, or dirty insulation probably won't.

It certainly seems wise, then, to avoid the risk of using a rewind insulation intended for anything less than full nameplate voltage. How sure can you be, after all, of the power system design, or of how long it may be before the motor ends up on a different system entirely? Offering long, satisfactory motor life through adequate design, materials, and workmanship is enough of a challenge without taking over responsibility for power system conditions, too.

Standard voltage ranges

The relation between motor operating voltage and insulation system voltage rating is often taken for granted, so a little explanation is in order here.

Once you get out of the fractional horsepower sizes, a winding operating at 155, 208, or 230 volts will normally use the same basic insulation system as one operating at 460 or 575 volts.

This system carries a 600-volt *system rating*, meaning that coils insulated with this system are suitable for long, trouble-free life at operating voltages within plus or minus 10% of 600. If the motor must be wound for 600 nameplate volts (not uncommon in Europe), the same insulation system is used. Final test voltage would then be 2(600) + 1000, or 2200 volts, winding to core.

Why not have a 120-volt system? Why not a 240-volt system, to take care of the 208- to 240-volt operating range, with adequate strength for use at 240 + 10%, or 264 volts? One big answer is the wide use of dual voltage motor ratings. A 230/460-volt machine must obviously be insulated for the highest voltage at which it could be used, even though it often may operate at 230 volts only. Another answer is the economy of standardization. A variety of system voltage ratings could require different parts, with different stator slot sizes, for each operating voltage, thus limiting flexibility of both original design and rewind.

So the 600-volt system is a basic, universally used compromise. Above 600 volts, it's not so simple. The principle of a relatively few standard system ratings to cover the operating voltage ranges still applies, though.

What determines these ranges? In working with electrical machinery, we hear a great deal about insulation *temperature* classes—such as A, B, F, H, 105°C, 155°C, etc. But we don't hear much about insulation *voltage* classes—which are equally important, though much less standardized.

The temperature class of an insulation system is derived from thermal endurance tests. These establish how long insulation will hold up under increasing temperature, the voltage being held constant. Voltage class, on the other hand, derives from voltage endurance tests, which show, at a given temperature, how long insulation lasts at various voltage levels.

How are such tests made? The procedure is to determine from test samples the applied voltage per mil (.001 inch) of insulation thickness which will yield acceptably long life without dielectric breakdown. The finished insulation system for use on motor coils is then built up from enough mils of thickness to handle the total motor voltage. Total space allowed for insulation in the winding includes those mils built up of integral layers of tape or sheet material, plus a margin for manufacturing variations, material thickness tolerances, and so forth.

In random windings, the total thickness is made up of slot liners, phase pieces, and connection sleeving or tape. For formed coils, the basic ground wall is usually layers of tape or sheet.

In IEEE standards, power system nominal voltage classes are:

600 v. optional; 480 preferred
2400 v. optional; 4160 preferred
4800 v. optional
6900 v. optional

Maximum system voltages for each of the above classes are, respectively: 635; 2540; 4400; 5080; 7260.

From that, it is easy to see how the motor design field has come to employ these values as common insulation system voltage classes: 600; 2500; 4400; 5000; and 7000.

Some manufacturers omit the 2500 and 4400 volt values, using 3000 instead.

Generally, both original manufacturers and service shops will use systems in accordance with the above, though many variations may be found. Common ground wall thickness ranges for some of the higher voltage classes appear in Table 6-I. These figures will be valid for either Class B or Class F temperature ratings; Class H, often using silicone materials, may be somewhat thicker.

This illustration offers another answer to the question, "Why not a separate insulation system voltage class for every operating voltage?" Tape is made only in certain common thicknesses. Each turn or layer of tape (or sheet wrapper) adds a fixed amount of ground wall. You can use three layers of tape for 2500 volts, and five layers for 5000 volts. Therefore, it would be possible to use four layers for some voltage in between. But you cannot use 3½ or 4¼ layers of tape and get the same thickness at all points around the coil.

So further intermediate voltage classes become impractical.

As mentioned already, the steady increase in required ground wall thickness as voltage increases causes a direct decrease in available copper space in the slot. The smaller the motor, the more drastic this effect, because total slot area is smaller, whereas the required insulation thickness remains the same. In Figure 6-6, we can see what this means for a typical slot having a cross-section of 0.4 by 2.50 inches.

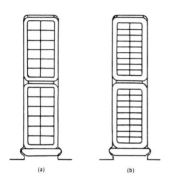

Figure 6-6: Difference between typical 4000-volt motor winding in slot (a) and 6600-volt winding (b) is apparent in these cross-section views. Winding at (b) normally has more turns, hence more slot space used up by wire insulation, and even when this can be equalized the ground wall is thicker, reducing still further the space available for copper.

Based on the typical insulation requirements of Table 6-I (with the addition of lamination "stagger," tolerances, etc.), the 4000-volt coil wound into this slot will look like Figure 6-6a. If the same slots are then used for a 6600-volt winding, the turns per coil will typically increase, so the amount of slot space taken up by wire covering alone will jump. But since this can perhaps be prevented by changing parallel circuits in the winding, we will ignore it.

Considering only the ground wall increase, then, we find that the net space available for copper in this slot at 6600 volts (Figure 6-6b) has been cut by 23%. The resulting motor performance suffers two ways. First, copper loss and winding temperature go up because less copper can be used. Second, the thicker ground wall causes poorer conduction of heat from winding to core and to cooling air.

The net result will be a stator winding temperature rise anywhere from 12% to 25% higher for 6600 volts than for 4000. (Some of this can often be compensated for in a completely new design, by making the slots larger or decreasing the number of slots. But that isn't possible for repair or rewind.)

An 80°C rise, 4000-volt motor will therefore run 90°C to 100°C rise at 6600 volts—maybe more. To bring that rise back down, it may be necessary to drop at least one horsepower rating.

Besides temperature rise, another effect tends to make the higher voltage motors larger in size. That is the added stator end coil length. Both coil and lead insulation are thicker, cutting down on the clearance between end coils so they become difficult or impossible to wind into the slots. To compensate, coils must be shaped with longer or more flared extensions, which take up more space and may force use of a larger frame or shorter core stack in the same frame.

In addition, higher voltage between coils and core or frame means more clearance is needed between the winding and the housing structure, bearing brackets, etc. This too may force the rating onto a larger frame.

The smaller the motor, the more drastic these effects all become. For this reason, few induction motors used to be designed above 4000 volts except in the largest sizes—1000 hp or more. Today, however, power system economics are dictating 6600 volt design down to as low as 300 hp, 2 and 4 poles.

Clearly, then, there are compelling reasons for trying to reduce insulation wall thickness at the higher voltages. Not only is loss of motor output a factor—so is cost.

Certainly one of the areas where insulation improvement may be sought is in more detailed testing, better materials, or both, resulting in safer operation at higher volts-per-mil. This permits motor design with thinner insulation for the same voltage. But such reduction must be made safely, without needlessly sacrificing motor life through use of an insulation wall inadequate for the voltages to be encountered in service.

Choices of insulation material

In choosing the proper insulation wall thickness for any motor application, the designer should be free to select the type of materials used.

For three quarters of a century, mica-based sheets, or "wrappers," have been used to insulate formed motor coils. No matter what the type of insulation (epoxy or polyester, vacuum-pressure or resin-rich) or the voltage rating (460 to 13,000 volts), wrappers remain in common use both by manufacturers and by electrical service centers. But the same basic materials may also be applied to coils in tape form—the so-called "continuous tape" method.

The choice between tape and wrapper depends on the particular insulation system design and on the economics of shop operation. Many service facilities and some motor manufacturers have found wrapper less costly than tape, especially for voltages below 7000. But either method produces good coils. During the 1960s, one manufacturer made extensive studies of six 4400-volt insulation systems, two using wrappers, the other four using tape. They all scored about the same in high temperature life tests.

Does it make any difference to total insulation wall thickness whether tape or wrapper is used? Generally not. However, for practical reasons this may not be true for some voltage ratings. Figure 6-7 shows typical rated or line voltage per mil of total insulation thickness. Note its rapid decrease at the lower voltages. This is because it's not possible to reduce wall thickness in direct proportion to decreasing voltage. You can't put on less than one thickness of material, nor can either tape or wrapper be made less than 4 or 5 mils thick.

As we shall see, taping permits finer adjustment of total wall thickness than does wrapper, which is of particular importance at high voltage. Wrapper

Insulation and Windings

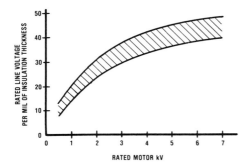

Figure 6-7: How voltage stress varies with total insulation thickness on formed motor coils for various voltage ratings. Figures include one layer of the standard covering for the wire itself. Both wrapper and tape systems are included, from motor manufacturers and from the service industry.

tends to do a better job at low voltage. But how true that is will vary greatly with shop practices, materials, and coil configuration. Here are some comparisons to consider:

(1) Tape offers flexibility in varying both wall thickness (for tightness of coils in slots) and effective insulation build. The two dimensions are the same with wrapper, but not with tape, because the latter can vary in amount of lap as well as in material thickness.

In Figure 6-8, for example, both lap arrangements yield exactly the same physical wall thickness or total buildup on the coil. But the half-lap shown at (a) produces a dielectric wall or insulation barrier equal to only two thicknesses of tape at some points. The quarter-lap at (b), which produces a minimum of three thicknesses at all points, is usable at a higher working voltage.

Figure 6-8: How four layers of tape on a coil can produce the same measured buildup, yet not give equal dielectric strength.

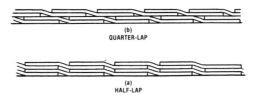

Figure 6-9 illustrates some other lapping variations. Note how both overall thickness and insulating ability can change by as much as three to one in a single wrapping of tape.

The relatively small differences between, say, 6600 and 6900 volts, or between 11,000 and 12,400 volts, can be accommodated by changing the tape lapping alone rather than by adding layers—as would usually be the only option with wrapper.

(2) On the other hand, when voltage is relatively low, it can become difficult to reduce a tape wall thickness as much as might be possible with wrapper. A single turn of wrapper can be safely used because no circumferential joints exist along the coil—no "edges" separating one turn from the next. But if tape is used, even a single layer must overlap from one turn to the next to make sure the insulation wall is unbroken.

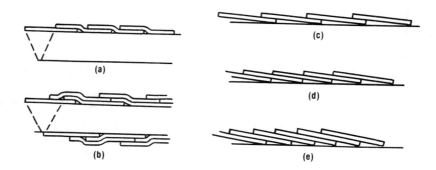

Figure 6-9: These tape lapping variations show how the number and arrangement of layers can produce wide ranges of both dielectric strength and actual buildup. At (a) one half-lapped layer provides two thicknesses of tape—but only one actual thickness at some points. Adding a third layer as at (b) still gives only a two-layer wall electrically. One wrapping of tape can furnish one thickness electrically, at (c), or two thicknesses (d); increasing the amount of overlap, as at (e), increases the buildup but still gives only two thicknesses at the thinnest points.

Hence, the minimum thickness possible with a layer of tape will exceed the minimum amount possible with wrapper. This means less room for copper. In many low voltage machines space may not be at a premium.

(3) Processing and curing time, and the steps in a vacuum-pressure impregnation (VPI) cycle, may vary considerably between tape and wrapper systems.

In general, impregnating resins or varnishes do not flow "through" the layers of insulation on a coil. Whether or not vacuum is used, the flow progresses edgewise between the layers, the material entering under the edges of tape or wrapper turns. Capillary action, or the VPI cycle if one is used, pulls the liquid between the layers from such edges, then down under the next layer in similar fashion, and so on into the body of the entire coil. This migration takes time. If treatment does not continue long enough, impregnation will not be thorough, so that internal voids will result. The higher the operating voltage, the more harm these will eventually cause.

Manufacturers have found it necessary to lengthen VPI processing time when using wrapper rather than tape. This is to be expected, because the layers of wrapper lack the many overlapping joints through which the VPI resin can penetrate from layer to layer in a tape system. Material suppliers concede that, "In general, tapes impregnate better than wrappers."

Why won't the resin or varnish flow through the pores of the sheet itself? Many insulating materials are faced with polyester or similar films which are impervious to impregnants. The liquid cannot pass through that facing. Others have porous backings, but are themselves already impregnated with resins that block any flow of other material through the sheet.

Some materials now on the market are claimed to enhance the migration of impregnant into a coil. One such tape, says its maker, is designed for the bonding resin to mix with the impregnating resin so that the latter can flow through the tape or sheet rather than only edgewise between the layers. In a test of this, some dummy coil slot sections were wrapped with 14 layers of half-lapped tape, after which the coil ends were completely encapsulated to seal them off. The coils were then put through a VPI cycle. "The results were outstanding," according to the manufacturer.

However, although it is true that no resin could enter the tape layers at the ends because of the encapsulation, it was still possible for the resin to work its way between tape laps at the coil surface. Although some improvement is to be expected with the new material, then, such a test cannot fully establish the degree to which direct "flow-through" can occur.

(4) Tape may be more difficult to wrap uniformly and tightly so that calculated wall thickness is actually achieved. If some turns are looser than others, or if the degree of lap is non-uniform from layer to layer, either voids or excessive buildup will result.

Machine taping is recommended whenever possible. This is especially important for high voltage coils. Some shops may be able to machine tape an entire coil. (See Figure 6-10.) Others can do only the slot sections, and even then only with coils of certain sizes or shapes. The economics of hand versus machine taping must depend on available tooling.

Figure 6-10: Machine-taping a high voltage coil.

Applying tape or wrapper

Taping machine manufacturers and insulation experts agree that tape should be applied to the slot section of a coil in one continuous pass, without stopping the machine. Non-uniform lapping or tension may otherwise result. "Thinning out" of tape layers is undesirable because it reduces the dielectric wall, but "double lapping" and undue thickening of the build can also cause trouble by creating a pressure point at which the insulation can be damaged during handling or crushed when the coil is tightened in the slot.

Figure 6-11 shows how wrappers are put on. To avoid abrupt discontinuities or "bumps" on coil surfaces, the end of a wrapper layer is always placed at a coil corner. There must be some overlap, as with tape, to ensure the needed thickness at all points, and this produces an "extra" wrapper thickness on one

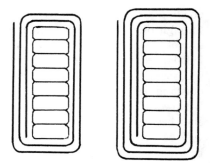

Figure 6-11: Coil cross-section showing how wrappers are used. The left-hand arrangement is "2½ turns"; the right-hand coil has "3½ turns."

side of each slot section of the coil. Wrapper orientation relative to coil forming should be as shown in Figure 6-12. The friction drag as a coil enters the slot will then have no tendency to loosen or unwind the wrapper. This can happen with a tight slot fit even though the wrapper is covered with the usual layer of protective outer binder tape.

The wrapper is liable to crease or wrinkle as it is being put on. And it's hard to hold uniform tension from end to end of the sheet while wrapping a long coil. There are a couple of ways to minimize these difficulties. The sheet can be pre-coiled around a stiff support from which it is unrolled onto the coil. (See Figure 6-13.) But a support rigid enough for wide wrappers can be hard to handle. Another method is the loose or "cigarette wrapping" technique (Figure 6-14), with sponging to work each turn down tight in a twisting process. At the same time, each turn is smoothed or worked endwise from the center of the coil to avoid either puffiness or wrinkles in each layer. Some authorities claim wrapper has a tendency to entrap air between layers as it is applied. Such "bubbles" must be removed.

(5) The most common difficulty with wrapper systems is making a proper transition between the end of the wrapper, where the slot section of the coil bends into the diamond, and the taping which must always be used to cover the end turn portion.

Describing insulation developments within his company, one major motor manufacturer once said that, based on extensive testing, "maximum voltage life is obtained through the use of wrappers. However, with wrappers we must also use tape on the end turn and ... we have a problem with the joint between the tape and wrapper." The higher the voltage, the greater this problem will be.

If the wrapper is square-cut, so that the edge of each turn falls directly over the edge of the layer beneath, there will be an abrupt joint at that point. (See Figure 6-15.) The end turn taping must overlap that joint so the dielectric barrier is maintained. This results in extra buildup that can make it more difficult to wind coils into the stator, promote stress damage when coils are flexed during winding, or even extend into the slot to prevent proper fit.

Insulation and Windings

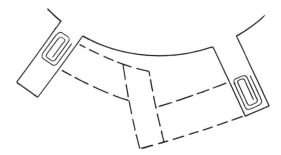

Figure 6-12: How wrappers should be placed so that no loosening takes place during winding. Pressing these coils down into the slots tends to tighten the layers.

Figure 6-13: Right: One way to keep a wrapper smooth and taut during application is to pre-roll it around a stiff support like that shown here.

Figure 6-14: Left: "Working" turns of wrapper around the coil for a smooth, snug fit.

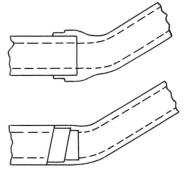

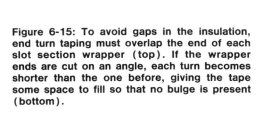

Figure 6-15: To avoid gaps in the insulation, end turn taping must overlap the end of each slot section wrapper (top). If the wrapper ends are cut on an angle, each turn becomes shorter than the one before, giving the tape some space to fill so that no bulge is present (bottom).

This can be avoided by cutting all or part of each edge of the wrapper on a taper. Then, as each turn is wrapped on, the portion of coil covered by wrapper becomes progressively shorter. (See Figure 6-15.) When the layers of end turn tape are placed over those areas, the first layer of tape can fill the space outside the inner layer of wrapper, and successive tape layers then build up to no more total thickness than the slot portion of the wrapper.

(6) A wrapper can generally be somewhat thicker than is possible for machine-applied tape. This may permit a wrapper system to accumulate a higher voltage rating—more wall thickness—with fewer layers.

Depending on coil cross-section, a twisting is imposed on tape as it progresses around the coil because of the lap from turn to turn. Tape must be flexible or conformable enough to form a tight layer despite this tendency. Therefore, it is seldom workable outside the 5- to 7-mil thickness range. But wrappers with a thickness of 10 mils or more are not uncommon.

Whether a continuous tape system or a wrapper system is "best" will depend on how you evaluate each of these many comparisons. Some experimentation may be necessary—as in the West Coast repair shop that had to make new coils some years ago for a 10,000-hp motor. "We didn't know that you couldn't wrap coils of this size," said the owner, "so we went ahead and wrapped them." No single method is always right.

Corona

One of the more mysterious and destructive conditions to attack high voltage motor and generator windings is corona. One authority supplies this definition: "Corona is a form of electrical discharge occurring between conductors when the breakdown voltage of the intervening gas is exceeded. When this occurs, the conductors are surrounded by a visible and frequently audible layer of ionized gas...." In a rotating machine winding, the gas is normally air.

How does this differ from an ordinary arc or flashover? The adjacent solid insulation can withstand indefinitely voltages far above that which will ionize the air. Because that insulation remains intact, no continuous arc joins the two conductors involved.

Energy released by the violent pulling apart of air molecules manifests itself as visible light—a characteristic purplish glow or "halo" effect—and a crackling, frying, or buzzing noise. More important, the process converts some of the air's oxygen to ozone. This unstable, chemically reactive form of oxygen will destroy many insulation materials. In an air-filled void completely enclosed by insulation, similarly destructive nitrogen oxides may be formed. Insulation surfaces are "eroded" or eaten away by the bombardment of ionized air particles, just as a metal surface might be sandblasted.

Corona, then, is soon followed by complete insulation failure. Because it involves a partial, rather than complete, breakdown of an air space, the phenomenon is perhaps better described as "partial discharge," a term used today by most insulation authorities.

Insulation and Windings 261

There are two kinds of partial discharge involving the coil surfaces in machine windings. One is the "slot discharge" effect. Here the two conductors are the steel laminations along the slot wall, normally grounded, and the copper coils connected to line voltage. Between these two is the ground wall insulation on the coils—plus some unavoidable small gaps where the uneven coil surface does not touch the also somewhat uneven surface of the slot. Even when coils are die-pressed, the surface is not completely flat.

Also making the slot area vulnerable to discharges between core and coil are the occasional "high laminations" and the abrupt discontinuities at air vent ducts. Any sharp corners or sudden surface changes produce high voltage stress points.

Slot discharges could theoretically occur even in 2300-volt machines. In practice, though, windings rated less than 5000 volts need not be considered susceptible to trouble. As one large motor manufacturer puts it, "Corona generally occurs at potential levels of about 10 kilovolts." Since a 4000-volt motor winding is factory tested at no more than 9 kilovolts, and since it should not be maintenance-tested in the field at more than 75% of that, corona is not a problem.

The advent of vacuum-pressure impregnated (VPI) insulation systems has raised the threshold of possible slot discharge damage to 7000 volts (rated). The reason is that VPI fills the small air voids between coils and slot, as well as between insulation layers on the coils themselves, where corona could otherwise start. Remember that the discharge takes place only in air—not within solid insulation.

Higher voltage machines, however, will be tested at far higher voltages. Insulation is thicker, of course. But the air along a coil surface has no more dielectric strength in a high voltage motor than in one of a lower voltage. Hence, the higher the motor operating (and especially test) voltage, the more necessary it is to supply some kind of protection against corona. Here is what a typical 13,800-volt winding will see during motor manufacture:

Short time test voltage on each coil after insertion in slot: 63 kV
Coils unwedged and temporarily connected: 61 kV
Completed winding: 56 kV
Final motor test: 29 kV

Above 7000 volts rated, then, whether or not VPI is used, slot discharges require treatment. Since the slot wall, the laminations themselves, is at ground potential, a solution to in-slot corona is to make sure the entire external coil surface is also at ground potential. This is done by applying a conductive coating to that coil surface.

Types of coatings

There are several types of coating. One is a "slot paint," a brushed-on varnish containing finely powdered graphite in suspension. (See Figure 6-16.)

Figure 6-16: A typical set of 13,200-volt stator coils ready for winding. The dark finish on the slot portions is the conducting coating used to prevent partial discharges in the slot.

Another is a glass or Dacron (one of several trade names for "polyethyleneterephthalate") type tape impregnated with such a suspension, wrapped over the slot portions of each coil as a final layer in the insulating process (partly conductive ferrous asbestos tape is even better). At least two European manufacturers use a graphitic paper.

Alkyd or certain polyester resins are common vehicles for conducting surface coatings. Some epoxies, however, have shown a tendency to "track." Tracking, the arcing along coil surfaces that rapidly destroys insulation, is normally avoided by keeping those surfaces clean and free of conducting contamination—water, chemicals, metal dust, etc. If the surface is deliberately made conducting, its chemical composition must prevent, rather than promote, destructive tracking. Hence the proper formation of a corona treatment compound involves more than just coming up with the right surface resistivity.

Most coatings air-dry within a few hours and some in as little as 30 minutes. They must be thoroughly mixed before use. These materials are suspensions, not solutions, so the conductive particles tend to settle out.

It is fortunate that VPI has rendered unnecessary the conductive slot paint below 7000 volts, which is where most industrial motors operate. Most manufacturers have found it impossible to use such coatings on coils prior to VPI treating. The problem is chemical. The coating tends to "leach out" or dissolve into the impregnating resin while the winding is in the tank. Contaminating the resin with a conducting material will eventually make impregnated coil insulation conductive as well, which would be intolerable. At the same time, the winding being treated would lose the necessary surface conductivity.

In recent years, several solutions to this problem have appeared. One is a VPI system in which the resin is carried in the tape itself, only the catalyst or "hardener" being in the dip tank. A conducting surface layer of coil tape does not react with the catalyst.

Although this surface finish, like the uncoated coil itself, will not completely

Insulation and Windings

contact the slot wall everywhere, there is only a short distance from any non-contact area to one of contact. If the surface resistance is low enough, the non-contact area will still be near enough to ground potential to prevent discharge. (With VPI, contact irregularities are harmless anyway, because resin completely fills the spaces. Only an air void can produce corona. In non-VPI systems, the slots must contain conductive side fillers or "side packing" to force the coils tightly against the opposite sides of the slot, for good electrical contact.)

It has been found that coating surface resistivity in the slot should be between a few hundred and perhaps 100,000 ohms per unit of surface area, to be effective as "corona treatment." (See Figure 6-17.) The practical lower limit depends on the extent to which short-circuiting of the stator laminations by the coating will increase machine losses. This could become significant at about one ohm. Various U.S. and European manufacturers have used the values shown in Figure 6-18.

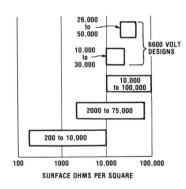

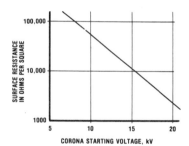

Figure 6-17: Right: The coil-to-ground voltage at which corona or slot discharge commences will be lowered if coil surface coating resistance is reduced. This graph shows the relationship as observed by one manufacturer.

Figure 6-18: Left: Various U.S. and European manufacturers have used these ranges of conducting coating resistivity for high voltage coils.

Note the term "ohms per square." You may ask, ohms per square what? The answer is that it doesn't matter—call it ohms per square centimeter, per square inch, or whatever other unit you wish. The answer will still be the same. Therefore there is no need to state any unit.

Figure 6-19 shows why this is so. Imagine a square sheet of material, of length L and width W. Dimensions L and W are equal—that is the definition of a square. We connect an ohmmeter M as shown, to measure the surface resistance across the square from one edge to the opposite edge.

Now suppose we double both L and W to form a larger square. When W alone is made twice as large, the resistance measured by the meter becomes

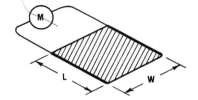

Figure 6-19: The significance of "ohms per square." The meter M reads resistance in ohms across the surface of the sheet of material having equal length and width.

twice as great as before. But if we also make L twice as large, that resistance is reduced by half, exactly cancelling out the increase caused by the larger W. So no matter what size the sheet is, so long as it remains a square, its resistance remains exactly the same.

When winding coils into the slots, filler strips on both width and depth are needed for various reasons. Depending on the voltage, for example, a "center strip" from 1/32 to 3/16 inch thick lets the two coil sides in each slot seat firmly against each other without the pressure being taken by the "bump" or enlargement in coil depth at the end of the slot insulation. There can be some air voids between these strips and the coil or slot. To maintain surface conductance at such voids, the filler strips themselves are given conducting surfaces at voltages above 7000. Graphite may be molded into the plastic itself, or the strips can be brushed with the same paint as the coils.

Slot paint normally extends ½ to 1 inch beyond the core stack. Remember that its purpose is to prevent breakdown of air space between coil and laminations. As the coil emerges from the slot into the air, the air along the coil surface will be subjected to voltage stress which can produce discharges, especially at the end of the stack itself where there is an abrupt change in shape of the electric field. So the protective surface coating must continue beyond the point where normal operating or test voltage can result in air breakdown between the coil and the end of the core, or the end finger supports for the lamination teeth.

End turn treatment

The second form of surface partial discharge encountered in rotating machine windings can occur on end coils outside the slot. This is a complex phenomenon. Unlike the slot discharge situation, where everyone uses the same basic remedy with only minor differences in material, end coil corona treatments vary widely and are not even universally used (as with the coils of Figure 6-16).

Figure 6-20 shows how voltage stresses along end coil surfaces can produce partial discharges in the adjacent air. Through any solid insulation wall there is always a flow of small "leakage currents," represented by the arrows in Figure 6-20. These will pass from the coil conductor to the surface coating throughout the length of that coating. Hence, the total leakage current passing along the coating at point S_1—including all the current to the right of that

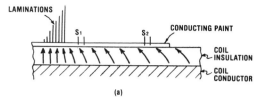

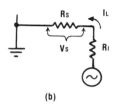

Figure 6-20: At (a), the arrows show paths of leakage current flowing through coil insulation from the copper to the conducting surface coating. Across the short length of coating S_1 the accumulated leakage current is higher than across an equal length further out at S_2. Coating resistance is uniform throughout. So voltage drop across the coating surface—which stresses the adjacent air—is higher at S_1 than at S_2. The coating actually forms many series/parallel branches like that of (b).

R_1 = constant insulation resistance

R_s = surface coating resistance

I_L = leakage current varying with length along end turn

V_s = varying voltage drop across portions of surface

point—will be greater than the leakage current passing through point S_2. The coating, of uniform composition and thickness, has a constant resistance at both S_1 and S_2.

If we then plot a graph of voltage to ground along the surface, as a function of length from the core, we get a picture like Figure 6-21a. If this voltage is rising steeply enough over a short distance, the air adjacent to those parts of the surface will break down, and corona results.

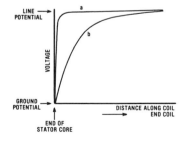

Figure 6-21: The difference between voltage gradient (rate of voltage rise with distance along the end coil surface) with no semicon paint (a), and (b) with semicon paint added to reduce concentrated voltage stress. Note that curve "b" rises less steeply than curve "a."

Coils are protected against that by adding a second coating, of higher surface resistance than the slot paint (more ohms per square), to the end turns beyond the end of the slot paint. If the coating were not used, the leakage currents reaching the slot paint from the end coil region would have to flow along the surface of the insulation itself, which has a far higher resistance than any conducting paint. Hence, the voltage gradient would be very high. A concentrated voltage stress would exist at the end of the slot paint as seen in Figure 6-21b—just as would happen at the end of the core stack if there were no slot paint. Because the object of the end turn coating is to reduce that stress by making the voltage rise less steeply along the coil surface, the added layer of protection is often term "stress grading" or "gradient" paint. To distinguish it from the lower resistance "conducting" or slot paint, it is also known as "semi-conducting" or simply "semicon" paint.

To make sure the maximum voltage across any unit length of coating stays below the discharge level, the length of this semicon paint must be based on the winding test voltage. The higher that voltage is, the further the paint must extend. One manufacturer uses this rule:

Coating length, inches = 0.4 [(Test kilovolts − 10)/3]
Example: 13,200-volt motor; test kilovolts = 27.4.
Coating length = 0.4(17.4/3) = 2.3 inches

Note that it is not the rated voltage V_o, but the test voltage V_t that is the criterion. Discharges during test could damage the coating so that it would not function properly in service. Normally, $V_t = 2V_o + 1000$ volts.

Some engineers claim about 2½ inches is the usable limit for semicon paint length. Beyond that no further benefit is gained. Others, however, claim that as much as six inches may be needed. It will apparently depend on the type of paint, the insulation system, and the voltage rating.

What should be the resistivity of semi-conducting paint? Typically it will be 100 to 1,000 times higher than that of slot paint. This is achieved in several ways, one being to use the same graphite suspension material as in the slot portion, but to dilute it further. Figure 6-22 indicates typical properties of a commonly available material diluted with polyester varnish. Semi-conducting tapes are also available.

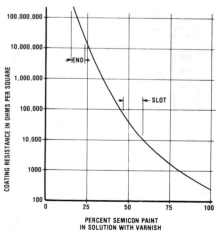

Figure 6-22: Proper surface coating resistance is achieved by varying the mixture of conducting paint in a varnish solution. Here the same paint in different proportions is used for both the lot and end turn coatings of the same winding.

Any end turn coating of constant resistance will "grade" or slope off the more extreme voltage stress, as Figure 6-23a indicates, in a non-uniform way. There is still a fairly sharp rise in voltage, even though the greatest concentration has been eased.

Ideally, the coating would have a resistance varying inversely with voltage. Where the voltage gradient tended to be steepest, the coating resistance would decrease to lower the voltage drop across that portion of the surface. Such a material does exist, both as a liquid suspension of fine grains of silicon carbide, and as tape carrying the same mixture. It greatly improves end turn voltage gradient as shown in Figure 6-23a.

Experimental attempts to get such results with constant resistance paints have occasionally been successful. Coils were painted with separate short lengths of different resistivity paints, each successive length outward from the core having higher resistance than the one before. Such a complex procedure is practical above 16,000 volts, where normally only large generators are involved.

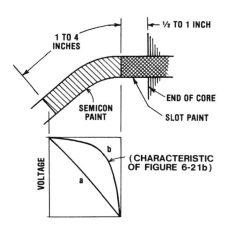

Figure 6-23: How coil surface voltage stress varies with a "variable resistivity" semicon coating (a) compared to the constant resistance type as in Figure 6-21b. The voltage gradient has an almost constant slope, minimizing the stress at any one point along the coil. For clarity, the overlap between slot paint and semicon paint is not shown.

Figure 6-23 shows the paint extending "around the bend" from the straight coil side into the diamond portion of the end turn. In the 11- to 15-kV range, it is common practice to press or "consolidate" insulation on the diamonds by using inflatable rubber molds. However, that is a compromise between an unpressed condition and the more effective die pressure used to squeeze the straight parts of the coil. So end turn insulation can't be quite as solid (void-free) as are the slot sections.

Because the surface coating is a conductor, partial discharges might occur within coil voids present in the less tightly pressed diamonds. Therefore, some manufacturers extend the coil straight section far enough beyond the core so that all the semicon paint covers only straight sections, rather than going around the bend into the diamonds. All protective surface coatings are thus applied only to die-pressed insulation. In any event, semicon paint should overlap the slot coating by half an inch to assure good electrical contact between the two types of conducting material.

What about corona treatment at the lower voltages? Here is a recent public utility specification for 4000-volt machines: "The stator coils shall receive a special shielding treatment to prevent the occurrence of corona discharge. ... The complete insulation system shall receive a final corona inhibiting treatment overall...."

Such a specification is neither practical nor necessary. In fact, its meaning isn't even clear. Certainly no manufacturer can ever apply conductive coil coatings "overall," because such treatment can't be used safely at coil connections (though this is possible with high voltage turbogenerator half coil designs). Normal corona protection as outlined earlier in this chapter can be used, of course, but is never necessary below 5000 volts with present-day insulation systems.

Another recent utility spec says "Anti-corona protection is desired and preferred on all 4000 volt windings and is mandatory on 4 kV motors rated 2000 hp and larger." This, too, is unjustified. There is no magic in the number 2000, especially since physical size, repair cost, and importance of the drive to plant operations are not determined by horsepower alone.

At 6600 or 6900 volts, corona treatment is recommended for any non-VPI winding. In rewinding a machine at those voltages, service centers should not rely on the fact that such treatment was not apparent on the original coils. If they were VPI-treated, they didn't need treatment. But the non-VPI replacements will.

Testing insulation

Is there any single "proof test" for motor insulation that will invariably reject a "bad" winding but accept a "good" one? Unfortunately not. But this is no reason to omit such testing. Rather, it is an argument for doing more testing, to compare the results of various procedures, and to make a reasoned judgment based on all of them.

Insulation testing has two purposes: (1) to determine if a winding is "healthy"—in proper condition for service under the stresses for which it was designed; and (2) to give some basis for predicting whether or not that healthy condition will remain, or if deterioration is under way which may result in abnormally short life.

The simple over-potential or "hipot" test serves only the first purpose. A winding that passes hipot is in good condition at the moment. But it is not necessarily due for a long life. This test is a "go no go" or "pass-fail" procedure that can tell the user nothing about the future. Such testing, therefore, is useful (and standard according to NEMA) for manufacturers of new motors, but it is not particularly useful in the field as a predictive maintenance tool. What's needed there is a test that enables the maintenance man to decide

Insulation and Windings

if the motor needs early repair or replacement, so an unscheduled and possibly catastrophic shutdown can be avoided.

Other types of testing have therefore become more common in the field. There is even argument over whether hipot testing should be used at all once a motor has left the factory. Some years ago, a group of experts was asked, "Is it advisable to use hipot tests on motor windings?" Their answers ranged from "Yes, it is a safe and recommended practice" to "I would not hipot any motors in service because of the chance for damage."

When making that choice, you have to realize just what such a test does. "Good" insulation should easily withstand voltage far above the rated or operating voltage of the motor. Hence, any test that evaluates insulation by seeing what it takes to "break it down" must push voltage stress well above the rated value.

Also, the test voltage must be applied for some reasonable period of time. Any material under stress can carry the load for a short time more easily than for a longer period.

Finally, there is frequency to consider. An a-c machine in the U.S. sees 60 Hz power throughout its lifetime. Therefore, it is logical that test overvoltage be at "power frequency," that is, 60 Hz.

When rated voltage is quite low, it is recognized that insulation thickness cannot decrease proportionately. Tape and sheet materials are typically the same at 200 volts as at 575 volts. So there is a lower limit for test voltage.

Putting all these conditions together results in the well-known hipot value of twice rated voltage, plus 1000 volts, for one minute, a-c. NEMA standards require any completed machine to receive this test before shipment from the factory.

'Acceptance' tests

Many users, particularly public utilities, specify their motors to get an "acceptance" test at this same hipot value, when motors arrive at their plants for installation. This unsound practice should be discouraged. The reason is that any over-potential test, as we have seen, does its job by over-stressing the insulation. To do that often enough with any material will break it down. If the over-stress is great enough, failure may occur after only a few applications—even if the material was in perfect condition to begin with.

Hence, acceptance tests (sometimes called "proof" tests because they are supposed to "prove" the insulation is still as good as it was when the motor left the factory) should never be performed at the original factory hipot voltage. Never use more than 80 percent of that factory value.

For "maintenance" testing—typically during a motor's annual checkup in service—use a hipot voltage of from 125% to 150% of rated motor voltage. This is typically about 60% of the original factory hipot voltage. (See Table 6-II.) This is also recommended for "reconditioned" machines. Such motors may have been thoroughly cleaned, dried, perhaps given a varnish treatment, but

Rated motor voltage	New or rewound motor	Acceptance or proof test	In-service or reconditioned test
460	1920	1550	1300
2300	5600	4500	3500
4000	9000	7200	5400
6600	14,200	11,400	8500

Table 6-II. 60 Hz for one minute, a-c hipot voltages recommended for motor windings. For d-c testing, multiply all values by 1.6.

they still contain the original wire and insulation—so lower that hipot voltage! A completely rewound motor, of course, is subject to the factory hipot level.

If d-c hipot testing is used, the proof test and maintenance test values should be reduced in the same proportions as for a-c testing. (Again, see Table 6-II.)

'Destructive' test

It is sometimes said that hipot is a "destructive" test. What does this mean? The a-c test set needs a relatively high current output capability to test large size windings. A-c current is thought of as "flowing through" a capacitor, whereas d-c current is "blocked" by a capacitor. Actually, during one half cycle the a-c current flows toward one side of the capacitor and away from the other side, then during the other half cycle both of these flows are reversed. The effect, to a meter in either side of the capacitive circuit, is a continuous "flow" of alternating current.

Insulated windings being hipot tested form a capacitor, the insulation being the dielectric, the copper being one plate of the capacitor, the grounded stator core being the other plate. Large motor windings have high capacitance, because of the large surface area of these "plates." High capacitance means relatively low impedance. Hence, the a-c current "flow" in that circuit is large. For example, a typical 6000-hp, 720-rpm motor with 108 slots in a 42-inch long core has a total area of slot contact between coils and winding of nearly 39,000 square inches, or 27 square feet. The impedance (to alternating current) per phase of the "capacitor" thus formed is 20,000 ohms with insulation in perfect condition. That is only 1/50 of a megohm! An a-c hipot tester, being used on such a winding at 9000 volts, would have to furnish 1.34 amperes for all three phases in parallel, or 12 kVA. This would be too much for even a 10 kVA set. To handle very large windings, some manufacturers have a-c hipot sets rated at 100 kVA. Cost, size, and weight rule these out for many shops.

How can you tell when you may run out of tester capacity? During routine maintenance or repair work, apply standard a-c hipot tests to a variety of motor sizes, making a record of the applied voltage, the tester output current, and the total "slot contact area" for each. Slot contact area is simply length of iron times number of slots times the sum of slot width plus twice slot depth. Impedance between winding and ground is test voltage divided by test amps.

Insulation and Windings

Then, when a larger motor is to be tested, compare its slot contact area with that for the largest similar machine already tested. The probable test amperes for the bigger unit, at the same voltage, will be:

$$\text{original test amperes} \left(\frac{\text{new motor slot contact area}}{\text{original motor slot contact area}} \right)$$

If this current at the proper test voltage results in a kVA beyond the tester rating, then test only one phase of the winding at a time, or go to d-c testing.

Locating faults

Smaller motors, of course, don't need high tester output. Sets are available in 1, 2, or 5 kVA ratings. But at any size, because of the capacitance of motor windings, tester output current available must be relatively high. If a fault occurs during test, high available current means that substantial arcing and burning are likely at the fault location before the test set is tripped off the line. A tiny crack or pinhole in one part of a coil may thus result in widespread damage, perhaps to several coils.

There is the advantage that it is usually easy to locate the faulted area by inspection after the test. But the a-c hipot is properly termed a "destructive test" because of this behavior.

On the other hand, the d-c test set need not supply such a high current, but only the small "leakage current" that can pass to ground through the insulation itself which has millions of ohms resistance even when in poor condition. Test set output can therefore be much lower than for a-c, so that destructive burning at faulted areas cannot occur. Damage remains so well confined to that small area that it is sometimes difficult to locate for repair.

Hence a d-c hipot should not be considered a destructive test. This, plus the much smaller tester rating needed, is one of the great advantages of such a test in maintenance work. The other advantage is that d-c testing can serve as a predictive tool, as a basis for estimating whether good insulation is going to remain good, or is getting bad. There are several specific d-c tests, other than hipot, each with its benefits and limitations.

Probably the most useful procedure (covered in detail by IEEE Standard No. 43) is the insulation resistance test. Insulation is, after all, a current conductor; it's just an extremely poor one, having very high resistance to passage of d-c current. Applying a d-c voltage for 60 seconds between a winding and ground generates a small leakage current which enables this resistance to be determined by Ohm's Law. The direct-reading megohmmeter is usually employed. Maximum test voltage ought to be:

Motor rated voltage	D-c test voltage
460, 575	500 to 1000
2300, 2400	1000 to 2500
4160-6900	1000 to 5000

A single reading may not mean much, for good insulation can be highly variable. A reading as low as 10 megohms may be satisfactory, especially on low voltage windings, though on new motors the reading is more commonly well above 100 megohms. Old, brittle insulation may show a high reading and yet be ready to "blow" under the mechanical stress of the next motor start. Table 6-III may serve as a rough guide.

Single or "spot" readings, taken once or twice a year on a given motor, must all be corrected to a common temperature for comparison with one another. Figure 6-24 shows how this is done, and Figure 6-25 shows typical results of a series of tests taken over a period of time. This long-term trend is the important thing to watch.

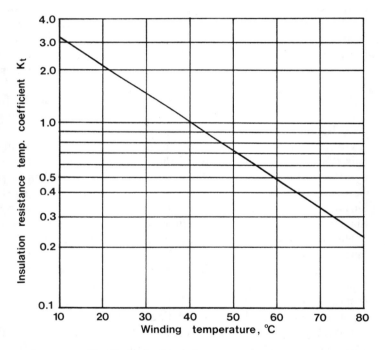

Figure 6-24: Temperature correction factor for insulation resistance. Take the insulation resistance at actual winding (not room) temperature T. Find the correction factor K_t at that temperature from the graph. Divide the actual insulation resistance by this value of K_t to get the equivalent insulation resistance at 40°C.

If the trend is downward, look for causes such as buildup of dirt or moisture in the windings. A clean, dry winding with steadily dropping insulation resistance is deteriorating and in need of overhaul at the earliest opportunity.

Insulation resistance should always be checked before taking any hipot test. If resistance is dangerously low, find and correct the cause before overstressing the insulation that has indicated weakness already.

Insulation and Windings

Motor voltage	Minimum acceptable insulation resistance, ohms
230-575	100,000
1000	1 megohm
2400	3 megohms
4160	5 megohms
6600-6900	10 megohms

Table 6-III. Suggested minimum insulation resistance values. Windings in good condition will normally have values far above these . . . if insulation is dry, cool, and clean.

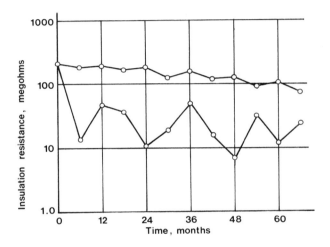

Figure 6-25: How "spot readings" of motor insulation resistance over a long period can show gradual deterioration. Lower graph is actual readings at various actual temperatures; upper graph is same data all corrected to 40°C. Note downward trend, especially after the second year.

Other useful d-c tests, with or without the megohmmeter, are the dielectric absorption and the step-voltage tests. In the dielectric absorption test, a single value of voltage is impressed on the winding for a period of time while insulation resistance is measured at regular intervals. As all the insulation components gradually realign their molecular structures under the applied voltage stress, the "absorption current" that flows will decrease, whereas the leakage component current flowing through the insulation resistance levels off at a constant value. Total current thus decreases, or—saying the same thing another way—total insulation resistance goes up, as time passes.

Dividing the insulation resistance value after the first minute into that after 10 minutes of test gives the so-called "polarization index," a measure of whether or not the absorption current does decrease with time. If it does not, so that the index is low, insulation is in poor condition. (See figure 6-26.) An index of 1 to 1.5 is a bad sign. From 2 to 4 is acceptable, although some users prefer a minimum of 3.

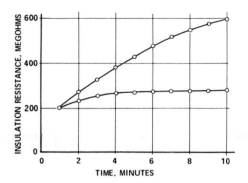

Figure 6-26: Dielectric absorption or "time-resistance" test results, with insulation resistance recorded minute by minute at a constant test voltage. Upper curve shows increase in resistance with good insulation; polarization index is 3. Lower curve is typical for insulation in poor conditions.

Applying test voltage for 10 minutes is difficult without a motor-driven megohmmeter or a rectifier test set. If only a hand-cranked megohmmeter is available, take readings at 30 to 60 seconds instead of 1 and 10 minutes. Here an index of less than 1.25 is a warning; more than 1.5 shows good insulation.

In the step-voltage or "voltage-resistance" test, time of application is held at 60 seconds for any given voltage, but the voltage itself is increased in successive steps, the insulation resistance being checked at the end of each 60-second step. The first step should be about one-third of the maximum d-c voltage to be employed. (See Table 6-II, page 270.)

The theory here is that wet or dirty insulation reveals itself easily under test voltages far below rated value. But the effects of aging or mechanical damage may not show up unless voltage is far higher than that. As test voltage is stepped up, such weak spots have an increasing effect on overall insulation resistance. Figure 6-27 shows typical results to be expected.

In the simple "two voltage" version of this method, two megohmmeters are used, each applied for 60 seconds. Figure 6-28 illustrates the result.

Figure 6-27: Step voltage testing produces graphs of this type. As voltage rises in increments, upper curve shows resistance of good insulation remaining constant or increasing slightly. The middle and lower curves indicate insulation is in trouble.

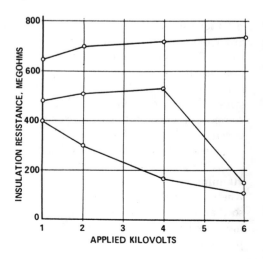

Insulation and Windings

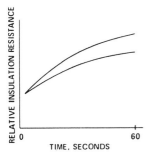

Figure 6-28: A simpler step voltage comparison using two megohmmeters of 500 volt rating (upper curve) and 2500 volts (lower curve). Drop in insulation resistance at the higher voltage shows insulation is weak.

A third means of using d-c testing diagnostically is to measure the leakage current itself as the test voltage is raised in a manner similar to the step voltage test. For this the test set must include a properly calibrated microammeter. If insulation resistance stays constant with increased voltage, of course, the current will rise in direct proportion to voltage, according to Ohm's Law for resistive circuits. But in weak or defective insulation the resistance will begin to drop, resulting in a more rapid rise of current. Such a sudden upward trend in current versus voltage (Figure 6-29) indicates potential trouble but is a non-destructive test because no actual breakdown has occurred.

Figure 6-29: How leakage current may vary with test voltage during a step voltage test. Note the sharp upward bend near the end of Curve C; when this warning of trouble appears, the test should be halted. Conditions applying these typical curves are:

A—Initial condition of motor; wet and dirty
B—After first cleaning and drying of winding
C—After a second cleaning and drying
D—After drying a third time
E—After dip and brake

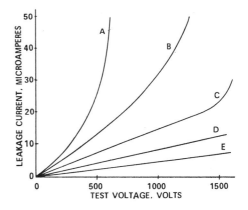

All these tests have the advantage that, unlike the spot check of insulation resistance, they need not be corrected for winding temperature at time of test. They are comparison rather than absolute measurements, and the temperature can be assumed constant throughout the test.

Ideally, each procedure used should be repeated at regular intervals over a motor's lifetime, because each is most useful as an indicator of long-term changes in winding condition. However, they can provide a great deal of insight into insulation condition even when the test is taken only once on a machine in the shop for the first time.

Insulation between turns

All the methods so far described are for testing only the insulation between motor and ground. Growing concern about transient surge or impulse voltages, especially in motors fed by solid-state power supplies, has focused greater attention in recent years on the quality of insulation between turns in a winding. This is because such voltage impulses act like pressure waves in a water piping system. Travelling rapidly along the low impedance cable circuit to the motor, they suddenly slow down upon entering the much higher impedance motor winding.

This tends to make the entire magnitude of the transient voltage pulse suddenly "pile up" across the first turns in the coil, where the leads enter the winding. (See Chapter 4.) The ground insulation in a 4000-volt motor may easily withstand a 7000-volt pulse, but if such a surge appears quickly between two adjacent coil turns, breakdown is quite possible. This is especially likely if such pulses are repeated hundreds or even thousands of times within a few days or weeks. Remember that windings in the best of condition that will sustain one or two applications of over-voltage will fail if the stress is repeated often enough.

So the integrity of turn insulation—which usually consists only of the wire covering itself—is as important as that of the ground insulation. But testing is much more difficult. Impulse testers, to supply the type of high voltage transient mentioned above, are complicated and expensive. There are no industry standards governing their use with rotating machinery.

Two methods have been worked out to meet this need. One is the so-called "surge comparison" test, usable on both random-wound and form-wound coils of almost any size. The tester applies a short duration, high frequency a-c voltage to individual pairs or groups of coils. A comparison is made between the wave shapes of the voltage surges applied to and reflected back from one coil or group of coils, and those simultaneously applied to a second identical coil or group. The reflected impulses are displayed on an oscilloscope screen. If they overlap exactly, this shows that both coils or groups are identical, and both are presumed good (assuming that identical turn-to-turn faults in both groups at the same time will not occur). If they do not overlap, one of the portions of winding under comparison contains a defect. Testing successively smaller coil groups pinpoints the trouble. Portable surge comparison testers are readily available.

Rylander test

Another way to check turn insulation, particularly in individual coils, is the Rylander high frequency test. Developed half a century ago, the Rylander test applies high frequency oscillations (up to 200 kiloHertz) to coil leads. Voltage can be high because at high frequency the coil impedance becomes large, limiting the current to very low values. Typically, 10-turn coils for a 4000-volt machine would be checked at 6000 volts high frequency (for glass-covered wire), or 12,000 volts for mica-taped wire. The voltage should be at least 75% of rated motor voltage and not more than 130 percent of the factory hipot ground test value. Somewhat lower voltages, and frequencies, are used when checking a complete stator winding.

The original Rylander set used a rotating, motor-driven spark gap to trigger the capacitor discharge which set up the high-frequency oscillation. Results were read from a calibrated "wave meter." A few years ago the equipment was refined to use solid-state switching instead of the spark gap and an oscilloscope display instead of the wave meter. The Rylander set is still a big piece of gear, however, unsuitable for use in the field, or for most repair shops unless they do a great deal of work on large apparatus.

Turn insulation testing requires long training and experimentation because of the variety of motor winding types and characteristics. Service people should be familiar with what can be done. But the average shop will seldom find it practical to test turn insulation in motors, either during maintenance checks or after repair.

Research continues into better ways of judging insulation condition, the goal being to be able to learn just what insulation will withstand without in any way impairing that capability through the testing process itself. Insulation experts currently contend that the greatest reliability results from making several different types of tests, rather than relying on only one. Says one service manager, "As there is no infallible test, it would be best if you use as many tests as is practical. It is in the 'battery' of tests that you may have more assurance. . . ."

Remember, too: "No electrical test will ever replace the inspection of the physical condition of the insulation by a competent and reliable man."

Winding temperature

Insulation in any motor winding deteriorates through exposure to chemicals, moisture, transient surges or corona, or physical stress. But the most widely known and universally understood measure of insulation "aging" is temperature. Hence, the various ways to measure winding temperature—and how to interpret the results—should be familiar to motor users.

"How hot is that motor running? Will it take more load without overheating?" Those are questions operators must often face. Answering them correctly requires understanding how and why winding heating is measured.

Why be concerned with temperature measurement? First, because to diagnose motor trouble properly you often must know the actual winding rise. Such testing is usually done in the field. It may also be done in a service shop, some of which are becoming equipped to load-test motors—which can be essential in deciding how far the rating can be upgraded through redesign. As motors of "tighter" thermal design become more common, you can't depend on rules of thumb or guesswork to uprate them. You can't assume the generous margin in temperature that was common some years ago when many motors ran much cooler than their nameplate ratings.

What do those temperature ratings mean? Older motors, of whatever size, carried a nameplate rise—such as 40°C, 55°C, or 60°C—which was always based on an ambient temperature of 40°C. Thus, total winding temperature limit was the rise plus ambient or 80°C, 95°C, or 100°C. Motor enclosure, service factor, and type of insulation determined just which rating was assigned.

Newer motors below 250 hp seldom carry a nameplate rise. Only the ambient and insulation class are given on the nameplate. Again, 40°C is the standard ambient.

That figure is arbitrary. Most industrial ambients—even outdoors—seldom get that high. Since it is total temperature, not just the rise, that determines winding life, an ambient below 40°C means that a higher-than-normal rise can be allowed.

So the first step in finding out if a motor is running within its rating is to check the actual temperature of its surroundings. In doing this, be sure:

(1) To measure the motor's true ambient. Temperature on the far side of the room, or over against the wall, isn't important. What is important is the temperature of the air directly surrounding the motor—the air that enters the enclosure intake passages. Certain motor designs show some "recirculation," meaning that the heated air discharging from motor outlets swirls around to mix slightly with the cool air being drawn in. That incoming air is therefore a few degrees (usually 3 to 5) above the "room" temperature a short distance away. Those few extra degrees add directly to total winding temperature.

(2) To avoid hanging a thermometer directly in the inrushing air stream at the intakes, with no shielding from air currents, because doing so can give a reading that is erroneously cool. Put up a cardboard baffle, or put the thermometer in a cardboard tube of fairly large diameter that can easily be slipped off for reading. (It can remain in place, of course, if a thermocouple is used instead of a thermometer.)

Measuring procedures

Next, how is winding temperature itself measured? Forget about core thermometers. To get useful results, either:

(1) Have thermocouples or resistance detectors right on, or embedded in, the winding—preferably in the slots, or

(2) Obtain the overall winding temperature by the change-of-resistance method. This is by far the most accurate measurement for the smaller motors, and is the most commonly used for almost any size.

In-slot detectors normally must be installed when the motor is built. You can't do a proper job of this in the field without a complete rewind. However, it is often possible to get thermocouples on the coils right next to the core, where windings emerge from the slots. Use at least two, one on each end of the stator, near the top of the winding (for horizontal motors). Fasten them in place with putty or an insulating pad to keep air from blowing right on the thermocouples. Be careful, though, not to tie them down with such a mass of tape or sheet material as to interfere with air flow through the winding.

For motors up to about 100 hp, especially of the totally enclosed variety, this will give readings fairly close to overall winding temperature. In larger machines you will have to add as much as 20°C to the reading of an end coil thermocouple, so that this is not a very trustworthy method.

Always keep in mind that "winding temperature" is not a single value throughout the entire motor, even in a small machine. It can vary greatly from one point to another. The "hottest spot" is seldom accessible, and can be in different places in different types of motors. Some kind of average is the most useful figure—which the change-of-resistance method produces.

Thermocouple output is a low d-c voltage, which you can convert into temperature in one of two ways:

(1) Read the voltage itself on a millivoltmeter. Consult standard thermocouple voltage tables to get the temperature of the "hot junction" where the thermocouple wires are attached to the motor winding. These tables can be obtained from the thermocouple supplier; the graph of Figure 6-30 is drawn from a portion of such a table.

(2) Use a direct-reading meter indicating actual hot junction temperature in degrees, on a dial or digital display. They are well worth the investment.

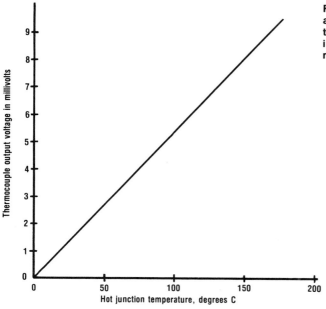

Figure 6-30: Output voltage versus hot junction temperature for standard iron-constantan thermocouple.

What type of thermocouple should you use? Of the many varieties of thermocouple wire, probably the easiest to use in motor work is either the iron-constantan or copper-constantan type. You can buy reels of duplex thermocouple wire of these types (two conductors paired in a single insulating jacket), usually 20- or 24-gauge.

It's easy to make test thermocouples from duplex wire. Cut pieces of the desired length to extend from the measurement area to a convenient, nearby location for the readout instrument; bare the ends of each strand for a short distance; then twist them together and fuse the junction with a torch to form a bead. This junction will be the temperature-responsive point, and it is for this reason that the thermocouple method is called "point sensitive." The temperature along other portions of the wire will have no effect on the reading.

The 'change-of-resistance' method

For the change-of-resistance method, you need only a resistance bridge and a few simple rules. You also need to be able to shut the motor down for a short time after it has run under normal load long enough to be "levelled off" or stabilized at its final temperature for that load (with one exception which will be mentioned later on).

How long does stabilization take? From a cold start, it may take three to eight hours. Usually a five hour run is enough. Here is where you can use a thermometer or thermocouple on the motor frame, to tell you when you've run long enough. Read it every half hour. When two successive readings are the same, stop the motor and take the hot resistance measurement between any two leads.

Watch that ambient, though. Read it every half hour also. If it is rising or falling, compare those changes with the motor thermometer changes. That way, you won't mistakenly assume the winding has levelled off when it was really just a dip in the ambient.

When you shut the motor down, be ready to attach the bridge quickly to the de-energized motor leads. This can be done at the starter if it is right next to the motor. For large, low voltage machines, however, the feeder cable resistance can introduce significant error because it will be at a lower temperature than the motor winding. Be sure the starter and any automatic controls on the motor circuit are locked out while the bridge is connected.

Call the hot resistance, between any two leads, R_h (in ohms). When the motor is "cold"—levelled off back down at ambient temperature—read the same resistance again; call this R_c. The corresponding temperature (use the housing thermometer again for this, because now both winding and housing will be at the same temperature) will be T_c.

Then, use this formula to find the "hot" winding temperature T_h:

$$T_h = \frac{R_h}{R_c} (K + T_c) - K$$

If the motor winding is copper, K = 234.5; if it is aluminum wire, use K = 225.

This will be total temperature, not rise. To get the winding rise, subtract from T_h the value of ambient temperature at the time of shutdown.

This value of T_h applies at the time the bridge is hooked up, which could be one or more minutes after shutdown. How do you know what the real T_h was before the motor stopped? To find out, make a cooling curve as in Figure 6-31. Read R_h, and calculate T_h, at one-minute intervals for 5 or 10 minutes. Plot these as shown below. Then, following the shape of the curve, extend it backwards from the first reading to time zero, to get the probable temperature at shutdown.

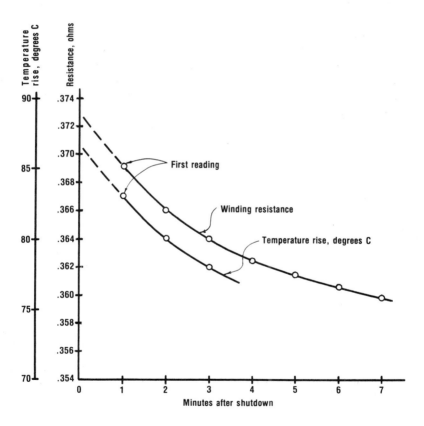

Figure 6-31: Cooling curve plotted by change-of-resistance method following shutdown of loaded motors. The dashed portion is the extension of the curve back to the time of shutdown, indicating what the resistance and temperature were before the drive was stopped.

Actually, if you can take the initial reading soon enough after shutdown, the IEEE test standard says you can ignore that extension back to time zero. Just use your first reading as the real T_h. To permit this, the elapsed time between shutdown and first reading must not exceed these values:

Motor size	Time, minutes
up to 50 hp	½
51 through 200 hp	1½
201 hp and up	2

What if the winding temperature does not start to decrease right away after the motor is stopped but keeps on rising for a short time? This happens in some smaller, older machines, especially when the rise is measured by winding thermocouple.

For that situation, the same IEEE test standard says this: "If successive measurements show increasing temperatures after shutdown, the highest value shall be taken." Actually, it would be more logical to extend the curve back to shutdown time no matter what pattern the readings take, though it may be a bit more difficult to be sure of the proper curve extension when temperatures rise after shutdown (as in Figure 6-32). It is more realistic to do that than to use a reading which is not reached at all by the running motor.

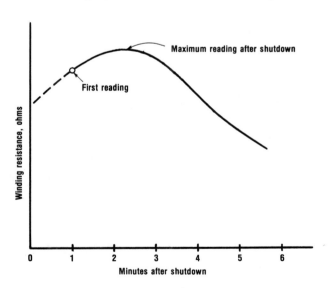

Figure 6-32: Cooling curve in which winding temperature rises for a short time after shutdown, instead of falling. Extending curve back to shutdown is still possible, but IEEE test standard says the maximum reading should be used instead.

Usually it's possible to read R_h within 45-60 seconds after shutdown, even for large machines—provided the insulated cable joints in the terminal box, if that's where the bridge will be connected, have previously been untaped for quick access. Smaller motors, especially if coupled to a loaded pump, conveyor, or mixer, will come to a quick stop when shut off. In some other drives mechanical brakes can be used. But a high inertia load, such as a large fan, may coast too long. You can't use plugging or dynamic braking to stop such a drive because these would cause extra winding heating, distorting the results.

If you take readings with a conventional bridge while the de-energized motor is still coasting at a good clip, the very slight residual magnetism in the core will induce small voltages at the leads which will prevent properly balancing the bridge. For motors rated 600 volts and below, a special test set—the Seely bridge—is available which permits monitoring winding resistance (and temperature) while the motor is on the line and running. No shutdown is necessary. However, this cannot be used with 2300- or 4000-volt motors.

How accurate is the change-of-resistance method? Ohmic measurement by bridge can be correct within a small fraction of a percent. But you have two values to deal with: R_h and R_c. There is also motor lead cable resistance in the circuit, and the resistance of joints between cable and coil groups within the winding. Manufacturing variations between motors presumed to be identical can cause differences in winding resistance, and much larger differences in apparent temperature, from one machine to the next. Core loss and air gap variation may have large effects.

This means that a temperature test taken at one time on one motor cannot be used to predict within 1 or 2 degrees the "true" rise on that motor, or another like it. For example, these winding rises were determined by the resistance method under laboratory conditions, for four identical motors all built at the same time:

Motor No.	Rise by resistance, °C	
	First test	Second test
1	58	—
2	42.4	—
3	47.8	—
4	65.9*	55.4

*Probably erroneous data; 2nd test is more likely correct.

If this can happen with such an accurate method, it is clear that you must leave some margin in deciding the possible excess thermal capacity a motor may have for desired overload or uprating.

RTD readings

If you do have occasion to make temperature measurements on any motor equipped with embedded resistance detectors (RTDs), make sure you know how to read the output from these useful sensors. Whereas the thermocouple reads at a single point, and the change-of-resistance method responds to the entire winding, the RTD sees temperature change averaged over a length of core slot—usually 10 to 20 inches. Rather than producing a voltage, as with the thermocouple, such temperature change in the RTD produces a change in resistance of the detector. You read this with a bridge circuit.

First, find out which basic type of RTD is involved. The two most common versions are:

(1) The "10-ohm" detector, using a copper wire resistance element. It measures exactly 10 ohms at a "reference temperature" of 25°C (77°F).

(2) The "120-ohm" element, using nickel wire. Its resistance is 120 ohms at 0°C (32°F).

Resistance of these two types of RTD varies with temperature as shown in Figure 6-33. (To get detailed calibration tables for them, request Tables 16-9 and 7-120 from MINCO Products, Inc., 7300 Commerce Lane, Minneapolis, Minn. 55432.)

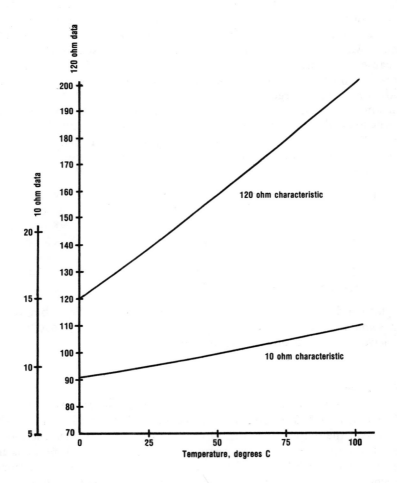

Figure 6-33: Resistance versus temperature curve for two standard types of RTD used in motors.

If you check the cold resistance of an RTD and find it is not one of these two types, ask the owner or manufacturer of the motor what the detectors are. There are several different platinum wire styles, for example, all having 100 ohms resistance at a reference temperature but each following a slightly different calibration table. Unfortunately, there are as yet no national or international standards governing the selection.

There is one major precaution to observe in reading RTD output. The 10-ohm detector normally has three leads—two white and one red. The two white leads are tied together at the detector itself. The purpose of the extra lead is to compensate in the measuring bridge circuit for the resistance of the lead wires themselves. These wires are small (No. 22 gauge is normal), and even a 5- or 10-foot length from winding to bridge can mean significant error in the results.

Figure 6-34 shows how lead compensation works. Be sure to hook up your bridge this way when reading 10-ohm RTDs. Three leads are also furnished with 100-ohm detectors, but for normal motor temperature checking the inaccuracy introduced by not using the lead compensation hookup is negligible.

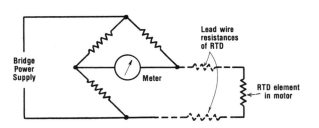

Figure 6-34a: Typical RTD bridge circuit, simplified, showing lead wire resistances between the RTD itself and the bridge. Note how both lead wires and the detector itself are in one leg of the bridge, so that the meter is influenced by the lead resistance as well as by the RTD itself.

Figure 6-34b: With this circuit, the added third lead permits connecting equal lead wire resistance in two legs of the bridge circuit, so they cancel each other out. The "lead-compensated" circuit wipes out the influence of lead resistance on meter reading. When reading 10-ohm RTDs, be sure to use this bridge circuit to avoid error.

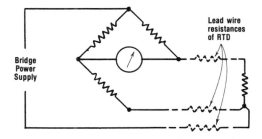

Make it a habit to measure the resistance of each RTD in the motor carefully at ambient temperature, with the motor cold. Don't use the readings from any RTD having a cold resistance that doesn't closely match the calibration table value for that ambient. If none of them match, you may have the wrong table for the type of RTD used; check further on just what was supplied in the motor.

The great advantage of RTDs is that, whatever the motor voltage, they permit continuous temperature monitoring while the motor is running. Read all the detectors, then compare the results—there may be 5°C difference between highest and lowest detector. One of the big causes of such variation is the difference in air flow around the various parts of the stator core. Average all the readings to get the overall figure.

How should RTD temperatures relate to values determined by the resistance method? NEMA standards for motors 1500 hp and below assume that RTD rise for Class B or F insulated machines will be 10°C above the resistance rise. This is equivalent to a "hot spot allowance" (though that term is not used or defined by NEMA today). That is, being located in the slots between coils, the RTDs tend to indicate a higher temperature than shown by the winding resistance change which includes the effect of cooler end coils. For a Class H machine, the standard difference is 15 degrees instead of 10.

In practice, you won't necessarily see those arbitrary figures. Extensive industry surveys have shown that rise by RTD may differ from resistance rise by anywhere from 0°C to 25°C. It's going to depend on the specific motor design. If the motor nameplate is not otherwise marked, the temperature rating is presumed to be by resistance. (Prior to 1964, NEMA standard rise was by "thermometer," which is no longer used because of its uncertainty.) The detector figure is simply a guide, then, and not necessarily usable to judge whether or not the motor rating has been exceeded. But on a large, high voltage machine, detector rise is so much easier to get than resistance rise that the former is always preferred.

Many engineers think RTDs are suited only to large motors and generators. It is true that motor manufacturers encourage this view by charging no price addition for RTDs in machines above 1500 hp. It is also true that the multichannel bridges, perhaps including a selector switch to transfer from one RTD to another in the winding, can be quite expensive. High monitoring equipment cost may be unjustified for smaller motors.

But RTDs are the most reliable hot spot temperature sensors available. They are being used in random-wound motors. Figure 6-35 illustrates this as commonly done in the 75- to 200-hp, 460-volt range.

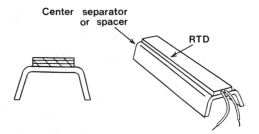

Figure 6-35: This is an easy way to install RTDs in a random winding. The detector element is epoxy-cemented to the top of the standard coil separator, about midway down the slot. If too short to extend full length of the core, plastic filler strip can be butted to it to fill the remaining space. Many motors have been built this way.

Insulation and Windings 287

The RTD is easily checked in the field. If its resistance is anywhere near the nominal value at room temperature, it is undoubtedly in working order. Only a zero ohmmeter reading (direct short) or infinity (open circuit) indicate a bad detector. Inasmuch as six are usually supplied—never less than three—one or even two bad RTDs in a winding are no cause for concern.

Some users prefer the thermocouple as a temperature sensor. It has some disadvantages, although it can easily be fitted into a plastic strip and embedded in a slot between coil sides just like the RTD—in either formed-coil or random windings. (See Figure 6-36.)

Because the thermocouple junction is somewhat bulkier than the flat RTD element, it always takes up a little more slot space than the thinnest RTD. Also, the thermocouple is "point sensitive"—that is, it responds only to tem-

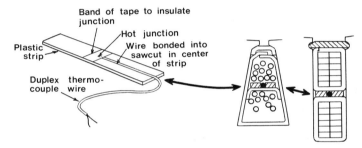

Figure 6-36: Slot thermcouple construction and use in either form-wound or random-wound stator slots.

perature at the single thermocouple junction point. The RTD, on the other hand, responds to the average temperature along the full length of its wire sensing element—usually at least 10 inches. Inasmuch as temperature variations along the length of the slot in any motor can be considerable, the RTD tends to give a truer picture of average conditions than the point-sensitive thermocouple.

Finally, the thermocouple circuit must be more carefully wired. Wire of the same special alloy as the thermocouple leads themselves must be used to extend the circuit, no matter how great the distance, to the control device that produces the actual temperature reading. Thermocouples produce a voltage output, not just a change in ohmage as does the RTD. This is amplified electronically to a readout or relay function.

Although normally used to indicate actual winding temperature on a meter or digital display device, either RTD or thermocouple circuits may also include a relay function to give an alarm signal on overheat, or to shut down the motor.

When either RTDs or thermocouples are used to trigger an alarm or shutdown function, the tripping point is variable and must be selected by the user—not the motor manufacturer. This is particularly true when both an "early warning" alarm and a shutdown are provided, one before the other. Here are the considerations affecting the choice of control circuit settings:

(1) How critical is process continuity in the plant? It has long been recognized throughout the motor industry that any machine, whether or not it has a service factor, can be safely overloaded by a considerable amount for an occasional short period—say a few hours. If a drive is essential to a process or downtime extremely expensive, a user would certainly want to let a motor overheat for a short while rather than shut it down automatically right at the insulation system rating limit.

(2) How well attended is the location? If trained operators are right there to respond quickly to any alarm and either diagnose the trouble or shut the drive down while investigating, an alarm point can safely be set much higher than if no one is around to react. In some installations, the alarm may sound miles away via dirt road.

(3) Can the motor be "unloaded" by closing valves, switching to backup equipment, and so forth? That also would justify setting a fairly high alarm temperature, because trouble can be compensated for in a hurry. But if such unloading is lengthy or difficult, then a lower alarm setting will provide an earlier warning.

(4) Alarm and shutdown temperatures too close together can lead to unexpected shutdown without prior warning. The motor manufacturer knows that thermostats, for example, may operate at any temperature from 5°C below to 5°C above their nominal settings. But only the user knows what tolerances may apply to his monitoring equipment in an RTD circuit.

(5) What range of motor ambient temperature may be expected? A setting suitable for an outdoor desert location might be much too high to furnish adequate protection for a motor in another location during another season.

(6) Will the motor be subject to suddenly applied overloads, or are they only likely to come on gradually? If a slowly rising winding temperature is more probable, a fairly "late" sounding alarm will still provide time for corrective action. But a suddenly applied overload will result in a high rate of increase in winding temperature, so thermal circuits should trip sooner (lower temperature setting).

It is evident that only the user can understand and weigh all the conditions involved in choosing temperature alarm or shutdown settings.

Other heat detectors

Though not usable for measuring actual winding temperature, several other types of winding thermal sensors should be mentioned here. Unlike the resistance detector or thermocouples, they can only furnish a "go or no go" indication that the temperature is either above or below some fixed set point. Usually that point cannot be adjusted in the field.

One is the inherent thermal protector. This is a thermostat that has a small heating element built into it, through which motor line current passes. Thus,

Insulation and Windings

the thermostat responds both to heat building up within the windings around it—which can result from blocked ventilation, gradual overload, or rising ambient air temperature—and to heat produced within the protector itself by the current-carrying heater. Stalling or severe starting will take some time to warm the entire protector from the outside, but the high current through the heater does the job quickly, taking the motor off the line in time.

This device is unsuited to motors above about 7½ hp. Thermostat size increases rapidly when its contacts and the heater have to handle full line current.

The other sensor, usable in larger motors, is the rate of rise or "anticipatory" protector, shown in Figure 6-37. Its inner rod and outer tube or shell will gradually heat up under steady overload, or loss of cooling air, each part expanding at about the same rate. It will take some time for the expansion to proceed far enough to operate the switch contacts.

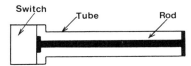

Figure 6-37: Simplified cross-section of a "rod-and-tube" thermal protector.

But if the winding is suddenly heated by a stall or severe start, the outer shell, in contact with the stator coils, heats up faster, expanding more quickly and pulling the relatively unexpanded inner rod with it. This operates the switch and gives an alarm sooner. In other words, the faster temperature is rising, the lower the temperature at which the switch operates.

Once the motor is above a certain size, however, it becomes quite difficult to properly match these variable trip times to actual differences in stator and rotor heating characteristics. Most difficult of all is getting the protector to trip at a stalled or acceleration time when the stator is not yet very hot, but the rotor is close to its safe temperature limit. With such a condition, there is a good chance of nuisance tripping under normal overload.

Also, proper heating of the outer tube is possible only when the protector is fully embedded in random-wound end coils. This sensor is not suited to formed-coil windings.

The other commonly used motor winding temperature sensors are thermostats. (See Figure 6-38.) Unlike the RTD or thermocouple, these cannot be embedded in slots where winding temperature is normally highest. Their bulk requires mounting on the end turns.

This results in several shortcomings for the thermostat. First, mounted against the end coil surface it can receive heat from only one side. The slot-embedded sensor is heated from both sides. Second, the thermostat is relatively large. It takes time to warm up. Third, though quite reliable, its snap-action switch is more prone to mechanical failure than a device with no moving parts.

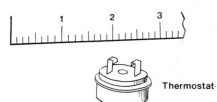

Figure 6-38: Relative size of thermostat for motor winding use. Scale is in inches.

Finally, it cannot produce any temperature readout or adjustable trip setting. It is strictly a factory-calibrated, non-adjustable, on or off device.

However, when considered as part of the total protection system, the thermostat is the least expensive sensor. No expensive bridges or electronic circuits are needed. In small motor sizes, the thermostat can directly make or break the starter circuit. For larger drives, only an auxiliary relay need be added, to control either starter or alarm circuit.

Thermistor systems

The thermistor overcomes some of the thermostat's drawbacks, since it is smaller and has no moving parts. Although an electronic controller is needed, the thermistor system is still relatively inexpensive.

What is a thermistor, anyway? How does it work? If a thermistor circuit fails, how can the trouble be located and corrected? These questions become increasingly important as more and more motor manufacturers adopt thermistor systems for built-in overheat protection in standard motor windings, in the U.S. and elsewhere.

The term "thermistor" stems from the two words "thermal" and "resistor." That is, the device has a resistance which changes with temperature. All ordinary conductors have that property, of course; the resistance of copper wire goes up when the wire gets hotter. However, the thermistor is a semiconductor which has been "doped" or mixed with certain chemicals which produce a drastic resistance change within a fairly narrow temperature range.

This change may be either a drop or a rise in ohms when the thermistor is heated. If the former is true, the device is called a "Negative Temperature Coefficient," or NTC thermistor. The other version is the "Positive Temperature Coefficient" (PTC) thermistor, the sort most often used for motor protection. (We will see why a little later on.) Figure 6-39 shows typical resistance-temperature curves for both types.

Thermistors are small. An actual semiconductor "bead" is about the size of a match head. Those used in motor temperature sensing may appear larger for several reasons. First, an epoxy overcoating is usually applied for mechanical protection and electrical insulation. Second, to connect the thermistor to its external circuit and survive shop handling during installation, insulated lead wires must be attached to the bead. Finally, to better "soak up" heat from the

Insulation and Windings   291

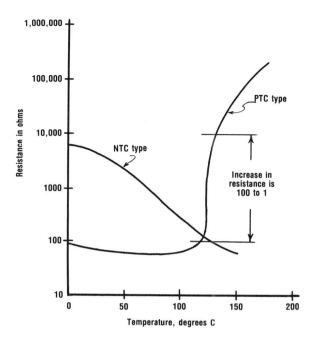

Figure 6-39: PTC and NTC thermistor resistance-temperature characteristics. Note the abrupt and massive rise in ohms at the "critical temperature" for the PTC type. Usual tolerance on this trip temperature is plus or minus 5 degrees. Some makes of thermistors are available in 5- or 10-degree steps from 60 to 180°C, but all will have about the same resistance at room temperature, or at the critical temperature. Note that resistance for the NTC type is much more gradual.

motor winding, some thermistors are sealed into copper or plastic foil "fins" extending some distance outward. (See Figure 6-40.)

Because it is so small, a thermistor alone cannot directly operate an alarm circuit or controller. The current-carrying capacity of the device is negligible. Hence, all motor overheat systems that use thermistors include one other element, the "control module," which converts the resistance change into useful output.

There are two types of modules, both performing the same task. One type is diagrammed in Figure 6-41. It uses a conventional relay, with a transformer and rectifier supply, to transform the thermistor resistance change into relay contact opening or closing.

This module is bulky. It is also subject to all the reliability problems—dirt, corrosion, and so forth—of electromechanical relays generally. But it does have the advantages of fairly high control current capability (governed by relay contact rating), of multiple contacts, and of easy conversion to different operating voltages. In fact, as Figure 6-42 shows, a typical 220-volt a-c module can readily be converted to work in a 48-volt, d-c control circuit.

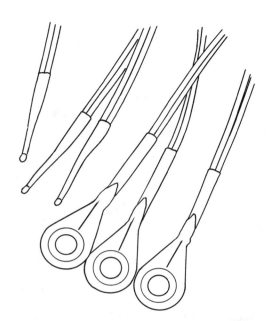

Figure 6-40: Typical thermistor sensors for motor winding use. Those at right have large heat collecting fins, but the actual thermistor bead itself is the same size as for the units at left.

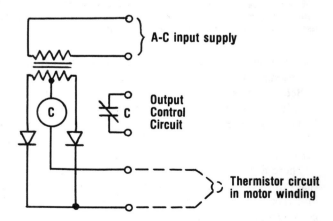

Figure 6-41: Basic circuit of simple thermistor control module using electromagnetic relay output.

Insulation and Windings   293

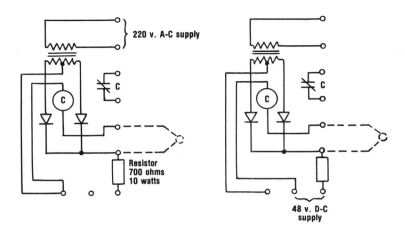

Figure 6-42: Extra terminals and leads have been added in the field to operate this module from either 220 volt a-c, at (a), or, as at (b), from an alternative source of 48 volt d-c control power.

The other type of module is all solid-state, replacing the relay with electronic switching circuits. In any event, what the module does is use the large swing in thermistor ohms at the critical temperature to trigger a switching action which opens or closes an external circuit. This in turn may operate an alarm or shut the motor down. Depending on sensor characteristics and module design, a motor overheat protection circuit may use one thermistor or several. Some circuits require three thermistors in series. Others use anywhere from one to six in parallel.

Parallel hookups have the advantage that the failure of one thermistor does not require that it be replaced in the motor. It need only be cut free from the circuit. The remaining thermistors in the winding will then continue to operate the module properly.

System trip temperature for a typical Class B motor winding may range from 115° to 130°C; for Class F, from 130° to 150°; and for Class H, 160° to 180°. Thermistors have not only a temperature sensing trip point, but also a thermal limit. Get them too hot, and they will be damaged. This limit is typically 200°C, so the sensors should not be harmed by normal dip and bake cycles even with Class H insulation systems.

Location

Where are the thermistors located in the motor winding? Their size and shape will not permit their being embedded in slots as a conventional RTD would be. They are always mounted on the coil end turns. With a random winding, thermistor beads are worked into or "buried in" among the coil wires as far as possible, for faster response to overheating. In form-wound machines, the thermistor is tied flat against the surface of the coil knuckle or diamond as space permits. Sensors can be retrofitted to existing motors in the field, though it is not always possible to get them in among the coil wires on a small machine after dip and bake. The careful use of a heat gun to soften the varnish or resin may make retrofitting easier.

Are thermistors of different makes interchangeable in an existing circuit? No. The same is true of control modules. The specific resistance-temperature characteristics of one make of thermistor are tied to the operating mode of the module designed for it. If you interchange either the thermistors or the module with some other make, you may render the circuit inoperative or destroy the components. Thermistor systems are complete systems and must be treated as such.

Control modules may be mounted in a terminal box on the motor, or remotely. They may include as options:

(1) A multiple-contact output. (With completely solid-state devices, the most you can usually get is one normally open and one normally closed output.)

(2) An alarm light.

(3) A manual reset button.

Control power to energize the module circuit is supplied normally at 110/120 or 220/240 volts a-c, 50 or 60 Hz, and except for those changes to electromechanical modules described earlier, this can't be altered in the field.

Selection

What governs the choice between PTC and NTC thermistors? Preference for the former rests on its greater degree of "fail safety." The most common defect that develops in service is an open circuit in the thermistor wiring. To the NTC thermistor circuit, an open wire looks like a cold winding. Hence, the module will not trip, so there is no indication of trouble. Yet overheat protection has been partly or entirely lost.

In a PTC thermistor circuit, on the other hand, an open wire—an infinite resistance—looks like a very hot winding and thus causes an immediate trip. The resulting alarm or shutdown alerts the operator that something is wrong.

Short circuits can occur, however, as well as open circuits, so one further refinement appears in some recent PTC systems. Normally, the modules are not designed to respond to unusually low thermistor resistance. The circuit is activated only by the sudden rise in sensor ohms at the critical temperature. But modules are now available which work within both upper and lower thermistor resistance limits. If sensor ohms falls too low, as would happen if the thermistor leads became short-circuited, the module trips just as it would on overheat. Again, the object is fail-safety—to warn the operator that a defect needs correction.

It is important to know if any of these thermal sensors ever do fail. Normally, at least two or three thermostats or thermistors are provided in any motor winding. In some thermistor schemes, however, the circuit will not operate properly if any of the sensors is defective or missing; in other systems, the sensors are all in parallel and only one working sensor will keep the system functioning.

Thermostats can be checked by continuity or by applying heat, although finding the exact tripping temperature can be quite difficult. Thermocouples can be checked with a millivoltmeter and standard calibration data.

How can thermistor circuits be checked to see if they are working properly? One obvious way is by simply overheating the motor windings, while monitoring temperature with a thermocouple, to see if the module operates properly at the right temperature. There are several drawbacks to this, however. Doing it by running the motor at overload is possible only if you have the loading device available. A locked rotor test is possible, but it may damage the rotor or harm the winding mechanically through sustained high current before the module trips. Remember that the thermistors sense only end turn temperature and are for emergency protection; the slot portion of the winding may overheat and jeopardize insulation before the sensors do their job. You're using up insulation life while the motor is still on the test bench.

Thermistors themselves can usually be checked simply by resistance measurement. Unless the manufacturer of the specific thermistor has approved it, however, don't use a conventional ohmmeter to do this. The semiconductors will safely pass only very small currents. Ohmmeter current alone may be high enough to damage the thermistor. Therefore, use a high impedance instrument.

Don't expect the same resistance from all thermistors in a winding, or throughout a set of new ones that are supposedly identical. Remember, at the critical point the resistance changes by hundreds of times, so at room temperature a range of two or three to one in ohms is meaningless. Typically, the thermistor manufacturer will specify broad limits, such as 550 ohms maximum at 25°C. One sensor might measure 127, the next 83, and the next 204. This is perfectly normal.

If the module itself must be checked, the usual procedure is to connect it up with pre-checked thermistors duplicating those in the motor winding (or, if the test is done before the winding is completed, use the ones that will be installed). Immerse the thermistors in a controlled temperature oil bath containing a reliable thermometer. Then observe where the module operates on rising temperature, and where it "resets" as the bath cools down. If the module doesn't work properly, replace it. Even if it is not of the completely solid-state type, repair components for the module itself are seldom readily available, nor is the repair cost worth it.

If modules do develop defects in service, check out two possible causes. The first is excessive ambient temperature. The typical top limit for solid-state module temperature is anywhere from 40°C to 70°C, depending on the module's make. In some control housings this may be exceeded, resulting in circuit damage, erroneous operating points, failure to reset, and so on.

Another cause of damage is a control circuit exceeding the volt-ampere rating of the module output. Again, these ratings vary widely depending upon manufacturer. Three commonly-used versions are limited to 2.3, 5.5, and 6.7 amps respectively, all at 120 volts a-c. If in doubt, ask the motor manufacturer to confirm the rating.

Why not consult the thermistor or module supplier directly? His application data or knowledge may not reflect the motor designer's adaptation of the parts to motor protection. For example, some catalogs give particular thermistor trip settings which are valid only if the thermistors are connected to standard catalog modules. The motor manufacturer may be using a different module to achieve more desirable characteristics, such as a greater spread between trip and reset temperatures. The special module contains different circuitry to produce a different trip point for the same thermistor, such as 120°C instead of 105°, for example.

So the combination must be dealt with as a system rather than as separate components. The final system test, performed after motor repair is complete or a new thermistor circuit has been installed in an old motor, is to test the wiring to ground. Most thermistors have momentary insulation test ratings of 3000 to 5000 volts. However, the control modules may have close terminal spacing and only be in the 300- to 600-volt dielectric class. To play it safe, avoid testing thermistor circuits to ground at more than 1100 volts a-c.

System variations

Most module designs today permit up to six thermistor inputs, tempting some users to ask, "Why not use one module to protect two different motors, three thermistors in each winding?" or, "Why not use the one module with 105°C thermistors to protect the winding, and 85°C thermistors to protect the bearings?" (The same thermistor beads are available in metal probes for insertion into bearing housings.)

There is no reason why it won't work, but it's not normally advisable. When the module trips, there is no way to tell which motor of the two is overheated, or whether the winding temperature or bearing heating caused the trip. When any protective device operates, the location of the trouble should be clear at once. Otherwise, proper corrective measures aren't possible without shutting down everything involved until it can be examined in detail.

It is not uncommon, however, to provide two sets of thermistors—each with its own module—in one motor, to give an "alarm" trip at one temperature plus a "shutdown" trip at a higher level. Some users wire the lower-temperature circuit to lock out a restart, after shutdown, until the motor has cooled below the "alarm" level. The circuits shown in Figure 6-43 are typical. Keep these points in mind:

(1) Avoid the mistake of interchanging the modules or thermistors. The lower temperature sensors will look just like the higher temperature sensors and will have the same range of ohms at room temperature—even though they trip at an entirely different figure.

Insulation and Windings

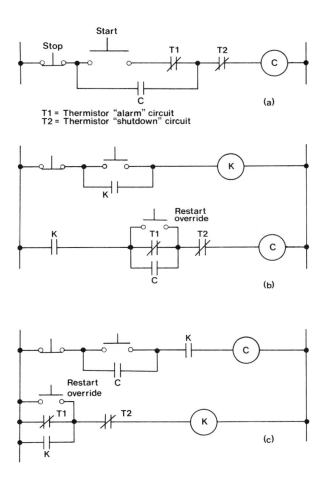

Figure 6-43: These are typical schemes for using both high temperature ("shutdown") and low temperature ("alarm") thermistors in a single motor winding. At (a), the motor is cold. As the winding begins to overheat, T1 will open, but the motor will keep running because the contactor circuit is still closed. After the winding rises above the high temperature limit, T2 opens and the motor stops. It cannot be restarted until the winding cools enough for T1 to reclose.

At (b), a "restart override" button will restart the motor directly, by-passing the normal start button.

At (c), the restart override simply restores the original starting circuit, permitting the motor to be restarted while still warm, using the normal start button.

(2) If you mix up a low temperature thermistor with a high temperature device in a parallel-channel circuit, the low temperature will operate first every time. In a series thermistor loop, this probably won't happen.

(3) Checking out individual thermistors or modules is done in exactly the same way already described.

In a two-winding multispeed motor, using a single module with two sets of thermistors, one set in each winding, may be all right. Diagnosing drive trouble can be more complicated, though, because after a trip there's no way to be sure which set of thermistors sensed the overheating. It may not be possible to tell at which speed the motor was operating when it overheated, because thermistors in the de-energized winding can pick up heat from those in the energized coils. The motor is protected, all right, but getting the drive operation back to normal may require more investigation because the trouble source isn't so easily pinpointed.

If a motor is rewound Class F but still carries a Class B temperature rise on its nameplate, what temperature limit should the thermistor system use? Some engineers claim that "The selected sensor temperature should be based on the original design temperature rise of the rewound motor." But this isn't necessarily correct. The user should be consulted before thermistors are selected.

It has been proven that the life span of a Class F system at its 155°C limit is at least as great as that of a Class B system at 130°C. So using thermistors with a Class F tripping point in a Class F winding, allowing full use of the thermal capability built into the motor, will not necessarily reduce winding life.

Of course, if higher temperature rewinds are made without considering bearings and lubricant life, the motor may suffer from operation at the new temperature limit. And if the user has no intention of ever operating the motor at the higher temperature, sticking with a lower thermistor trip setting would be reasonable.

Like any other stator winding temperature sensors, thermistors cannot be expected to respond in time to prevent rotor damage during a stall, or from starting too frequently. The temperature changes so rapidly at locked rotor that a correlation between actual end coil temperature and rotor heating is impossible to establish. This is particularly true in a large machine, and—as a relatively inexpensive and more reliable version of the sealed thermostat—thermistors have been used in many motors in the 1000- to 2000-hp range.

Users often have a thermistor protection scheme already in operation, or they have thermistor control units on hand. When buying a new motor, they request thermistors in the winding, expecting to use them with their existing controller. They can do so only if the controller was designed to work with that particular style of thermistor. Remember that the controller is an electronic circuit in which the exact variation of resistance with temperature in the thermistor itself is used to trigger a circuit response, translated into relay action. The circuit may be biased to take account of thermistor resistance below, as well as above the trip temperature. Most PTC thermistors change resistance very suddenly at the critical temperature; the NTC type may change much more gradually.

Furthermore, the thermistors and controller form an operating system. Who will assume responsibility for proper functioning of a system that uses components never designed to go with one another? Certainly not the motor manufacturer. A user could install a protection system of his own choice, using components as he wished, if he were willing to warrant the operation of the combination. Otherwise, he should not attempt it.

Similarly, RTDs of one resistance cannot be used as input to a monitoring unit designed for a different RTD resistance. Nor can thermocouples of one type be used to replace another.

Keep in mind also that whatever the sensor used, its purpose is only to protect the winding insulation system—not to react when temperature rise happens to exceed the nameplate figure. Many motors have service factors as built-in overload capacity; others are bought with unusually low nameplate rise, well under the insulation system limit, for long winding life. Any temperature sensor installed in such a machine will, nevertheless, have a tripping point coordinated with the actual insulation system limit, not with the rise that happens to be on the nameplate.

For example, suppose a 140°C thermostat is normal for use in a Class F winding, as is the case with some manufacturers. A motor ordered as an "80 degree rise" machine, using the Class F as added safety or life margin, will still use that same 140°C thermostat.

Of course, at times the operator will want an "advance warning" that the winding is heating up. He will want to know trouble is brewing ahead of time, perhaps because he may have to keep running as long as possible or because the location is unattended and it may take some time to find the trouble. This calls for a second set of temperature sensors to trigger the alarm before the other set operates to shut the drive down.

Thermostats and thermistors alike generally have a plus or minus 5°C tolerance on their tripping points. This means that if separate alarm and shutdown sensors are chosen with tripping points only 10 degrees apart, the tolerances could result in both operating at the same time. There would then be no "advance warning" at all. The motor designer selects the shutdown device for the insulation system, as noted above. If the user or application engineer then specifies the alarm temperature, it should be at least 15°C below the shutdown device temperature, which means the designer should be consulted about that point before the choice is made.

Interpreting temperature tests

Recognizing the importance of actual winding temperature in evaluating a motor redesign, or judging a complaint about overheating, we sometimes face this question: Can we predict temperature rise at one load by tests taken at some other load? Many shops cannot test a large machine at its full horsepower rating. Neither loading devices nor power system capacity may be large enough, or the demand penalty from the utility may be too great. A full load heat run on the average 3-phase induction motor may take several hours before temperatures stabilize and the machine can be stopped.

Therefore, it may be useful to apply the load for only 15 minutes or half an hour. Can the eventual "continuous" winding temperature at such a load be predicted from the short-time test?

Perhaps the winding presently runs well below the design temperature limit. How hot will it run if overloaded? This question may take one of two forms:

(1) If the motor is subjected to a steady overload indefinitely, how hot will it get?

(2) If the overload is applied for a short time only, how quickly will winding temperature go up?

What will be said in answer is not necessarily true for any synchronous, wound-rotor, or d-c machine. Their rotor and stator heating patterns are quite different from what exists within a squirrel-cage motor.

To begin, consider what happens when a steady load is applied to a motor that has been sitting idle and "cold." Winding temperature—however it is measured—goes up rapidly at first. Then the rate of rise tapers off, as a balance is reached between heat entering the coils from I^2R loss generated within the copper, heat produced in the lamination steel, and heat lost to the outside cooling air.

For a simple heat-producing device like a single wire or coil in undisturbed air, this temperature-time curve follows a simple mathematical formula. If a short-time test were taken, with accurate measurement of temperature after 15 minutes or half an hour, one could calculate the part's ultimate temperature accurately.

But heating in an actual motor winding is not that simple. The curve doesn't fit simple mathematics. Heat from several sources is flowing back and forth between copper and steel, with considerable non-uniformity in the different directions. Calculating ultimate temperature rise from a short-time test isn't normally possible in the field.

For a rough estimate, however, assume that 30 minutes after a cold start the winding rise (if measured by slot-embedded detectors or by resistance) will reach about 55% to 60% of its eventual value. A 15-minute test is too short to permit prediction. Taking readings at two or more points during this test will help establish the shape of the curve (Figure 6-44) more accurately, but with the resistance method the motor must be stopped for several minutes each time to get readings. These interruptions distort the shape of the temperature/time curve. With low voltage machines, that can be avoided by using a Seely bridge which can measure winding resistance while the motor is energized.

This drawback can also be overcome by monitoring thermocouples on the end coils. But that can be misleading for many designs. Figure 6-45 shows what can happen. The initial rate of rise for even an hour or two looks deceptively low, because heat is not reaching the end coil surfaces. But, in

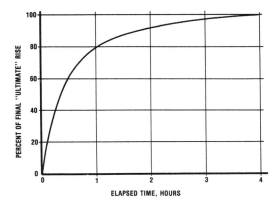

Figure 6-44: this time-temperature curve is typical of induction motors carrying constant load from a cold start. It applies only if the winding temperature being measured accounts for the hottest part of the coils, using the resistance method or embedded detectors.

Figure 6-45: If motor temperature is taken from surface thermocouples, on either core or end coils, the initial rate or rise may be deceptively slow, as shown by curve (a). From a short-time test, this motor looks better than the more normal machine represented by curve (b), which has more efficient cooling. But with a longer test the trouble becomes apparent.

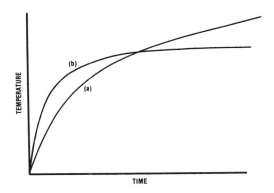

fact, the machine's ventilating system is unable to level out the overall winding temperature at all. A short-time resistance measurement would show much higher values, but even that would not prove that the ultimate rise would ever level off.

So be careful in making predictions from short-time tests unless your measurements involve the hottest part of the winding—the slot portion.

Load changes

Given the temperature rise of a motor at one loading, how much does temperature go up when the load is increased? For many years, the standard assumption was that an increase of 15% in horsepower meant 10°C higher winding temperature.

The same assumption is often made today. When a 1.15 service factor is furnished, it is still presumed that the 80°C rise (for a standard Class B machine) goes up 10° to become 90°C rise at the service factor load.

But that is an arbitrary basis for rating, not an expression of actual temperature change with load. When we take tests to find out what a motor will do at these two loads, we find what is shown in Figure 6-46. The average difference in temperature for a 15% load swing is not 10°C but 22°C. (See Chapter 1.)

There are ways to calculate the effect of overload on rise. A simple equation won't work for more than small changes in load, though, because just as with Figure 6-44 the mathematics aren't simple. Winding temperature rise at zero horsepower is not zero. There is an idle or unloaded rise produced by heat generated in the core laminations. As load is applied, this is augmented by an added rise increasing with load. (See Figure 6-47.) The added rise itself has several components.

The I^2R losses in the stator coils, for example, vary with both current I and resistance R; the resistance goes up as the temperature does, so it is not constant with load. Current does not vary directly with load, either, though for high speed motors near rated output the difference may be slight. The same kind of thing is occurring in the rotor, except that it cools itself much better.

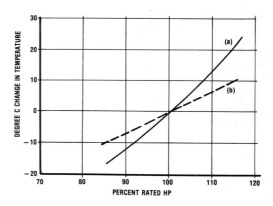

Figure 6-46: Service factor overloads seem to have a far greater effect on winding rise today, (a), than the old NEMA standards implied, (b). But in terms of percentage temperature change there is no contradiction; today's rated load rise with Class B insulation is itself twice as high as yesterday's Class A standard.

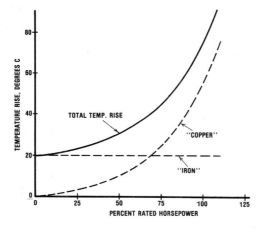

Figure 6-47: Motor winding temperature is not negligible at no load, because of core heating. This component of temperature, the "iron" portion, changes very little with load. The variable component, or "copper" temperature, does change rapidly with load. Total rise is the sum of these two components.

The "copper" line is not actually zero at no load, but very close to it.

Stray load loss has an effect also, some parts of it varying with load, slip, or rpm, and other parts varying with the magnetization in the core which will go down somewhat as horsepower output rises.

With synchronous machines, there is a way of separately measuring the "iron" and "copper" rise components so that the curves of Figure 6-47 can be plotted and total rise predicted at any load. One heat run is taken with open-circuited armature to get the iron loss temperature using full field excitation. A second run is made with the armature short-circuited, excitation being reduced to a very low value so iron loss is negligible, while rated load current flows in the armature coils. Adding the temperature rises from these two tests gives very nearly the correct total armature rise under full load conditions.

But this won't work for an induction motor. One reason is that the only way to circulate load current in the winding is to demand the right amount of horsepower at the shaft. If applied voltage is lowered enough to make the iron loss effectively disappear, the motor will be unable to develop enough torque to carry that horsepower.

Is it possible to predict winding temperature at full load from a single no-load test? You can make only a rough guess. In typical motors the no-load iron or core loss will be 25% to 30% of the total heat-producing watts at full load. This loss heats the winding, but some of it is carried away by the air surrounding the core. So one might expect that the winding rise at no load would be 20% or 25% of the temperature rise at full load—and this is about right. Multiply no-load rise by 4 for a crude estimate of full load rise.

We can refine such a prediction by taking two tests, one at no load and the other at 75% to 100% of rated load. The result can permit estimating rise at loads up to 150% of rated. Figure 6-48 illustrates how the data are plotted. When the ampere values, squared, are multiplied by the temperature correction ratios as shown, the prediction accounts for the extra winding heating produced by increased copper resistance at the higher temperatures.

This method has limited value outside the factory. If a test at or near rated horsepower is possible, a good enough overload temperature prediction may be made from that one test by one of the simpler methods described below. And if a test cannot be made at well above half load, the extrapolation or extension of the data becomes too great for accuracy. An error of only a couple of degrees in measured rise may cause a 10% to 15% uncertainty in the predicted rise at the higher load.

To get around the complications of such work, here is how to predict effectively how hot an overloaded winding will get:

(1) Calculate total temperature (rise plus ambient) as proportional to (horsepower)$^{1.5}$. For a 25% increase in horsepower, as an example, $(1.15)^{1.5} =$ 1.233. The temperature should go up 23%. Example: 80° rise plus 40 ambient = 120; 1.233 times 120 = 148; 148 minus the ambient = 108° rise, or 28° above the original test value. This is somewhat on the high side, but conservative.

(2) Calculate rise alone as proportional to line current squared (not horse-

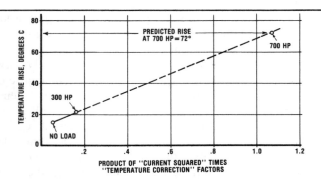

Figure 6-48: this graph illustrates a method proposed almost 40 years ago for using reduced-load temperature tests to calculate temperature rise at full load or overload. In theory this accounts reasonably well for two effects: (1) the two separate "iron" and "copper" temperature components of Figure 6-47, and (2) the extra loss and temperature in the "copper" portion caused by increased resistance in the hot winding.

This example is a 700-hp motor. Two temperature tests were taken, one at no load and the other at 300 hp. The results were:

	No load	300 hp	700 hp
Per unit current	.25	.43	1.0
Measured rise	17°C	21°C	?
Ambient temperature	25°C	25°C	assumed 25°C

For the idle test, the "current squared" factor is $(.25)^2$, or .0625. The "temperature correction factor" for winding resistance is:

$$\frac{234.5 + \text{ambient of } 25 + \text{rise of } 17}{234.5 + 75°} = .892$$

Multiplying these together gives .0557. This is plotted on the horizontal axis, and the measured rise on the vertical axis. For the 300 hp test, the horizontal axis value becomes .168. Next, a 700-hp rise is assumed, a horizontal axis value calculated, and the result plotted to see if it falls on the dotted line drawn by projecting or extending the line connecting the other two plotted points. If it does not, a new 700-hp temperature is assumed and the process is repeated.

The shortcoming of this method for such a large projection in load is evident: very small errors in either test can make a big difference in the result.

---

power squared). For a motor in which 15% overload means 13% or 14% higher amperes, this would mean that temperature rise would go up about 29%; 1.29 times 80% = 103, or 23° higher rise. This is a little closer to the tested average, and still conservative.

(3) If you know the actual losses in the motor at any load for which temperature has been measured, then use the method in Table 6-IV. This is within 90, so for a class B insulated motor a service factor of 1.20 could be allowed. (Method 1 would give a T_2 of about 94; Method 2 about 88.)

At the test condition	At the overload condition
Add: CU_1 = 1.15 times stator I^2R	(overload)^2 times CU_1
plus	plus
CU_2 = rotor I^2R	(overload)^2 times CU_2
plus	plus
FE = .75 times core loss	FE
plus	plus
STR = stray load loss	(overload)^2 times STR
Sum = original effective heating watts	Sum = effective heating watts at overload
Measured winding rise = T_1	New winding rise = T_2

The new winding rise $T_2 = T_1 \left(\dfrac{\text{new effective watts}}{\text{original effective watts}} \right)$

Example (200-hp, 2-pole rating):
  Under test, 200 hp

CU_1 = 1.15 times 4000 watts
     = 4600
CU_2 = 2000
.75 FE = 2500
STR = 1500
Total watts = 10,600

Loaded to 240 hp
(20% overload)

(1.2)^2 times 4600 = 6620
(1.2)^2 times 2000 = 2880
.75 FE = 2500
(1.2)^2 times 1500 = 2160
Total watts = 14,160

Test rise $T_1 = 64°C$

New rise $T_2 = 64 \; \dfrac{14,160}{10,600} = 85.5°C$

Table 6-IV. Method for predicting how hot an overloaded winding can get.

This procedure allows for several possible inaccuracies in the first two methods. One is that the heat produced within the stator coils themselves has a greater effect on their rise than heat produced elsewhere in the motor. Another is the lack of variation of iron loss with load.

Besides permitting judgments about overload capability, tested variation of temperature with load and with time can reveal certain motor deficiencies. For example, in Figure 6-44 (page 301) the time scale for stable rise may stretch out to as long as 6-8 hours for some motors. If after that much time the temperature is still rising, the load is probably exceeding the motor's capacity. There are simply too many watts of heat loss being generated for the cooling system to throw off. (See Figure 6-49.)

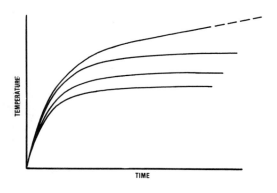

Figure 6-49: Each extra degree of temperature drives winding resistance up, in turn increasing loss, which raises temperature and resistance still more. This tail-chasing process becomes "thermal creep," "thermal piling," or "runaway"—the winding temperature never levels off, continuing to rise indefinitely until burnout. This occurs when the motor's cooling system is completely overloaded, as in the upper curve. The lower curves represent steadily increasing loads which are still within the motor's cooling capability.

It is then necessary to drop the load until temperatures do level off and then look for clues to possible defects in cooling. For example, it might be expected from similar machines that a curve of temperature versus load like Figure 6-50a might apply. Instead, tests show something like Figure 6-50b. This points to some blockage or restriction in the cooling air. Look for vent spaces plugged with varnish, incorrect or missing fans, and so forth. Make sure the direction of rotation is correct for proper fan action.

Figure 6-50: If tests at different loads show a pattern such as (b), when (a) was expected from similar designs or earlier test, motor ventilation is probably restricted. The motor runs hotter than it should regardless of horsepower. But if (c) is observed, the problem is more likely to be a heat load too great for the parts—in other words, the motor is being "worked too hard" at the higher loads.

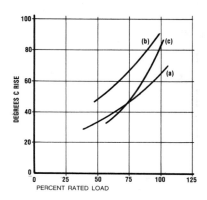

Another clue is the difference in temperature between air entering and leaving the motor. This "air rise" will typically be about one-third of the winding rise. If the air is getting much hotter than that passing through the machine, it is a warning that air flow is insufficient to do the job. Thus, knowing how to interpret temperature tests and make predictions from them is often just as important as knowing how to take the readings correctly.

7 Bearings and Lubrication

WHEN APPLYING MOTORS, many engineers understandably think of them only as electrical machines. Because a motor has major moving parts, however, it is also subject to the strictly mechanical problems of all rotating machinery. Most such problems involve what is widely agreed to be the single most common cause of motor failure: the bearings.

So much has been written about bearings that a comprehensive treatment here is neither possible nor necessary. Literature, from maintenance bulletins to complete textbooks, is readily available from several bearing manufacturers.

However, that literature is general. Bearing selection and use in electric motors imposes specific requirements not always clearly understood. Remember that motor bearings are unlike many others in common use, such as automotive wheel bearings. Rotating at 700 or 800 rpm, the wheel bearing may see 30 million revolutions during 50,000 miles of driving. In a 4-pole industrial motor, running two shifts daily five days a week, the bearings may reach two *billion* revolutions at 1780 rpm in less than five years.

Bearing standards

What determines how long a bearing lasts? Why are there so many different types of bearings, and how should each be used? In the NEMA integral horsepower motor sizes, standard horizontal machines use ball bearings. Sleeve bearings are sometimes an optional alternative, but motor delivery will be long and the price high. From about 400 hp up to at least 2000 hp at 1800 rpm, machines can be furnished with either ball or sleeve bearings. Still larger ratings use sleeve bearings as standard, with optional anti-friction (ball or roller) bearings available when end thrust or belt pull cause high bearing loading.

The reason for these practices is primarily economic. Small ball bearings are inexpensive, quickly replaceable, and highly reliable. As bearing size increases, anti-friction bearing cost rises rapidly. Large sleeve bearings are less expensive. They are also less sensitive to dirt, or to loose tolerances in shaft and housing machining (except at high speeds or very heavy loads).

"Standard" bearing size selection in horizontal motors varies from one motor manufacturer to the next. Table 7-I is typical of one supplier's practice.

2 pole	4 pole	6 pole
3: 203-205	2: 203-205	1: 203-205
7½: 205-206	5: 205-206	2: 205-206
15: 206-207	10: 206-207	5: 206-207
25: 207-309	20: 207-309	10: 207-309
40: 207-311	30: 207-311	20: 207-311
60: 210-312	50: 210-312	30: 210-312
100: 213	75: 213	50: 213
150: 214	125: 214	75: 214
400: 216	400: 216, 318	300: 216
800: 316	900: 316	500: 316
1500: 3½×4	1500: 318	1750: 326
3000: 3½×4½	2500: 320	3500: 330
5000: 4×5 φ	4000: 322	6000: 6×7
7000: 5×7 φ	8000: 6×7 φ	

φ requires forced-oil lubrication

Table 7-I. For standard open motors, one manufacturer normally offers these bearing sizes for direct-coupled service. Motors designed for belting, flange-mounting, or other mechanical specialties will exhibit many variations. In each column, the numbers at the left are the maximum horsepowers corresponding to the bearing sizes at right. Where two bearing sizes are given, the first is usually at the motor's front or outboard end; the second, at the shaft extension or drive end.

The question of which type of bearing should be used, sleeve or anti-friction, has no universal answer.

If the surroundings are fairly clean, or the operating speed high, with little or no load applied to shaft and bearings, then sleeve bearings are the likely choice. But if large shaft loads are present, especially of a cyclic or vibratory nature, then anti-friction bearings are preferred.

Vibratory loads are troublesome for sleeve bearings because of the large internal clearances necessary to maintain the oil film. These clearances will be from two to four times the "looseness" present in anti-friction bearings. As the shaft is moved back and forth by an external vibrating load, then, the sleeve bearing tends to fail by being "pounded out."

How often the motor starts is also important. Sleeve bearings require a film of oil between bearing surface and shaft journal for wear-free operation. That film is not present when the motor is standing idle. At each start, the film is re-developed by the rotating journal, and for a short time until that occurs there will be wear in the bearing. Ball or roller bearings are not subject to that kind of wear, so are the better choice for frequent starting.

A great many different kinds of ball bearings exist for specialized machine applications. In motors, the most common version is the Conrad-type "deep-groove" radial bearing, shown in Figure 7-1. The deep ball groove provides sizeable thrust capacity, typically ⅔ to ¾ of the bearing's radial load rating.

What do we mean by "radial" and "thrust" loads? Radial means at right angles to the shaft. If the motor shaft is horizontal, the load may be up, down, left, or right; as long as it acts at right angles to the shaft centerline, it is still considered radial.

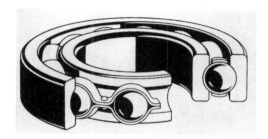

Figure 7-1: A typical Conrad ball bearing. One basic size range includes several different outside diameters for a single shaft diameter—the largest outside diameter being a "heavy" series; a somewhat smaller diameter, a "medium" series, with less load-carrying capacity; etc. A second grouping includes different inside diameters—again, heavy, medium, light, etc.—for the same outside diameter (O.D.). Grease lubrication is normal.

A thrust load acts parallel to the shaft. In a horizontal motor, any pushing or pulling force along the shaft axis is a thrust load. In a vertical motor, only upward or downward force—including the weight of the rotating parts themselves—is a thrust load.

Ball bearings and some types of roller bearing will withstand loads made up of both radial and thrust force components. It's wise to remember this: A ball bearing designed for thrust loading will normally carry some radial load safely. The reverse is not necessarily true, certainly not to the same degree.

Carrying thrust loads

When either radial or thrust load exceeds the capacity of a Conrad bearing, either a double-row ball bearing or a roller bearing may be supplied instead. The straight roller version, shown in Figure 7-2, is common for belt or gear loading. The spherical roller bearing, shown in Figure 7-3, is inherently self-aligning to accommodate some shaft bending, so it is best for the highest belt loads which cause such deflection. It is not suitable for 3600 rpm (which is too high a speed for belted motors above about 30 hp, anyway), nor is it suitable for 1800 rpm in the larger sizes.

Vertical shaft motors, often used on deep well pumps, almost always include ball or spherical roller thrust bearings. (See Figure 7-4.) When thrust load or motor rpm exceed the capability of such an assembly, the Kingsbury or plate-type thrust bearing, shown in Figure 7-5, is used. High cost and complex construction rule it out for integral horsepower (NEMA frame size) machines.

This is basically a sliding bearing. Think of it as a sleeve bearing that has been rolled out flat instead of being cylindrical. Because of its large load-carrying surfaces generating extra heat, the plate-type bearing almost always requires external cooling of its lubricating oil. This usually dictates an oil-to-water heat exchanger consisting of water cooling coils immersed in the bearing oil chamber. When that isn't possible, a circulating coolant/pump assembly must dissipate the heat either to outside water or outside air. (See Figure 7-6.)

Figure 7-2: Left: Available in similar size progressions, having higher load capacity because of greater contact area between rolling and stationary elements, this cylindrical roller bearing is popular in belted motors. Either oil or grease lubrication is common.

Figure 7-3: Right: The next step up in radial load capacity, with tolerance for shaft deflection, is the oil-lubricated spherical roller bearing.

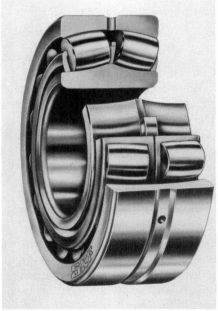

Figure 7-4: Left: The spherical roller principle adapted to vertical-shaft motor bearings that will sustain a high thrust load.

Bearings and Lubrication

Figure 7-5: Right: Available in sizes from about 7 inches to several feet in diameter, this plate-type thrust bearing contains a circle of separate highly-finished stationary "thrust shoes" to which load is applied by a rotating runner, or "thrust collar," attached to the motor shaft. In this view, two of the shoes have been removed. Each is supported by a swivelling mounting structure, allowing the shoe to tilt so a wedge-shaped oil film may develop during rotation as in the horizontal sleeve bearing.

Figure 7-6: Left: Bearing oil cooling was necessary for these high thrust vertical motors in a power plant (foreground). But a circulating water supply was not available. So a closed cooling system was installed, using a separate pump, fan, and radiator, to dissipate bearing heat to the surrounding air.

This is costly, lowers overall motor reliability, and may require some special means of furnishing cooling water where none would otherwise be needed. (Spherical roller thrust bearings also may require cooling, especially at 1800 rpm.)

It is not uncommon for users or consultants to specify the plate-type bearing, not because it is needed to carry the load, but only to yield unusually long bearing life. That is a commendable objective. But the choice may result in a bearing far outlasting other motor components, and the energy cost of operating the bearing can be surprisingly high.

A 1979 IEEE paper gave an example of such an operating cost penalty. Assume a 1500-hp 6-pole motor, having a normal full load efficiency of 95% as a horizontal unit, but built vertical-shaft for 25,000 pounds continuous downthrust load. A spherical roller bearing suitable for that thrust would reduce full load motor efficiency to .947. If a 6-shoe plate bearing were used instead, the efficiency would become .942. Difference in loss between those two bearings is about 5 kilowatts. (See Figure 7-7.) If the motor runs fully loaded only half the time, on power costing 5 cents per kilowatthour, the extra operating cost for the plate bearing design is over $1,000 annually. Depending upon interest, inflation, and tax rates, over the expected motor life span that could amount to equivalent capital investment of $20,000 or more. (See Chapter 2.) Besides that, the plate bearing's first cost is substantially higher.

Here's another example. Consider an 800-hp, 1800-rpm motor subjected to 10,000 pounds downward thrust. A spherical roller thrust bearing under that thrust load would develop 4.7 kilowatts loss; a plate-type bearing, 14.1 kilowatts. The motor runs 7000 hours per year. The difference of 9.4 kilowatts, with power cost at 4 cents per kilowatthour, would raise annual operating cost by $2,640. So specifying more bearing than you need can be expensive.

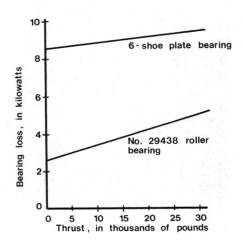

Figure 7-7: How bearing heat loss, caused by internal friction in the oil, varies with thrust load for two types of bearing usable in a 1500-hp, 1200-rpm vertical motor.

The smaller vertical motors will use the angular contact thrust bearing as standard. Its outer race has a high shoulder on one side of the ball path or raceway, a low shoulder on the other. (See Figure 7-8a.) The normal line of contact between balls and races forms a large "contact angle" with the race centerline. That permits the bearing to carry a large thrust load. When that load becomes too great for a single bearing, others may be added to it, two or even three high. (See Figure 7-8c.) Total thrust capacity of such a "stacked bearing" increases as $N^{0.7}$, N being the number of bearings in the stack.

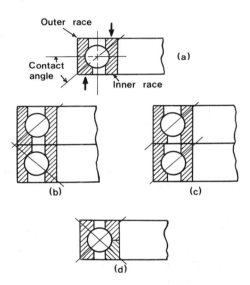

Figure 7-8: Various types of angular-contact ball thrust bearings. Basic bearing construction, at (a), permits thrust in the direction of the arrows because of the high outer race shoulder below the raceway. At (b), two such bearings are assembled together with the high shoulders adjacent. This "back-to-back" assembly will carry thrust load of the same amount in either direction. At (c), one bearing is turned around in a "duplex tandem" assembly to carry increased thrust in one direction only. The bearing at (d) has a split inner race, allowing it to be assembled despite high shoulders on both sides of the raceway. It will carry thrust in both directions without the need for a second bearing (unless the load is too high).

Normally, whatever its type, the thrust bearing is at the upper end of a vertical motor. This makes it easy to get at for servicing; the thrust bearing, carrying the heaviest load, is likely to need the most maintenance. Its larger size also favors assembly at the top. To keep the rotor-shaft assembly properly centered within the stator, a lower "guide bearing" is provided. With a plate-type thrust bearing, there will also be an upper guide bearing, because otherwise the top end of the shaft could wobble from side to side.

Guide bearings normally don't carry much load. However, pump operation subjects many vertical motors to upward, as well as downward, shaft loading. Because such "upthrust" is often present only momentarily, rather than continuously, and seldom exceeds one-third of the downthrust, a radial ball lower guide bearing can take the upthrust if locked to the shaft.

Another method of carrying upthrust eliminates the extra parts needed to lock the guide bearing. Some motor manufacturers use the "split inner race" thrust bearing, shown in Figure 7-8d, with high shoulders on either side of the raceway in both races. This provides the high contact angle, and thrust load capacity, in both directions. (A similar arrangement uses two angular contact bearings stacked "back to back," shown in Figure 7-8b.)

Such a bearing can be assembled only if the inner race is split in the middle. Like the other type of angular contact thrust bearing, this may be used singly or stacked in combination with standard angular contact bearings. Stacking two or more split-race bearings to increase the thrust capacity in both directions is not recommended, however, because doing so presents lubrication problems.

When ordering a vertical pump motor, let the motor manufacturer know if the pump thrust is to be "balanced" for more than a few seconds at a time. That can happen with condensate booster pumps, for example, which may run lightly loaded for long periods. If neither upthrust nor downthrust is present, the motor shaft can "float." Running that way can cause a split-race thrust bearing to fail quickly due to ball skidding.

Statistical life prediction

Whatever the anti-friction bearing type or the nature of its loading, its useful life span is determined by a combination of operating speed (rpm versus size) and the load. Bearing operation depends upon rolling contact between ball or roller and the raceway. The result is continual alternating stress in those parts, leading eventually to fatigue failure of the steel—regardless of lubrication (or the lack of it). The heavier the load, the faster the bearing parts move, and the sooner something will break.

"Bearing life" is the operating time prior to occurrence of such failure. There is no way it can be predicted except as a statistical average for a large number of similar bearings.

The prediction takes two forms. First, the "minimum life" for any bearing is the time at least 90% of a group of such bearings will run before showing evidence of fatigue breakdown. This is the so-called B-10 or "L_{10}" life. It is not a minimum in the true sense of the word, because any bearing in the group could in theory fail during the first minute of its operating life—though that is highly improbable. (See Figure 7-9.)

Says one bearing manufacturer, "It is not possible to predict the life of an individual bearing, nor the minimum life which every member of a group of bearings will equal or exceed under a given load." This means that under the best of conditions it is to be expected that one bearing out of every ten will fail before the L_{10} life. Every bearing user should understand this. It is an unavoidable consequence of the nature of bearing manufacture and the mechanism of bearing failure.

Another way of looking at it is in terms of "reliability." We could say that any given bearing is "90% reliable" at its L_{10} life. As Figure 7-10 shows, the reliability becomes very much greater for that same bearing at an operating time well below its L_{10} life.

Remember: That degree of reliability, and the L_{10} life itself, are valid only if

Bearings and Lubrication 317

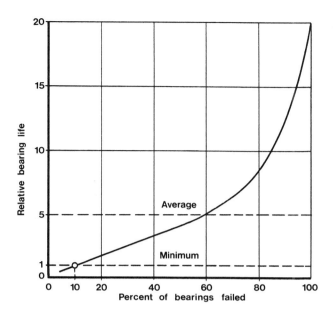

Figure 7-9: This "life dispersion curve" illustrates the statistical nature of L_{10} bearing life. For the required "large group" of bearings of the same design, 10% may fail within the L_{10} minimum life span, plotted as relatively one unit of time on the vertical scale. When the "average" life span has been reached, five units in duration, as many as 60% of the group may have failed. A small number of survivors may last 15 to 20 times the L_{10} life. Remember: There is no way to predict where on this curve the life of any one bearing in the group will fall.

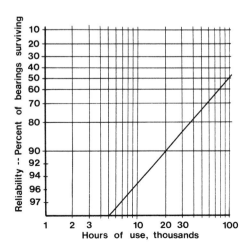

Figure 7-10: A reliability plot like this is a way to express individual bearing life. Any one bearing has a 90% probability of lasting out the L_{10} period, which for this bearing design happens to be 20,000 hours. For only 5000 hours of use, any individual bearing is 98% reliable—that is, it has only a 2% chance of failing by that time.

the basis for the L_{10} life prediction is accurate. If actual bearing load is greater than predicted, the L_{10} life will be reduced. Even a 10% increase in load will reduce life by nearly one-third. Loss of lubrication, improper mounting, or dirt in the bearing—all these can invalidate any life span forecast.

A second predictable bearing life is the "average life." This is the average time-to-failure for each bearing in the statistical group, and is taken as five times the minimum life. (Testing has shown the ratio is probably closer to seven times.)

Both minimum and average bearing life vary predictably with the amount and type of load (radial, thrust, or combination of the two). The prediction data, found in bearing manufacturers' catalogs or handbooks, are based on a national standard developed by the Anti-Friction Bearing Manufacturers Association (AFBMA) and adopted by ANSI as Standard B3.15-1978.

The methods are complex. For a typical ball bearing, they involve calculation of a theoretical basic load rating in pounds, C, which a "group of apparently identical bearings" will endure for one million revolutions. Then, the L_{10} life of such a bearing is expressed by $(C/P)^3$ in which P is the "equivalent radial load" and is equal to $(XF_r + YF_a)$. The factors X and Y vary with bearing type and dimensions; F_r and F_a are actual radial and thrust loads on the bearing in pounds. (See Figure 7-11.)

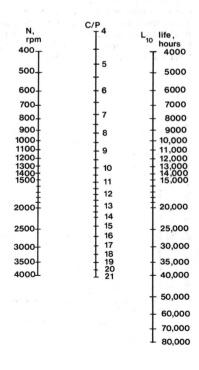

Figure 7-11: Typical of the guides found in bearing handbooks, this nomograph relates bearing rpm, N; basic load rating, C/P; and the resulting L_{10} life. Suppose you want a 20,000 hour life for a bearing carrying no thrust load, but a radial load of 1000 lbs. Speed is 1200 rpm; shaft diameter, about 3 inches. A straight line connecting 20,000 hours with 1200 rpm on this graph intersects the C/P line at about 11. For conventional motor service (bearing outer race stationary), radial load P = 1.0 times the 1,000 lbs. Therefore, C must be C/P times P, or 11,000. A 216 bearing would do the job.

Because C is a constant for a given bearing, it is evident that the basic relation between actual load and bearing life is:

$$\frac{\text{Life at load } P_1}{\text{Life at load } P_2} = \frac{(P_2)^3}{(P_1)^3}$$

Just as with insulation and its rapid deterioration when a winding overheats, overloading a ball bearing reduces its life expectancy quite sharply. Similar relationships apply to the various types of roller bearing.

Unless a horizontal motor is ordered for some specified bearing loading condition, no stated bearing life is expected to apply. Rigidity and strength of the shaft itself lead to bearing sizes large enough to ensure many years of bearing life. But when the application imposes special radial or thrust loading, the user should understand what bearing life the manufacturer will use in selecting bearings. A common practice is to furnish an L_{10} capability of either one or two years, larger or more critical motors getting the higher figure.

Average life, therefore, will be 5 to 10 years at the specified loading. If the motor runs only half the time, service life may average 10 to 20 years. Some motor specifications call for L_{10} life of 100,000 hours or more (at least 12 years). If the motor is a vertical unit carrying high downthrust (especially at 1800 rpm), an angular contact ball bearing must usually give way to a water-cooled spherical roller thrust bearing at a much higher cost. The result is likely to be a bearing life far longer than that expected of the stator winding. This is a further example of how it may be not only costly, but also irrelevant to motor reliability, to insist on extremely long bearing life.

Whereas bearing *rating* methods are standardized, bearing *usage* is not. One motor supplier may use cylindrical roller bearings for heavy belt loads; another may prefer double-row ball bearings. Lubrication methods may be different for the two. One manufacturer may clamp bearing races with snap rings; another, with threaded locknuts. One may use the 200 series ("light") ball bearing in a given application; another, a 300 series ("medium"). No standards govern such decisions.

What about the "new developments" in anti-friction bearings that have been widely advertised? In general these are improvements in the metallurgy of the steel used for races, balls, and rollers. The first one of modern times, about 1965, was the vacuum-degassing process. In a vacuum, entrained contaminating gases in the molten steel can be drawn off, reducing chemical impurities that caused weak points in the bearing structure. The result is greatly increased bearing life.

ANSI B3.15-1978 includes a "life adjustment factor for material" to reflect such benefits. When the advantages of vacuum-degassed bearing steel became clear, for example, the standard used a factor of 3 to account for the increased life. More recently, such new metallurgical techniques as "precipitation deoxidation" and "ladle refining" have further improved the "cleanness" of bearing steel, offering still greater increase in bearing C/P rating.

Theoretically, longer life at a given load allows a bearing size reduction for many motor applications. However, because in standard horizontal machines the limiting factor is usually shaft diameter, the tendency has been to retain the same bearing sizes but offer L_{10} bearing life several times greater (though not all bearing suppliers recommend the practice, which wasn't used in the example of Figure 7-11).

Oil or grease?

For any bearing construction, when we speak of "oil" versus "grease" lubrication, we aren't really talking about two entirely different materials. Only oil can lubricate a conventional motor bearing (we're not concerned with those very special designs using water-lubricated bearings). Grease is simply a convenient means of supplying oil in a non-liquid form. In either form, the lubricant serves these purposes:

(1) To maintain a film of oil between the rotating and stationary surfaces within the bearing, thus minimizing friction between them.

(2) To cool those surfaces so that friction heat causes no damage to the parts or to the lubricant itself.

(3) To help flush out microscopic particles broken away from bearing parts by wear or high surface stress.

(4) To prevent corrosion.

(5) To seal out dirt or chemical contamination.

The amount of lubricant necessary to maintain an oil film on the bearing parts is extremely small. Less than one-thousandth of one drop of oil at the right viscosity can properly lubricate a ball bearing on a 2-inch shaft running at 3600 rpm. (That's why an oil mist system, as described later in this chapter, is adequate for most motor bearings, although not always affordable.)

Because of its convenience, grease is the choice for most industrial motors. Grease lubrication has the further advantage of providing an inherent seal against bearing contamination. The close-running clearance spaces between shaft and bearing cap or housing are filled with grease at initial assembly. At each subsequent regreasing, some grease is forced outward into those same areas, and "purging" of excess grease from the bearing while the motor runs also produces this effect. That movement pushes dirt out so that it cannot enter the bearing.

Wherever motor operators congregate, the three most persistent questions about grease lubrication are:

(1) How often should bearings be greased?
(2) How much grease should we add?
(3) How should we add it?

Disturbing though it may be in this scientific age, the answer to each question is the same: No exact answer is possible. What works in my motor, in my plant, in my machine, may not work as well in yours.

To understand why, consider what conditions affect each of these three situations. A bearing needs grease addition—better described as grease replacement—when the original lubricant is no longer doing its job. We say that the grease has reached "the end of its useful life." That can occur in many ways, some of which are:

(1) Hardening of grease such that it no longer freely "feeds" oil to bearing surfaces. This can result from absorption of dirt, for example.

(2) Chemical breakdown because of excessive heat. High winding temperature, especially in totally enclosed motors, can accelerate the process.

(3) High bearing load, as in some belt drives.

Most such degrading influences are present only while the motor is running. To reach proper conclusions, then, you should know the operating cycle. Does the machine run fully loaded about the same number of hours each week, each quarter, every six months? Will your lubrication schedule, including some tolerance, result in a bearing being re-greased more than once with only a few hours of operation intervening—or in a long period of hard service with no grease addition at all?

Like most other physical deterioration—like the expiration of bearing life itself—loss of grease life is a gradual process, and it is a statistical process. Grease does not cease abruptly to be an effective lubricant at the same instant in all "identical" bearings.

When to add grease

Given all the operating conditions, grease life expectancy can be calculated. But defining all those conditions for each motor in a plant isn't practical. Many lubrication specialists have tried to do that, for "typical" operation, with widely varying results. To list just two recommendations:

(1) NFPA Standard 70B, "Electrical Equipment Maintenance," sums it up this way: "The normal frequency range [for grease relubrication] is from one year to five years depending on the environment and shaft speed." Where to work within that range depends on whether the bearing load is "moderate" or "heavy," and whether the environment is "typical industrial" or involves "heavy dirt."

(2) A leading bearing manufacturer's "Guide to Better Bearing Lubrication" gives relubrication interval in hours as a function of bearing diameter and speed, provided that "loading conditions are normal." These published intervals should be halved for every 15°C by which bearing outer race temperature exceeds 70°C (what if it is 75°?). But, according to the text, "...relubrication intervals may vary significantly even where apparently similar greases are used" and "when there is a definite risk of the grease becoming contaminated the relubrication intervals should be reduced." (The difficulty of measuring bearing temperature in the field will be touched on later.)

Motors in extremely wet, dirty areas need frequent relubrication simply to keep damaging moisture and contamination from building up in the bearings. Salt, for example, draws moisture out of the air, resulting in rapid ball bearing corrosion. The electrical maintenance supervisor in a salt processing plant recently said, "One 400-hp motor needs lubrication every week to keep the water out, but most of them only once or twice a year.... We have standardized on a polyurea/lithium grease. When we get a new motor in, we take it apart, remove the grease that's in there, and substitute our own, filling the bearing cavity about 40%. During the off-season, when we pull motors out to dip and bake them, we change the bearings ... mostly every five years.... It depends on how critical the drive is."

All such recommendations concur that the higher the speed and larger the bearing, and the more dirt or moisture is present, the more often a bearing needs grease. Figure 7-12 is an attempt to combine numerous published suggestions. It's evident that no "magic numbers" exist. There is no substitute for keeping careful, complete records of bearing and grease condition versus relubrication interval until it's clear that your interval works for your situation. Use running time, not calendar days, to measure that interval.

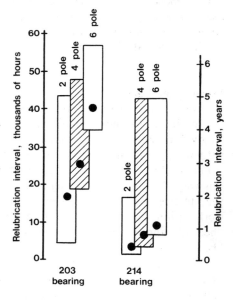

Figure 7-12: For just two frequently used motor bearing sizes, this chart indicates the range of re-greasing frequency recommended by a number of published sources. The black circles represent a "weighted average"—the period most often suggested. (The 203 bearing would be typical for a 1-hp, 1800-rpm open motor; the 214, for a 125 hp motor.)

Do not be afraid to set the interval "too long" rather than too short. About 40 years ago, one user said, "we were greasing motors every four months regardless of horsepower rating and speed. Later the greasing frequency was extended to six months and a few years later extended to one year." Finally, the interval was made three years up to 7½ hp, two years for 10 to 25 hp, one year for 30 to 100 hp, and twice a year for 125 hp or larger. It appeared likely still further increases were possible.

Bearings and Lubrication 323

Re-greasing methods

Authorities concur that more bearings fail because of over-lubrication than from under-lubrication. There are a couple of reasons for that. One is that pumping new grease into a bearing, even under the best conditions, all too often pumps in some outside dirt as well. Another frequent source of trouble is either the addition of too much grease, or leaving too much of the old grease in the bearing. That leads to escape of grease into the motor interior, rapid contamination or breakdown of the new grease, or excessive churning (and heating) of grease in an over-filled bearing.

This introduces the second of our three common questions: How much grease should be added? Again, many rules of thumb exist. One standard recommends various grease amounts by weight for each direct-coupled motor frame size up through the 449. But for larger frames, or belted motors, you're on your own. Others suggest a grease weight based on motor shaft extension diameter.

If we combine these recommendations for the two typical bearing sizes of Figure 7-12, we find:

Bearing number	Ounces of grease to add
203 (I.D. 0.67 in.)	0.1 to 0.5
214 (I.D. 2.76 in.)	0.4 to 1.5

The variation of 4 or 5 to 1 in these figures is not consistent. That is, the authority prescribing the lowest quantity for the small bearing is not recommending the lowest quantity for the larger bearing.

So the "right" amount—like the proper interval—must be determined by experiment. Find out what works for you. One of the first steps is to find out what weight of grease is discharged by each stroke of the grease guns used. These tools are not weight-calibrated. How the grease is added can definitely affect the amount involved, which brings up the third common question. Most often meant by "How should I add grease?" is really:

(1) Must I stop the motor to grease the bearings?

(2) Do I need to open a drain to let the old grease out? When I do, it doesn't seem to work, so what's the use?

The reason users and repairmen ask question (1) so often is that no single answer is always correct. Why? Because the bearing assembly itself, the seals, the grease path through a bearing chamber when grease gun pressure is applied, the grease chemistry itself, and numerous other details, vary widely from one motor to the next. For every manufacturer's instruction manual that says "Stop the motor when re-greasing," you will find at least one satisfied operator (and perhaps at least one other instruction book) that says you don't have to stop the motor.

This contradictory advice may seem frustrating. But consider how grease behaves in a confined space. The small clearances between moving and stationary bearing parts, discs, seals, shaft surfaces, and so forth can form what have been called "viscosity pumps." Relative motion between parts acts to push grease through such spaces. Internal bearing cap fins, vanes, or even rotating shaft impellers may be present to aid in distributing new grease and purging the old. (See Figure 7-13.) Stop the motion, and the pumping action stops with it.

Similar behavior has been put to good use in some oil-lubricated bearing assemblies, wherein lubricant is "lifted" through the bearing by nothing more sophisticated than a spinning disc. This is common in some large European motors. The principle of oil ring operation in a simple sleeve bearing is closely related.

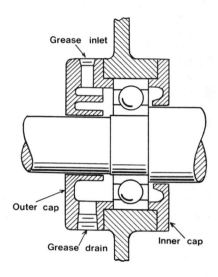

Figure 7-13: Most greasable bearing assemblies, like this one, have both inlet and drain on the same side of the bearing. Here, the outer cap (left) contains a "directing passage" under the inlet plus a "grease control ring" around the shaft to improve grease distribution during relubrication and normal running.

So if the bearing assembly in your motor happens to provide the right sort of viscosity pumping action, the proper flow of grease to purge old lubricant and make way for new may occur effectively with the motor running. Stopping the machine may hinder this action.

But in other assemblies, the running motor may push grease rapidly along the shaft and into the windings. If stopped, the motor loses this pumping action, avoiding the leakage. Here's how one maintenance engineer described the problem: "In one case grease being slowly pumped into a ball bearing housing passed to the inside [of the motor] even though a plug had been pulled from a one-inch diameter opening in the center of the bearing housing." With motors running, old grease simply would not drain from the bearing chamber in preference to flowing along the shaft into the motor interior. The only cure was to shut down the drive while lubricating.

In some motors, of course, proper use of grease fittings for relubrication exposes the worker to moving parts, or may even require some parts to be removed from the motor. (See Figure 7-14.) But most designs avoid such problems.

Bearings and Lubrication 325

Figure 7-14: Removing a grease inlet plug preparatory to relubricating a small fan-cooled motor. To get at the fittings, the fan guard had to be removed. Such construction requires stopping the motor to re-grease bearings.

(Always stop an oil-lubricated motor to add oil to a bearing chamber. Bringing the oil up to the indicated level in a sight glass (shown in Figure 7-15) may result in overfilling if done when the motor is running, because of level changes produced by air pressure differences or movement of oil within the chamber.)

Part (2) of the "how to add grease" question is more straightforward. Both to avoid overgreasing and to allow deteriorated grease to escape, the housing drain should always be open when a motor is re-greased. If the grease is so hard it will not escape from the opening, then clean it out with a rod or flush the housing with oil.

Do not rely on automatic "pressure relief" drain plugs.

A ball or roller bearing chamber should not be filled more than 30% to 40% full of grease. When close-fitting seals prevent the bearing's purging, a more complete filling will result in too much churning of the grease. The lubricant will overheat and oxidize. When the atmosphere is extremely wet or dirty,

Figure 7-15: This sleeve bearing assembly is equipped with both an external sight glass and a constant-level oiler (right). The arrow mounted on the bearing bracket shows where the correct oil level should appear in the glass.

seals are looser to allow escape of excess grease, and re-greasing is frequent, the bearing chamber can be more completely filled. This is especially desirable, to exclude contamination, when a motor must be shut down in dirty surroundings.

Choosing a lubricant

Beware of mixing greases. Beware also of claims that a single "universal" grease is usable everywhere. Petroleum producers use complex additives in modern lubricants to enhance lubrication properties, chemical/moisture resistance, temperature resistance, etc. These may be incompatible between different makes. Mixing such greases may eventually thin or thicken the mixture such that it can't function properly. But simply mixing samples together and observing the result, without a long period of operating time in bearings, isn't a reliable test.

If different greases must be mixed in a motor bearing, one handbook states, "the application should be closely observed to permit correction of any lubrication problem before damage to the bearing." This can be risky. It's better to simply purge the old lubricant, put the new in, and increase relubrication frequency temporarily.

The motor manufacturer's lubricant recommendation is based on general conditions. If your experience indicates that it isn't suitable, discuss this with the manufacturer to seek a solution. Among the operating conditions important to such a discussion are:

(1) The nature of the surroundings—dirt, chemicals, steam, etc.
(2) The kind of bearing loading.
(3) The actual operating temperature.

Many users have found the often-recommended "EP" (Extreme Pressure) lithium-base petroleum grease works well. As the usage of Class F winding temperature ratings has increased, however, others have adopted synthetic greases to withstand higher bearing temperature.

Plant operators have been told for years about the economy of purchasing lubricants in large quantities. Considering only the material cost seems to justify such purchases. However, it has been estimated that the cost of applying grease to bearings is typically ten times the cost of the grease itself. One plant took that into account in a lubrication study which showed that the lowest overall cost involved 14½-ounce cartridges as the basic unit of grease supply. Yet the cost of the grease itself in cartridges was double that of 400-pound drum lots. The material cost differential was more than overcome by the saving in time when grease guns did not have to be taken back and forth to a central supply for refilling.

More important is assurance of grease cleanliness. Said one bearing expert, "I never saw one of those big drums with the lid on it." Each time it is used, some dirt probably gets into it, and contamination may be general long before the bottom of the drum is reached.

Shields versus seals

Should shielded bearings be used? What is a shield? Which way should it face in the assembly?

A shield or "grease plate" is merely a stationary disc attached to the bearing outer race that leaves only a small annular opening around the inner race for access to the bearing interior. The bearing balls act as miniature centrifugal pumps, able to draw grease in through that annulus and expel it at the outer race on the opposite side of the bearing. Note the arrows in Figure 7-16.

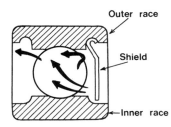

Figure 7-16: A "single-shielded" ball bearing. The shield (right) is attached to the stationary outer race. As the bearing revolves, the action of the balls tends to move grease in the direction indicated by the arrows.

Such a bearing construction serves a number of purposes. First, new grease can be forced into the bearing, past the shield, during relubrication. Second, the bearing will purge itself, by means of the pumping action, of excess grease which can escape through the unshielded side. Third, the shield minimizes entrance of contaminants. Fourth, the bearing can be mounted so that the shield serves in place of a bearing cap, to prevent grease leakage into the motor interior.

Not all those functions, of course, can be performed by the same type of bearing assembly. One argument is to put the shielded side towards the motor interior, to minimize grease leakage in that direction. A counter-argument is that this promotes overfilling of the bearing during relubrication. If single-shielded bearings are furnished, the motor manufacturer adopts one method or the other, and unless user operating experience clearly supports a change, the original arrangement should be preserved. But there is no universal standard assembly.

Here are some common arrangements and the reasons for them:

(1) Single-shielded bearing (as in Figure 7-16)—shield facing the outside of the motor. This assembly is common when an inner bearing cap is used (as in Figure 7-13) to hold the bearing in place. That cap acts as the grease seal to prevent leakage, and that side of the bearing is left open so the bearing can purge excess grease into the cap reservoir.

(2) Single-shielded bearing—shield facing the motor interior. This may be expected when there is no inner cap. The shield is needed to prevent internal leakage.

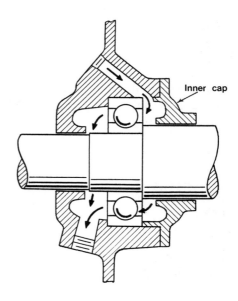

Figure 7-17: A "flow through" bearing lubrication system, in which grease enters one side of the bearing and leaves from the other. This arrangement is claimed to aid removal of old grease from the bearing during relubrication. Wide cap fit around shaft (right) is needed to keep grease gun pressure from forcing lubricant into motor.

(3) Unshielded bearings may be used when the lubricant system is arranged for "flow through" greasing as in Figure 7-17.

Also available are double-shielded bearings; bearings having a shield on one side and seals on the other; etc. The type selected must suit the lubrication system.

"Sealed" and "shielded" ball bearings may look the same on the outside, but they are not. A sealed bearing is made to prevent the inflow of lubricant (and dirt) as well as its escape. Such is the construction of the "prelubricated" bearing widely used years ago in NEMA-frame machines. It is still available, but to most users the advantages of being able to supplement or replace grease over the years outweigh the disadvantages.

In a typical "four way seal" bearing of this kind, there is a pair of sealing rings on each side of the bearing. Each is formed and attached like the shield of Figure 7-16. However, the inner ones are stationary and attached to the outer race; the outer seals rotate and are mounted on the inner race. The spaces between the two seals, and between seals and races, are quite small and grease-filled, so that the interior of the bearing is inaccessible to either contamination or relubrication. A "double sealed" bearing uses only one seal on each side, having a turned-in lip fitting closely over "seal grooves" cut in the inner race.

Today, such bearings are almost never used in motors manufactured by U.S. firms. They are standard, however, on a number of foreign makes, in the smaller frame diameters. Prelubricated bearings may be standard up to the 286 frame, for instance, with re-greasable types as standard above that size.

Bearing speed limits

Although some engineers have advocated ball bearings for 2-pole motors up to several thousand horsepower, 3600 rpm is generally accepted as too high a speed for ball bearings large enough for the shafts of such machines. Even sleeve bearings need lubrication system changes at high journal surface speeds.

Bearing "speed limits" are of several types. First, there are the general "published limits" for anti-friction bearings. The faster bearing parts move, the more pronounced become several damaging effects. For example, even assuming such ideal lubrication as a steady flow of cooled oil directed at every contact surface, too rapid a rotation will fling the oil aside so that it can't perform its function. Fluid friction within the oil increases with agitation or churning of the oil, which raises its temperature, making the oil "thinner" so it clings less readily to bearing surfaces. Furthermore, centrifugal force increases stress on balls and races to reduce the bearing's load capacity.

Therefore, even with ideal lubrication anti-friction bearings have speed limits, often expressed as a "DN" number, "D" being the inner race I.D. in millimeters and "N" the inner race rpm. Such general catalog speed limits, to quote one supplier, "serve as a useful guide for the majority of normal bearing applications."

There are also *thermal, lubricant,* and *design* speed limits for specific conditions. The user may impose a thermal limit—for example, because his experience shows it's needed to avoid frequent relubrication. Different lubricants have different viscosities, manner or rate of breakdown, etc. So for the particular lubricant used, an operating speed limit either above or below the catalog figure may be justified.

The design speed limit marks the point beyond which the bearing will destroy itself, and is of little interest to anyone but a bearing engineer. It is normally well above the catalog value.

What concerns the average user most is the lubricant speed limit. Certain combinations of bearing load and rpm will require a change from one type of grease to another; from grease to oil; from ring or sump "static oiling" to a circulating system; or from oil bath to oil mist. Yet the bearing itself may be basically the same in all cases. Some users have found that under their operating conditions a different lubrication system reduces maintenance, lowers the risk of failure, or decreases noise and vibration.

For motor bearings the typical DN value ranges from 250,000 to 400,000. In a 2-pole (3600-rpm) motor, then, the largest practical ball bearing is a size 316 (DN = 288,000) or 318 (DN = 325,000), the 316 being the largest for which grease lubrication is practical at such a speed.

What should bearing temperature limits be? This has become increasingly a subject for negotiation between manufacturers and major users of motors of all sizes. As with winding temperature, there is understandably a belief that "cooler is better." But a cooler-running bearing does not necessarily mean a longer-running bearing. If unduly low temperatures are required, the result may often be a reduction in overall motor reliability because of the cooling

Bearing temperature rise, degrees Centigrade

Bearing journal diameter, inches	Motor rpm			
	3600	1800	1200	900 or less
2.75	35	35	30	30
3	45	40-45	35	
3.5	45		35	
4	50		35-40	
5	—	45-50	40	40

Table 7-II. This table gives typical temperature *rise* of sleeve bearings in induction motors. Figures will vary somewhat with type of motor enclosure, and bearing length. Temperature rise of the oil itself, measured within the bearing housing, will be 15 to 20 degrees lower than shown. Maintaining the bearing rises shown at 3600 rpm for bearings larger than 3 inch diameter requires pressure lubrication, without which bearings would get at least 10 degrees hotter.

means needed. (See Table 7-II for typical bearing operating temperature ranges.)

The presence of fluid oil in a bearing chamber permits the use of bearing temperature sensors to reliably monitor bearing heating. Sleeve bearings also have thick walls, so there's room to mount a sensor in the bearing metal itself.

Like the winding temperature sensors described in the preceding chapter, bearing overheat detectors are commonly of the resistance, thermocouple, or thermistor type. Thermostats are also used, but are usually not of the simple on-off switch type. (See Figure 7-18.) Instead, they are more often "vapor-filled system" devices. A hollow metal tube is closely fitted into a hole drilled in the bearing shell. A smaller-diameter "capillary" tubing extension of the sensing bulb connects it to a dial gage mechanism, on which a moving pointer is rotated by pressure in the bulb-and-tube system increasing as the bulb is heated. The dial may be fitted with electrical contacts to sound an alarm when the pointer reaches a certain limit. (See Figure 7-19.)

Figure 7-18: Using a vapor-filled bulb sensor, the bearing temperature detectors on this motor operate contacts only (no thermometer reading is provided). (Photo courtesy Louis Allis Div., MagneTek)

Bearings and Lubrication 331

Figure 7-19: A typical "bearing thermometer" installed on a vertical motor. Again, a vapor-filled sensing bulb in the bearing transmits temperature-related pressure changes through the tube shown to operate the device. Adjustable contact arms on the dial face cause alarm circuits to function at pre-set temperatures.

These sensors are not conveniently added in the field unless the bearings and bearing housings have been properly machined at the factory to accept them. Therefore, remember these points when considering the use of bearing temperature detectors:

(1) If they will be installed later, in the field, give enough information when ordering the motor so the mounting holes will be properly drilled. Different makes and types of sensor require different mounting holes, threads, etc. There is no "industry standard" device.

(2) If the detectors are to be supplied with the motor, make it clear what's wanted: whether dial thermometer only, or alarm contacts; whether contacts are to be open or closed.

(3) If "filled system" bulb-and-tube sensors are required, and the relay or thermometer unit is to be located away from the motor, be sure the desired distance does not exceed the limitations of the sensing system. Also, be sure the motor manufacturer knows how much tubing to provide.

Grease-lubricated ball bearings can overheat too, but when failure is imminent the temperature rise is extremely rapid. The best possible temperature detector is a tip-sensitive device such as a thermocouple touching the bearing outer race. But even that is not considered a sufficiently fast-acting indication of impending bearing failure, so it is not recommended as anti-friction bearing protection. Sensing the bearing's high-frequency noise output, or vibration level, gives a much more reliable warning of trouble.

If a ball bearing does gradually overheat and stay that way for a long time without failing, the problem is overgreasing. Clean it out, then repack it with less grease. What should be of concern is not a temperature above some absolute limit, but one that becomes significantly higher than it had been, or than what is observed for identical bearings in the same service.

If ball bearing overheating occurs rapidly, following loss of bearing clearance or other potentially fatal damage, it will seldom be detectable in time. By the time you observe it, the damage will have been done.

Are the ball bearings used in motors made to special quality standards? Is there such a thing as an "Electric Motor Quality," or EMQ bearing? The answer, in general, is no. The existence of EMQ bearings is probably the most persistent myth in the motor industry.

Its origin is unknown. But it is fostered by supplier advertising containing statements like "Super Quiet electric motor quality (EMQ) bearings are available in 150 types and sizes." (What is "Super Quiet"?) Or one may read, "Both standard ABEC-1 bearings and EMQ (electric motor quality) bearings are available," or "Electric motor quality deep-groove bearings are available in .0625 to 4 inch I.D. to fit virtually any motor application."

When such offers are investigated, these facts emerge:

(1) No AFBMA, NEMA, ANSI or other industry standard defines any feature of bearing design or manufacture which could be termed "electric motor quality." For a specific bearing number, internal clearance, tolerances, seals, etc., are the same regardless of the potential uses of that bearing.

(2) What is often meant by "EMQ" in practice is that the manufacturer, or perhaps a major distributor dealing with certain user industries, selects from the general run of bearings those having operating noise level below some limit a particular user has specified. Normally these are very small bearings—seldom above a 203. Their load/life relationship, mounting, and internal fitup are not measurably different from bearings having a higher noise level. For normal industrial motor service the difference in noise is insignificant.

So if you go into the general market seeking EMQ ball bearings, there is no standard to which you can buy. The term will mean nothing to many suppliers.

Sleeve bearing loading

Little need be said about the sleeve bearing. In motors, it is normally a simple babbitt-lined steel cylinder. Riding on, and rotating with, the shaft will be one or more oil rings that carry oil up from a sump beneath the bearing, letting it then flow down onto the journal to spread out between journal and shaft surfaces. The babbitt is usually a lead or tin base alloy. Aluminum-base babbitts have been used under particularly corrosive conditions.

Sleeve bearings were once commonly made two to three times greater in length than in diameter. Modern practice, however, favors a length-to-diameter ratio from 1:1 to 1½:1. Slight shaft deflection, inevitable in larger motors, tends to intensify the load applied to the ends of the bearing, an imbalance minimized by shortening the bearing. Figure 7-20 shows a typical split sleeve bearing.

It is widely believed that sleeve bearings, unlike the anti-friction type, will "last forever"—that they enjoy unlimited life. This is untrue. The difference between ball and sleeve bearings is that for the latter, unlike the former, there

Figure 7-20: A typical sleeve bearing assembly in a horizontal motor.

is no standardized, statistical relation between load and life. Therefore, the sleeve bearing life span has no counterpart to the ball bearing's L_{10} figure.

But wear does take place in sleeve bearings, as we saw earlier in this chapter. Eventually increasing clearances break down the oil film, babbitt "wipes" or smears onto the journal, the bearing overheats, and failure ensues. If undertaken in time, repairs are not difficult or costly, which is another reason for the popularity of the sleeve bearing in large sizes.

Except for "thrust faces" at one or both ends, (see Chapter 9, Figure 9-32, page 466), the sleeve bearing has no continuous thrust load capacity. Special large sleeve bearings are available with oversize thrust faces capable of sustaining endwise loads of several hundred pounds. These have been used in motors driving Jordan refiners in paper mills.

Motor sleeve bearings will safely carry radial load up to about 150 pounds per square inch (psi) of projected bearing surface. For example, in a horizontal bearing 4 inches in diameter by 5 inches long, the actual surface area of the lower half of the journal is about 31 square inches. The "projected" area is simply diameter times length, or 20 square inches. Downward load—the gravity pull on rotating parts—is applied mostly to the bottom-most portion of the bearing surface, and little is carried by the sides of the bearing. Hence, the smaller projected area more nearly represents load-bearing capability than the total surface area. Theoretical bearing load limit, then, is 150 times 20, or 3000 pounds.

Actual capacity is reduced somewhat below that by cutouts in the surface which provide oil passages or assist in maintaining the oil film. Therefore, a more conservative limit of 60 to 100 psi usually applies. That's far below the 1000 psi limit for sleeve bearings in some non-electrical machinery. But it reflects the far longer life expected of most motor bearings despite shaft misalignment and fairly loose fits (compared to conditions in an auto engine, for example).

All sleeve bearings are lubricated by oil only. Unlike anti-friction bearings, they are not harmed by excess lubricant. But overfilling a bearing chamber is wasteful and invites oil leakage into the motor. Add oil only when the chamber level drops below its proper value. This shouldn't be necessary more than once or twice a year. To decide when a complete oil change is needed, drain off a little oil from time to time and check it for dirt, discoloration, and acidity.

Some motor users feel that maintaining the correct oil level, especially in locations difficult to get at or seldom visited for maintenance, is easier if the motor is fitted with constant-level oilers as in Figure 7-15. When the oil level drops slightly in the chamber, oil automatically feeds into the chamber from the bottle. If any oil at all is visible in the bottle, the bearing requires no addition.

Figure 7-21: Shaft-driven gear pump supplies oil pressure for this 3600-rpm motor. Piping is connected to header at lower left, which forms one end of cylindrical oil reservoir running inside motor enclosure to the other end. Reservoir is cooled by ventilating air stream entering through motor base. Feed line to other bearing runs through the reservoir itself for convenience.

Bearings and Lubrication 335

Keep in mind, however, that such an oiler bottle contains only a few ounces. It's intended to replace small quantities lost through normal leakage over a long period. Don't expect such a device to maintain oil level indefinitely on a large bearing housing containing several gallons of lubricant.

On motors not equipped with oilers, there may or may not be a separate sight glass (as in Figure 7-15) for observing the level. Small machines may have only fill plugs on the bearing housings. If it is not feasible to check oil level with the motor stopped, air pressure differentials may cause a loss of oil—and an incorrect level reading—if the fill plug is opened with the motor running. Such motors should be ordered equipped with sight gauges.

Sleeve bearings have speed limits determined by their temperature. The higher the journal surface speed, the hotter the bearing will run, and as both rpm and journal diameter increase, forced lubrication will become necessary to carry away that heat. At 3600 rpm, this happens when journal diameter exceeds a little over 3 inches; about 6 inches for 1800 rpm.

Pressure lubrication

Forced lubrication systems go far beyond the simple grease fitting or oil cup provided for smaller machines. How do they work, and what can go wrong with them?

The purpose of a pressure lubrication system (sometimes called "forced lube" or forced-flood) is to force oil to or into the motor bearings, balancing that inlet pressure against proper outflow drainage or venting. With sleeve or journal bearings, the object is to carry away excess heat produced by the "shearing" of the oil film as the shaft spins inside the bearing. With ball or roller bearings, the system serves to maintain the needed oil film despite rolling element speeds which might otherwise make that impossible.

All the system components may be mounted on the motor itself. In other instances, the pressure source and some controls may be with the driven machine, or in some separate location. Either way, the whole system must be kept in proper working order to avoid unexpected bearing failure. It's useless to repair such failures without making sure the lubrication system is functioning as it should.

In a sleeve bearing, oil rings alone will maintain the oil film needed for lubrication, in the bearing sizes and speeds found in industrial motors. So a pressure system serves only to keep the bearings cool. In a typical 3600-rpm motor, bearing temperature will be 10°C to 15°C lower when oil is circulating.

A force-lubricated motor will go on running safely for some time—often for hours—after a loss of oil pressure. But the oil and the bearings will eventually overheat, causing deterioration of lubricant, then wiping of the babbitt, which in turn will increase the heating still more.

If the system is working right with bearings in good condition, bearing temperatures should be well within the values of Table 7-II. Many pressure lubricated motors have bearing temperature sensors to monitor this. If there are none, the best you can do is get a reliable thermometer or thermocouple into the oil as close to the bearing liner as possible.

Regardless of motor speed, bearing operating temperatures quite normal for a horizontal machine may be far exceeded in high thrust vertical motors, where a roller thrust bearing may run well above 100°C. The oil used, whether externally cooled or not, must be a heavy grade to retain adequate viscosity at such a high temperature. Sometimes users don't believe it when told, for example, that an oil as heavy as SAE 40 or 50 is needed in a spherical roller thrust bearing. But at the fully loaded working temperature of such a bearing, lighter oils would virtually turn to water.

There are three common methods of pressure lubricating sleeve bearings:

(1) A shaft-driven pump, normally for 3600-rpm motors only. Figures 7-21, 7-22, and 7-23a show typical versions of this. The pump is usually a gear type, its own internal bearings lubricated directly from the oil being circulated.

(2) Inlet and drain connections only, including metering orifices or valves, on the motor. The user supplies his own oil piping to the motor, pump, and controls.

(3) A complete set of motor-mounted parts including oil tank, separate motor-driven pump, etc., as shown in Figures 7-23b, 7-24, and 7-25.

Any of these versions must include:

(a) An oil pump.

(b) A sump or reservoir.

(c) Some means of cooling the drained oil before it is returned to the bearing. This may be simply through exposure of a large volume sump to ambient air. For this to work, tank capacity should be at least ten times the total amount of oil flow per minute. Free air circulation around the tank must not be interfered with.

(d) Pressure and temperature controls for oil metering and system monitoring. The individual user preference, the nature of the process in which the drive is used, and the plant maintenance setup will dictate the details. However, there should be some kind of alarm/shutdown circuit that operates when

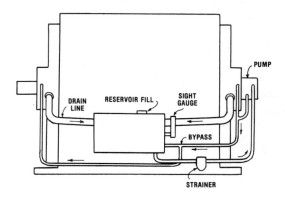

Figure 7-22: Principal parts of shaft-driven pump system are shown here, with large oil reservoir outside motor enclosure and cooled by surrounding ambient air. Proper slope of drain lines and cleanliness of oil are critical.

Bearings and Lubrication 337

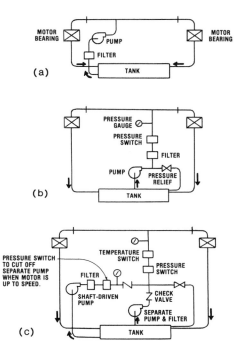

Figure 7-23: Typical lubrication systems for large high-speed motors. Version at (a) uses shaft-driven pump only; (b) is separate motor-driven unit; System at (c) uses both, for critical application. The two vertical lines at the top of (a) indicate metering orifices.

Figure 7-24: One version of separately-powered pressure lubrication system built up from individual components by the motor manufacturer.

oil pressure fails or temperatures rise too high. One control scheme is diagrammed in Figure 7-26.

(e) Some way of metering oil flow to each bearing, such as orifice plates in the piping at or near each bearing housing. There must also be adequate drainage. If oil inflow exceeds outflow, the excess will flood the bearing housing, get past shaft seals, and damage the windings.

Extending the idea of backup or safety factor a step further, some motors may have both a shaft-driven pump and a separately powered pressure system. (See Figures 7-23c & 7-27.)

Figure 7-25: Right: This complete oil system with motor-driven pump, for mounting on the side of a large motor enclosure, is sold as a package unit by a lubrication equipment supplier.

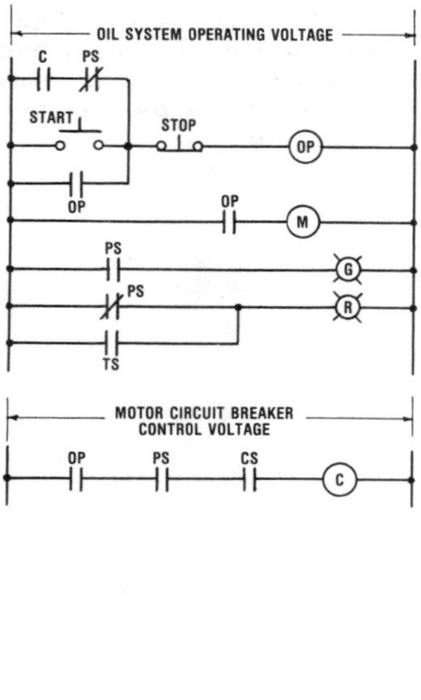

Figure 7-26: Left: One possible control scheme for an oil system as shown in Figure 2-23c. The oil pump must be started and pressure built up before the main motor breaker can be closed. Oil pump may then be stopped, and shaft-driven pump maintains pressure. Should that fail, separate pump is automatically restarted. Red and green lights alert operator to system condition. Many variations of this are possible.

PS = pressure switch; contacts shown in low pressure condition; CS = breaker closing contact; C = breaker closing coil; OP = oil system starting relay; M = oil pump motor; TS = temperature switch which closes on high oil temperature.

Bearings and Lubrication

Figure 7-27: This large motor uses both a shaft-driven pump (at left) and a separately-powered pressure system, mounted in the cubicle at right. Motor rating is 2500 hp, 1800 rpm.

Types of systems

Let's review the details of typical systems, starting with the simplest from the motor standpoint: the separate user pump, with only piping connections at the motor. How should you make sure this system is in proper working order?

First, check for adequate drainage. The drain line from the bearing housing should slope downward at least half an inch per foot all the way back to the reservoir. It is not enough simply to provide a drain hole at the proper level and then assume that the oil will naturally seek its own level regardless of supply rate, viscosity, or differential air pressure inside and outside the bearing.

One of the more common troubles with pressure lubrication systems is inadequate oil return piping from bearing housing to sump. The motor manufacturer usually specifies a minimum pipe size and a minimum down-slope of the drain line. Remember that these are only minimum figures. Operation in unusually cold surroundings may require increasing them considerably so the initially cold oil can flow properly.

Second, make sure metering orifices are in place and not blocked. When piping is disassembled for any reason, avoid interchanging orifices between bearings; sometimes (particularly with shaft-mounted pumps) the orifice in the line feeding the bearing nearest the pump will be smaller than the other because of the lower pressure drop in that shorter line.

These orifices may be quite small. Typical oil requirement for 3- to 5-inch bearing diameters ranges from ⅓ to ¾ gallon per minute per bearing, depending on the design. Figure 7-28 shows what the range may be. Supply pressure will vary from 3 to 25 psi, with 10 to 15 as a reasonable average. Check the motor manufacturer for specific values.

For a given supply pressure, P, in psi, the oil flow rate and orifice diameter D, inches to provide it are related by this formula:

$$\text{Flow in gals. per min.} = 24\ D^2 \sqrt{P}$$

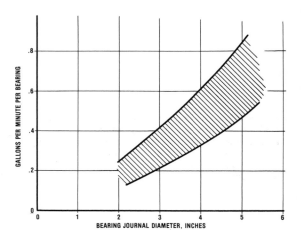

Figure 7-28: Pressurized oil flow needed for typical journal bearings, 1800 and 3600 rpm.

As an example, used for a 3600-rpm, 3-inch bearing: Desired flow was ¼ gallon per minute; user supply pressure 15 to 25 psi. Taking 20 as an average pressure:

$$D^2 = 0.25 \text{ gpm}/(24\sqrt{20}) = .0023$$
$$D = .047 \text{ inch.}$$

A No. 60 drill, .040 diameter, was used.

The orifice plate is normally bolted into a flange joint of the inlet piping to the bearing housing. Such small holes are easily plugged by sediment, or carbon particles produced in overheated oil. Hence a key element in these circulating systems is a strainer. These must be kept clean, and so located in the piping that it is impossible for sediment to drain back into the line.

Some manufacturers offer flow control valves instead of metering orifices. Whereas the latter work best only if supply pressure stays constant within narrow limits, the valve can be adjusted for wider pressure variations.

There are two types of valves: the manually adjusted needle valve (Figure 7-29), or the constant flow valve—either fixed rate or adjustable rate. The typical fixed rate valve can be purchased for any one of the following flows: 5 or 10 cubic inches of oil per minute; or ½ pint, 1 pint, or 8 pints per minute. The adjustable flow control valve can be set to maintain any flow from 3 ounces to one gallon per minute, while pressure ranges anywhere from 15 to 125 psi. Constant flow valves are preferable to the needle type because the latter must continually be reset if pressure varies, or else it will behave just like a fixed diameter orifice. (See Figure 7-30.)

Flow rates

Flow rates aren't hard to check. All you need is a large can and a watch. Pipe the metered oil line into the can for one minute. Then measure how much oil you collected. Repeating several times, then averaging, gives more accurate

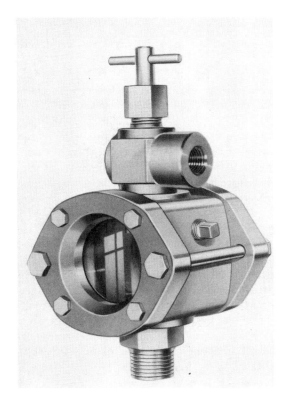

Figure 7-29: Typical needle valve, with flow sight glass. Before making any adjustments on such a valve, find out what type it is and how it is supposed to operate.

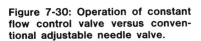

Figure 7-30: Operation of constant flow control valve versus conventional adjustable needle valve.

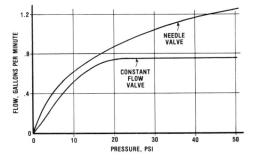

results. Be careful, though, not to run the system reservoir dry. This could ruin the pump.

Finally, examine the rest of the piping. Is it clean? Is it blocked anywhere? Above all, is it properly insulated if the motor bearing itself is insulated to interrupt shaft currents? In many motor designs, oil piping runs will "bridge" or short-circuit bearing insulation unless each line contains at least one insulating coupling or union. Like the bearing insulation itself, these can be tested with a megohmmeter to make sure they are in good condition. Watch out for accidental substitution of standard pipe or tube fittings for insulated connections.

Now consider the shaft-driven pump system of Figure 7-21. It needs the same plumbing checkout we've just described, plus some further review. Does the pump show evidence of leakage? Is it in fact able to pump at all?

That may seem a silly question. But when piping on one 2000-hp, 3600-rpm motor was disconnected from the pump for examination, a number of pump rotor gear teeth fell out in the workman's hand. Loss of suction, then lubrication failure, had "locked up" the pump, causing all the teeth to shear off. So be certain that all pump suction line fittings and joints are oil-tight and air-tight. Even a pinhole leak can break suction, and at 3600 rpm the pump can be destroyed in seconds.

In some systems, sight flow gauge glasses have vent lines connecting them to spaces inside the bearing housing, or within the motor frame. These are needed to properly balance air pressure across the gauge, preventing either an incorrect level reading or a siphoning of oil out of the bearing. If any such pipes are removed, be sure they are replaced tightly to prevent air leakage.

Pump shaft alignment can be important. The drive may be through a small stub shaft bolted to the end of the main motor shaft. Make sure that isn't bent or cocked. Some misalignment may be tolerable if the right type of coupling to the pump is used. If it's a gear coupling, though, that will need lubrication also, usually via a small hole drilled through the pump shaft. See that this hole is open. Examine all gaskets and seals for damage or leakage.

This kind of lubrication system of course works only when the motor is running; when the motor stops, the oil pump stops. Pressure and flow are controlled entirely by the piping and pump design.

Next, let's consider the operation of the separately driven pump/tank system, as in Figures 7-24 and 7-27. Why are these used? First of all, they suit motors at speeds other than 3600 rpm. Second, though it isn't normally a serious problem, some users are concerned about having an oil supply available at the bearings when the motor first starts to rotate. The shaft-driven pump doesn't do much work until the motor reaches full speed. The separate pump, however, may be arranged to start before the main motor is energized. Third, the system may be to some degree serviced, flow rates adjusted, oil easily added or removed, etc., with the motor in operation.

Cooling of the oil should be checked. Oil should not enter the bearings above 40°C. One arrangement for cooling uses a system reservoir or sump which stretches the length of the motor, located in the ventilating air intakes so that a large cool air flow is continually removing heat from the sump wall.

Sometimes, however, an oil-to-water heat exchanger ("oil cooler") is furnished. This is especially likely if the air ambient must be well above normal. Check coolers for cleanliness, for leaks, for free circulation of both oil and water, and for the correct water temperature.

Just as too high a temperature may require special lubrication system features, so too low an ambient may also dictate accessories which could be overlooked during maintenance. Bearing oil heaters, common for outdoor machines in cold climates, may also be needed in pressurized systems mounted on the motor. Immersion heaters may be provided in bearing housings. These should be of a special "low watt density" design (no more than 15 watts of heat developed per square inch of exposed heater surface) to avoid carbonizing the oil, and even then they should be thermostatically controlled so they're energized no more than necessary. If the wrong type of heater is substituted, or controls don't work, small particles of carbon can quickly cause bearing damage or block oil lines. Heaters have been known to build up a quarter-inch thick coating of carbon during a winter season.

If heaters don't work, on the other hand, oil may not flow, which may result in even greater damage. So all oil warming devices and controls should be checked out when servicing the system.

Many lubrication systems will include such accessories as pressure gauges, flow switches, or temperature alarms. Some user specs require systems to "be arranged with dual oil filters, switch-over valves and oil pressure gauges to indicate pressure differential across the filter." Each bearing outlet has an oil flow indicator plus dial thermometer. A pressure switch in the pump discharge line has two sets of contacts—one for alarm, the other to shut the motor down.

Control maintenance

All of the electrical controls involved, as in Figures 7-23 and 7-26, need the same maintenance as in any other electrical system critical to proper motor operation. The importance of this is emphasized by the recent destruction of a large 13,200-volt machine while it was shut down with its space heaters energized. Its separately-powered pressure lubrication system controls malfunctioned. Somehow, the system was started up, but normal oil circulation was hampered because the motor was not running. Oil overflowed the bearing housings, ran down inside the enclosure and contacted the hot space heaters, then ignited. The resulting fire ruined the stator winding.

Many users of large industrial machinery have become accustomed to one other type of system for lubricating numbers of bearings automatically—the oil-mist system. The mist is not a spray, but a fog of pressurized air containing very fine atomized oil droplets that stay suspended in the air so they will travel long distances at low speed through piping without settling out on the pipe walls. Injected into a bearing, this mist is quite effective in lubricating high speed ball or roller bearings that would otherwise overheat from grease or even oil splash lubrication. The air flow itself provides added cooling. Very little oil is used in maintaining the necessary film on bearing surfaces. A 3- to 4-inch I.D. bearing may use only an ounce of oil daily. (See Figure 7-31.)

At the user's request, motors are sometimes furnished with self-contained oil mist generators, and these too need to be serviced. Although adaptable to sleeve bearings, most oil mist lubricators for motors are used with ball bear-

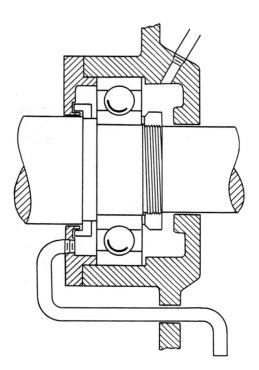

Figure 7-31: Typical assembly of an oil-mist lubricated ball bearing. Mist enters the bearing at top right. The vent line leaving the bearing chamber at lower left is needed to allow escape of the pressurized air that enters with the oil. A rotating "slinger" on the shaft, at left, prevents oil leakage into the motor.

Figure 7-32: Oil mist unit with 1 gallon reservoir, for mounting on 3600-rpm motor.

ings, at 3600 rpm. There are several different makes, but they have certain features in common.

Figure 7-32 shows a unit with these basic components:

(1) Solenoid air valve. This electrically triggers the flow of inlet air (typically at 35 to 65 psi).

(2) Next in line is the water separator/filter combination to make sure only dry, clean mist is produced.

(3) Third is the mist generator itself, usually with pressure gauge and regulator. This uses the high-pressure incoming air to atomize oil drawn from the reservoir beneath it.

Many accessories, plus a variety of control circuits, are possible for oil-mist systems. Figure 7-33 illustrates this. Oil and air heaters condition mist quality, prevent it from condensing in cold lines, and permit the use of various oil weights in varying ambient temperatures. Pressure switches provide for alarm, shutdown, or lockout of motor starting if the mist system is not working as intended.

Control circuits may be as shown in Figure 7-34. Since a high speed ball bearing might "run dry" and overheat quickly, it is common to interlock the system so the motor can't be started unless mist is flowing. Such controls may not be "fail safe," so they should be checked out completely if any bearing problems have occurred.

Ordinary system maintenance includes:

(1) Looking for blocked lines, particularly at the mist discharge fittings in the bearing housings.

(2) Making sure drain vents in those housings are open. Pressurized mist will not flow properly unless the pressure built up in the bearing is vented to the outside.

(3) Cleaning filters and separators.

(4) Cleaning sludge from oil reservoir and suction lines.

(5) Looking for leaks.

In operation, oil mist has two controllable properties: *volume* and *density*. Mist volume is controlled by the air pressure regulator, which typically will produce 8 to 10 psi. Mist density is regulated by a flow valve adjustment on

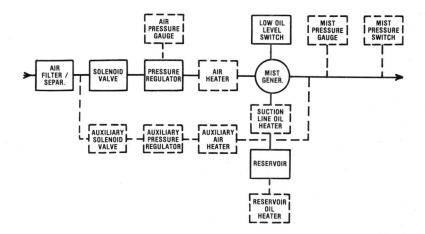

Figure 7-33: Oil mist system components. Solid lines are basic items; dashed lines show optional features. Arrows indicate air supply entering at left and mist being discharged at right.

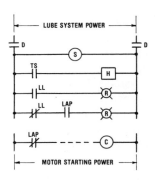

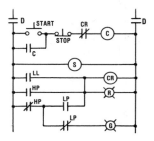

Figure 7-34: Typical control circuits for oil mist lubrication systems. D = line switch to energize system: C = motor starting contactor coil; S = mist solenoid valve; LL = low oil level switch on reservoir; LAP = low air pressure switch; TS = air temperature switch; H = heater in suction line; HP = mist high pressure switch; LP = mist low pressure switch.

the mist generator, which varies the oil feed from the reservoir. Proper use of these controls, and the amount of adjustment, will vary with the style and manufacturer of the unit.

Oil mist systems are best suited to two types of installation: (1) the large, important machine for which a self-contained mist generator may be economical; (2) large numbers of smaller motors grouped within a plant which can be conveniently supplied from one or more central systems. In the second category is at least one large refinery, where many motors even in the 50- to 200-hp range have been converted in the field to oil mist lubrication. (See Figure 7-35.)

Figure 7-35: A refinery pump motor equipped for oil mist lubrication from a central supply system serving a number of drives. Note oil supply tubing connecting to top of bearing chamber.

Bearing currents

Why are motor bearings (anti-friction or sleeve) sometimes insulated? Is it necessary to insulate both bearings in a motor?

The reason for electrically insulating a bearing or bearing chamber from the motor frame is to interrupt what would otherwise be a closed path for circulating currents that can damage bearings. Bearing surfaces are pitted by tiny electric arcs that form as the current jumps across gaps between the parts or breaks intermittently through the intervening oil film.

These currents originate from "dissymmetries" in the magnetic paths through stator and rotor iron. Here's what that means: As the rotor revolves within the stator, and as the stator's alternating magnetic field rotates, slight differences in the magnetic "reluctance" of the core parts (reluctance being the opposition to magnetization that is inherent in any steel) generate small voltages between the ends of the shaft. The magnetic dissymmetries result in a net amount of magnetic field or flux encircling the shaft. Because the shaft is a conductor, that looping flux produces voltages within it.

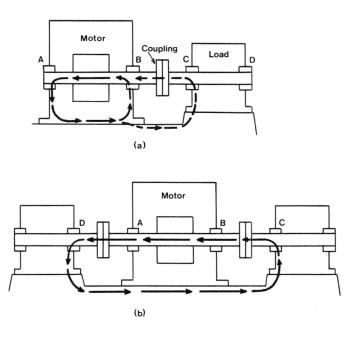

Figure 7-36: (a) Normal motor with single shaft extension—insulate bearing A to break current path (arrows) through entire circuit. If bearing B alone is insulated, current can still flow to damage bearings A and C. (b) For a double-shaft machine, insulating bearing A and/or B will not suffice; a circuit still exists through the common baseplate and bearings C/D. The solution: insulated couplings to interrupt the shaft circuit.

In most machines, especially those with ring or one-piece stator core laminations, this "shaft voltage" is too small to produce any damaging current flow through the possible path shown in Figure 7-36a. But with increasing motor size, the cost of a breakdown, the sensitivity of bearings to electrical current damage, and the likelihood of dangerous shaft voltages all increase as well. For that reason, many motor manufacturers insulate one bearing as standard practice above a certain motor size. Some insulate all vertical thrust bearings, because those heavily-loaded units are sensitive to damage which other bearings might tolerate. The same is true for 2-pole sleeve bearing machines, with their high journal speed.

Bearings may be insulated in several ways, one of which is by insulating the bearing itself by surrounding its outer surface with a ring or collar either pressed into the bearing chamber or built onto the bearing. Facings of similar material separate the bearing from any cap or retainer holding it in place.

In other machines, the bearing itself may not be insulated. Rather, the bearing chamber or bracket may be insulated from the motor frame. This requires insulating collars and washers on the mounting bolts, as well as insulating pads or rings at the mounting surfaces. Any temperature sensors, vibration probes, or oil piping attached to the bearing chamber must be insulated whether or not they touch the bearing.

When you are planning the motor installation, then, be sure you understand what type of insulation system is used.

In a conventional drive, there is no need to insulate both motor bearings. Opening the circuit at one point is sufficient to block current flow. However, when the bearing itself is insulated, it is likely that both bearings will be made the same way so that both assembly variations and spare parts stocking will be simplified.

If the motor has a double extended shaft for connection to other machines at both ends, both bearings *must* be insulated in some way, because, as Figure 7-36b shows, the possible current path cannot be interrupted by insulating one bearing alone. Current can still flow to the bearings of the other machines. If they are not protected, then one or more of the connecting shaft couplings must also be insulated.

Shop or field tests can seldom show the magnitude of the end-to-end shaft voltage for any motor. A rule of thumb is that anything over 100 millivolts is potentially harmful for ball bearings, and a 200 millivolt limit is reasonable for sleeve bearings.

Some old-timers have gauged the risk by observing the spark when a cable attached to a shaft-riding probe at one end of the motor is touched to the other end of the shaft, then drawn away. But measurement is quite difficult because the motor must be running at full speed. Because voltage is so low, a very high impedance instrument must be used, and good contact must be made with the rotating shaft inside the bearings. This can be impossible in some motors and dangerous in most. Figure 7-37 shows why measuring anywhere else is useless.

Figure 7-37: Lower diagram shows electrical circuit involved in "shaft current" flow through motor assembly in upper view. Bearing resistance R_B varies with bearing clearance, oil film, and type of bearing. Shaft resistance R_S is fixed by shaft size and material; may be much smaller than R_B. Using a voltmeter, V_2, to measure voltage across the bearing itself—from shaft to frame—is useless, because that voltage varies with both current flow (if any) and R_B. There is no correlation between observed voltage and possible current. Only true measure of potential damage is the generated shaft voltage, measurable by voltmeter V_1. Meter leads must contact shaft "inside" (towards the rotor) of the bearings. If a shaft grounding brush is used, it introduces resistance R_G. Unless that is very small, shaft voltage appearing across both R_B and R_G in parallel will still cause some bearing current to flow.

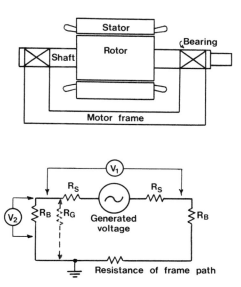

Will the "grounding brush" work to make insulation unnecessary, or "fix" a motor in which insulation was not provided and bearing damage later appears? Don't count on it. Shaft/brush contact voltage drop is uncertain, but generally high. If the resistance of that parallel or shunt path to ground is not extremely low, the voltage across the bearing path will remain high.

Grounding brushes are useful in such non-electrical machines as steam turbines to bypass much higher voltages, where contact drop is a much smaller fraction. But these are not "electromagnetic" voltages generated by induction as in a motor. They are "electrostatic" voltages produced by steam passing over turbine blades, much like the static voltage caused by the friction of walking across carpeting in a dry room.

This necessarily brief review of common concerns about motor bearing usage cannot suggest solutions to all operating problems. Motor manufacturers, bearing makers, and some large bearing distributors have capable engineering and service staffs. Don't hesitate to call on them when you have a bearing or lubrication question.

8 More about Optional Accessories

STATOR, ROTOR, bearings, shaft, frame. Put them all together and you have a complete motor—almost. Many industrial motors include at least one other component or assembly needed for control, protection, or coordination with other equipment.

Such extras we call "accessories." This list of a few common motor accessories indicates the variety involved:

Space heaters
Vibration sensors
Temperature detectors
Special connectors
Brakes
Ground pads
Surge protection
Oil/water/air flow or temperature controls

We have discussed several of these already. In this chapter, let's take a close look at some of the accessory items or functions which seem to generate the most questions for both equipment designers and users. The first of these, the speed sensor or motion indicator, is often supplied with the motor although it may be located elsewhere in the drive.

Speed sensing

How fast is the drive going? In which direction? Is it rotating at all? Those are the three questions a drive speed sensor may be required to answer. Most such devices answer only one, or perhaps two. Let's consider how they work and why they are used.

Probably the most familiar type of speed sensor is what's usually called a zero speed switch or "plugging switch." Directly coupled to a drive motor's special shaft extension, this switch uses centrifugal force or a magnetic field produced by its rotation to open or close one or more contacts in a control circuit at a predetermined speed.

Figure 8-1 illustrates the operation of three common kinds of speed switches. In the version shown at top left, centrifugal force moves a pair of pivoted weights outward. Links connected to each weight pull a contact arm which

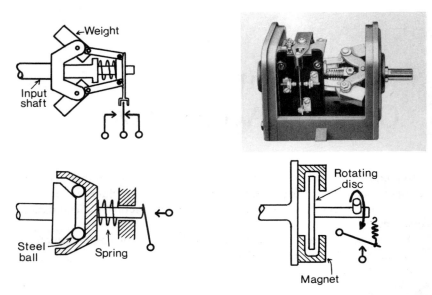

Figure 8-1: Operating principles of three common types of direct-coupled speed switches for electric motor drives. The two types at the left are centrifugal. The lower right version operates through eddy current torque. The photograph shows a centrifugal speed switch of the type in the top left diagram, with the cover removed to show the mechanism.

opens or closes contacts. At bottom left, steel balls pushed outward by rotation into the space between two discs will force the outer disc to move away from the inner so as to operate an electrical contact.

In the switch shown at the lower right in Figure 8-1, eddy currents induced in the inner member by rotating magnets in the outer member produce enough internal shaft torque to rotate a trip arm against a stationary contact. This is basically the same principle used in induction disc relays common in power system protection, or in the simple residential electric meter.

Five of the many uses for such devices will be discussed here. Only the first of these can properly be termed a "plugging" switch application.

When an induction motor is plugged to a stop for fast deceleration, its windings overheat. The usual reason for such a stop is to permit quick recycling of a batch process, as in a rubber mill drive. To minimize the overheating, as well as to keep the cycle time as short as possible, it's undesirable for the drive to re-accelerate at all in the reverse direction once it has reached a complete stop. What the plugging switch does is to sense the zero speed condition (or very nearly so), disconnecting the reversed or plugging power at the proper time so the motor will not restart in reverse.

Figure 8-2 shows several speed switch circuits used for that purpose. These schemes are intended for drives having only one normal direction of rotation ("forward") and therefore one direction of plug-stopping ("reverse").

More about Optional Accessories

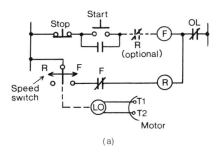

(a)

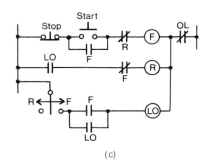

(c)

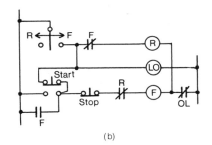

(b)

Figure 8-2: Several plugging switch circuits for bringing a motor to a stop from one direction only without allowing it to restart in the opposite direction.

Consider the arrangement in Figure 8-2a. When the motor is running at normal forward speed, the forward ("F") starter is closed, and the speed switch forward contact is also closed because of rotation. If the stop button is pressed, dropping out the "F" starter, a contact on that starter closes to energize the reverse ("R") or plugging contactor through the speed switch. The motor then slows down until at a low speed (typically no more than 50 to 100 rpm) the speed switch contact re-opens to drop out the "R" contactor. Normally the motor is loaded—typical of batch processing drives to which such switching is normally applied—so that from this low speed it quickly coasts to rest. The optional "R" contact in the forward circuit would prevent any possible attempt to restart while plug-stopping is in progress.

This sort of speed switch is generally obtainable with an optional magnetic coil "lockout" (LO) device wired to the motor terminals as shown. In some drives, such as machine tools, the operator may have occasion to rotate a shaft by hand with the power off. Should that rotation be sufficient to operate the "F" speed switch contact while the "F" starter is de-energized, the movement would start the motor without warning by picking up the "R" contactor. To prevent that, the lockout coil blocks the speed switch mechanism unless power has been applied to the motor terminals.

The circuits in Figures 8-2b and 8-2c perform the same basic function using the same kind of speed switch. There are also drives requiring a plug-stop capability from either direction of rotation. That uses the same switch, but with a more complex connection. (See Figure 8-3.)

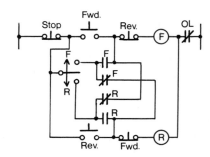

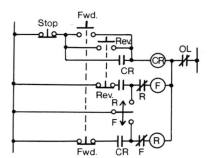

Figure 8-3: These circuits use the same type of speed switch for plug-stop control from either direction of rotation.

Anti-plugging

A second basic use of the speed switch is to prevent, rather than permit, motor plugging. This is often called "anti-plugging" service. Figure 8-4 shows a control scheme for a drive which is reversible, but not by plugging. The object of this arrangement is to make it impossible for the operator to plug-stop the motor, in situations where that is not necessary and may be dangerous to the windings.

In the switch itself, the "F" contact opens on forward rotation; the "R" contact opens on reverse rotation. Suppose the motor is running in the forward direction. An "F" contact then keeps the reverse rotation circuit from being energized when the reverse button is pressed. Pressing the stop button will drop out the "F" starter and close that "F" contact—but the open "F" speed switch contact still prevents application of reverse power. Similarly, the speed switch makes it impossible to apply forward power if the motor is coasting in the reverse direction.

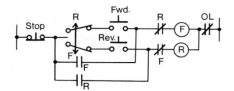

Figure 8-4: An anti-plugging circuit using a speed switch to allow reversal of a drive, but not by plugging.

More about Optional Accessories 355

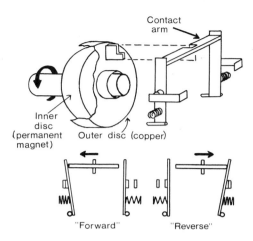

Figure 8-5: An eddy current "drag cup" speed switch which operates a different circuit for each direction of rotation. By adjustment of the contact springs, each contact may be made to operate at a different rpm.

One way that speed switches sense direction appears in Figure 8-5. When forward rotation in this "drag cup" device torques the disc in one direction, the forward contact operates. Rotation in the opposite direction will instead close the other contact.

A third function for a speed switch is to govern the transfer point between two drive speeds. There may be one two-speed motor (Figure 8-6) or, as in Figure 8-7, a pair of motors driving the same load but each wound for a different speed. Large fan drives often use two-speed driving power. One reason is that a single motor capable of bringing a high-inertia fan up to top speed may be unduly large, costly, and inefficient because of the high resistance rotor and oversize parts needed to withstand acceleration heating. That can be avoided by accelerating in two steps, switching from low speed to high at the proper time. Also, the lowest-cost operating cycle often requires the fan to run at reduced speed for long periods—and the lower speed is easily achieved with a second winding on the motor, or by using a second motor entirely.

Figure 8-6: A large tunnel ventilating fan drive using a two-speed motor with speed-sensing device coupled to a special shaft extension on the motor. (Photo courtesy Louis Allis Div., MagneTek.)

Figure 8-7: A power plant draft fan driven by two separate motors, each of a different speed. At maximum plant output, the high speed is needed; during periods of lighter load, the low speed option is much more efficient.

In such drives the higher speed should not be energized until the machine has accelerated up to the limit of the lower speed. Only then should it become possible to transfer to the second winding (or motor). The speed switch can be set to lock out that transfer until the proper rpm, as indicated in Figure 8-8.

Switch contacts can also be set to prevent reverting to the lower speed connection while the drive is coasting at a higher rpm, avoiding plugging of the high speed winding (or motor). In that situation, the switch contacts need not respond to direction of rotation, but only to actual rpm.

There are other similar applications throughout industry. For example, in a conveyor system, one conveyor may have to reach full speed before a second conveyor drive is started. A speed switch on the first conveyor drive, wired into the second conveyor motor control circuit, will do that job.

Motor protection

A fourth application for the speed switch is to protect motors that are subject to accelerations so severe that normal overcurrent relays would trip before the drive could reach full speed. To override relays for the required acceleration time, the speed switch senses the difference between normal rotation—for which the override can be permitted—and a stalled condition for which a trip-out should occur.

Figure 8-9 shows one way of doing this. Here, too, the switch operates the same way for either direction of rotation, responding only to rpm. The timer

More about Optional Accessories 357

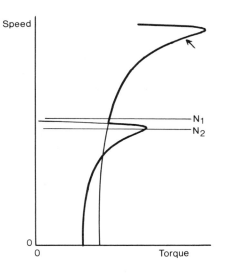

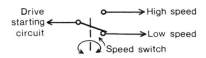

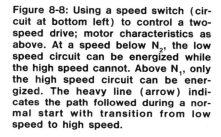

Figure 8-8: Using a speed switch (circuit at bottom left) to control a two-speed drive; motor characteristics as above. At a speed below N_2, the low speed circuit can be energized while the high speed cannot. Above N_1, only the high speed circuit can be energized. The heavy line (arrow) indicates the path followed during a normal start with transition from low speed to high speed.

Figure 8-9: How a speed switch and timing relay can be used to protect a motor during a severe acceleration (right). The top view shows the circuit prior to a start. The middle view illustrates a stalled condition—the shaft fails to rotate, and the speed switch closes the trip circuit after a set period of time. In the bottom view, rotation allows the speed switch to transfer the trip circuit to the overcurrent relay for running protection.

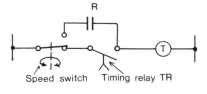

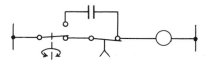

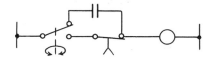

"TR" begins its cycle when the start button is pressed. The "trip" circuit (shown as a circuit breaker trip coil "T") will be closed, to take the motor off line, in one of two ways. If the start is normal, so that the speed switch senses rotation to a preset rpm, its upper contact closes to put the overcurrent relay contact "R" into the circuit. That relay takes over protection, once the motor is under way, against any subsequent overload.

If the start fails, so that the motor does not reach the preset rpm (typically 5% to 10% of full speed), the lower speed switch contact remains closed. After a short interval the timer energizes the trip circuit to remove power from the motor.

In Figure 8-10, a similar circuit uses two overcurrent relays of differing characteristics, without the timer. A normal start puts the R_1 relay into the circuit for "long time" response to overload current. Failure of the drive to start keeps the R_2 relay in circuit instead, resulting in tripout after a much shorter time.

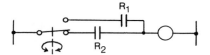

Figure 8-10: Another means of acceleration protection using a speed switch in conjunction with overcurrent relays of two different characteristics.

In either arrangement, the speed switch in effect "supervises" operation of the relays when it would otherwise be impossible to ensure normal starts without "nuisance tripping."

Speed switches may also monitor certain safety functions. In an oil company research center, for instance, a set of exhaust fans was used to remove toxic fumes from an experimental process. Undetected fan stoppage could allow dangerous fume accumulation. Simply having an alarm in the motor control circuit wasn't enough; actual fan shaft rotation had to be monitored. A circuit like Figure 8-11 will do this.

Figure 8-11: This circuit uses a speed sensing device to furnish an alarm if a critical drive comes to a stop. If there is rotation, the speed sensor "S" is energized, opening the circuit to the alarm light "A" and bridging the start button. If there is no rotation, S is de-energized, closing the alarm circuit. Assuming the motor contactor M to be de-energized, it will then be necessary for an operator to hold in the start button until rotation builds up.

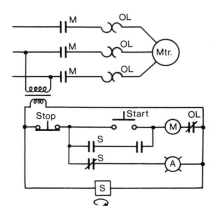

More about Optional Accessories

Because of the way it operates, the typical centrifugal speed switch is not precise. It need not be accurate within a few rpm. Depending on the make, adjustable contact settings from 10 rpm up to several thousand may be available (one switch spans the range from ½ to 5000 rpm). Common contact operating ranges are 15 to 60, 40 to 140, 140 to 750, and 150 to 900 rpm. Tolerance on the exact operating point within those ranges will typically be plus or minus 10% of the value selected.

Another kind of speed sensing switch recently introduced operates on an optical principle, yet is shaft driven. Within the switch housing is a light source and receiver combination, powered from 115 volts a-c. The light beam to the receiver is interrupted by a rotating, segmented disc driven by the switch shaft. An electronic circuit analyzes the frequency of interruption. When that frequency, corresponding to shaft rpm, exceeds a chosen limit, an output relay operates. Maximum running speed of the unit is 5000 rpm. Its working range, however, is from ½ to 100 rpm, so that it serves primarily as a zero speed device.

If a motor shaft extension isn't available for direct coupling, speed switches can be either belted to the motor or connected to another shaft in the drive train. Because that can change the switch operating rpm range, perhaps making it impossible to set the desired tripping points, such arrangements must be clear before the switch is ordered. A belt drive means extra maintenance, as well as possible belt breakage to render the switch inoperable.

Non-contact sensors

A new family of speed sensors appeared on the market in the early 1980s, having greater precision (sometimes operating within one rpm of the set point), multiple contact operating speeds, and no need for mounting or coupling to the shaft itself. Instead, they use inductive or magnetic pickups (Figure 8-12) to sense shaft rpm by "seeing" magnetically the rate at which one or more projections or "spots" on the shaft sweep past a sensing head mounted a short distance away. Electronic circuits translate that impulse rate into rpm, causing various kinds of switching according to internal programming. Some of these devices will distinguish whether or not speed is increasing or decreasing at a given rpm.

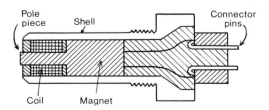

Figure 8-12: A non-contact inductive probe used with various types of electronic speed sensors.

One application in which these non-contact sensors excel is the switching of starting capacitors used with large motors. Capacitors boost feeder voltage during motor acceleration so inrush current doesn't cause too much voltage drop. But once the motor nears full speed (typically at 95% of rated rpm), the capacitors must be disconnected to avoid overvoltage on the circuit. A centrifugal speed switch usually has too great a set-point tolerance for that. A timing relay can't be used, because the exact time for acceleration varies too much. But the electronic speed sensor does the job nicely.

The "motion detector" of Figure 8-13 includes a "field effect" transducer that senses movement of metal across its sensing face. The transducer output signal is proportional to the speed of such movement. In use, the sensor face (either curved to suit motor shaft contour, as in the illustration, or flat for shaft diameters above 3 inches) is positioned within 2 millimeters of a rotating shaft. No parts need be mounted on the motor or drive machinery.

Figure 8-13: A non-contact "motion detector" that will sense both speed and direction of shaft rotation, usable for either plugging or anti-plugging applications. The sensing head is at left; control module, at right. (Photo courtesy Square D.)

The transducer causes relay operation in the accompanying electronic control module at shaft speeds from 5 to 1200 rpm, depending on shaft size and control adjustment. This device can discriminate between forward and reverse rotation, making it suitable for anti-plugging duty. It also can control up to a NEMA size 2 motor starter directly. Because the sensing head is encapsulated, dirty or corrosive environments present no operating problem. Nor is there any explosion hazard.

Other types of sensor work optically by responding to a light beam reflected from coded markings on a shaft surface. Whatever the principle, sensors requiring no direct attachment to a shaft have the advantage of not being subject to physical damage from overspeed, as well as not taking up space on the shaft or in the coupling area that may not be readily available.

Because of their sensitivity, these sensors can serve two purposes other than the simple switching described earlier. First, they can accurately indicate speeds all the way down to zero rpm. They are therefore suited to very low speed systems, or to systems in which some function must depend upon the occurrence of any rotation at all.

Second, they will operate as tachometers to read out actual shaft rpm. Though not needed in the typical induction motor drive, this can be important in variable speed applications. Tachometer output then provides feedback to the speed control system.

In one common type of tachometer (Figure 8-14), a permanent magnet rotor within a wound stator produces an alternating voltage proportional to rotor speed. A two-phase winding option lets the output circuit sense direction of rotation as well as rpm, because the sequence of the two phase voltages can be detected.

Figure 8-14: A typical direct-coupled tachometer generator used to monitor the exact rpm of a drive rather than simply to operate a switch at a predetermined speed.

That kind of generator must be shaft-driven. Others, however, use non-contact pickup schemes as in Figure 8-15. An electronic pulse-counter converts the signal from the shaft into a digital rpm display.

Figure 8-15: This non-contact speed sensor uses a disc attached to the end of the rotating shaft (left) or wrapped around it (right) having a number of magnetized spots on the periphery. The sensor generates an electrical signal each time a magnetic spot passes it. An electronic control circuit counts and times those impulses to determine shaft rpm within a tolerance that can be less than one rpm.

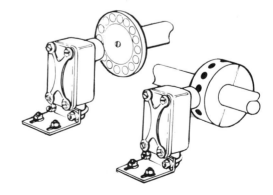

Still another kind of electronic speed sensor is the "zero motion detector," which actually derives no rpm signal as such from any rotating shaft. Rather, it responds to the extremely slight voltage generated by a de-energized motor winding with the rotor still turning. That voltage is produced by the residual magnetism remaining in the core after power is switched off, and cannot even be measured by the usual instruments. But it will disappear entirely only when rotation ceases entirely. The motion detector, connected as in Figure 8-16 (top), will sense the disappearance of that residual voltage, to permit reversing (an anti-plugging function), or to disconnect d-c dynamic braking power when that has served its purpose. (See Figure 8-16.)

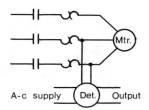

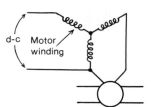

Figure 8-16: How the electronic motion detector can be used. There are no shaft or motor mounted devices, electrical or mechanical. Instead, the unit electrically senses the tiny voltage generated by rotation from residual magnetism in the de-energized motor's core lamination. Like the other forms of non-contact sensor in Figures 8-13 and 8-15, this is typically used for zero speed or plugging duty as shown at top. ("Output" represents the output control circuit.) Below, the same detector operates to disconnect d-c dynamic braking power when the drive has stopped.

One such circuit operates on residual terminal voltage in the 30- to 40- millivolt range. At this level the motor is considered stopped. The "voltage sensitive relay" then operates, indicating zero rpm to the external circuit in whatever way desired. For motors up to 600 rated volts, the direct connection of Figure 8-16 is possible; higher voltage motors can be coupled to the relay via current transformer. Isolation between relay and output circuit is either via mechanical relay or optical coupling. Between relay and 115-volt supply, there is transformer isolation.

The same relay can be made to trip at other speeds by using voltage dividers or control circuit adjustment to vary the residual voltage to which the circuit responds. This particular make is adjustable up to 400 millivolts, and the range from 40 to 400 millivolts corresponds to a wide range of rpm. That will naturally vary with the particular motor design, so the proper circuit must be worked out for each individual application.

For controlling dynamic braking, the relay must be "polarized"—that is, its internal circuitry will block the d-c braking voltage applied to the motor, while still allowing passage of the residual a-c ripple voltage being generated in the windings. Thus, although the Figure 8-16 diagram is for a wye-connected motor which will transmit no d-c braking voltage to the relay circuit, the device can be applied just as well to a delta-connected motor where the d-c would appear at the relay connections.

At zero speed, of course, the motor output is zero millivolts. But a relay set to respond there would probably give spurious indications (false tripping) caused by line noise. A low, positive voltage is close enough. Some devices of this sort work in the microvolt range.

"Residual voltage" as thought of here must be distinguished from the voltage remaining at a motor's terminals immediately after the power supply is interrupted. Consider a typical 50-hp, 4-pole standard open motor. From its initial value of 460 volts, what is measured across the open-circuited terminals will quickly decay to about 170 volts within half a second. That is the motor's "open circuit time constant," which depends only on internal inductance and resistance.

But that presumes little or no change in motor speed. As time goes on, whether the motor is coasting to a stop or being braked, the terminal voltage becomes more dependent on rpm as well as time. The relatively high value generated by the strong magnetic field imposed through the energized stator winding will disappear rapidly, but the far lower voltage produced by permanent residual magnetism is independent of that.

A typical voltage-sensing speed relay is accurate to within one millivolt out of 350, perhaps corresponding to only a few rpm. But because such accuracy comes only at the price of a "pre-calibration" of the circuit for every application, this type of device is used mostly for zero speed switching.

The circuit may have a "direction memory" option, permitting anti-plug use. That's done through a latch/reset double-acting relay that sets up a path for closing the motor starter in either direction only if the motor was last started in that same direction. But the path for opposite direction starting is routed through the speed sensing circuit so that closure is impossible unless the motor is essentially stopped.

It's evident that a wide variety of drive speed sensors is available today. One cannot properly specify, install, or use such devices without full knowledge of their operation and their intended function.

Friction braking

As we've seen, not only getting a motor drive safely started, but also properly monitoring its speed afterwards, can be an application challenge. But there are some applications in which the problem is getting the drive stopped. One of the oldest methods of stopping a motor, yet one involving a wide variety of circuits and devices, is the friction brake.

In d-c braking, a-c dynamic braking, or plugging, when the motor circuit itself is altered to produce braking torque, no external brake is needed (although the added control equipment may be considerable). But those methods, once they decelerate the load to zero speed, will not hold it there. They produce no torque at standstill. Such torque is necessary whenever the motor must be kept from rotating, as in a crane hoist drive which otherwise could not keep heavy loads safely suspended. That requires a mechanical friction brake.

Brakes are of two basic kinds: the disc type and the shoe type. The former produces stopping torque by clamping stationary pressure plates against rotating friction discs keyed to the motor shaft. Figures 8-17, 8-18, and 8-19 are typical. The shoe type tightens friction shoes against a drum which rotates with the shaft. (See Figures 8-20, 8-21, and 8-22.)

Shoe brakes add more inertia to the rotating system than do the disc type. To maintain stiffness and strength with ample capacity to absorb heat, the shoe brake wheel has a fairly heavy rim for the shoes to contact. But because all braking action takes place along that rim, which is furthest from the shaft center, this brake can produce more torque from the same spring or magnet force than the disc brake in which the torque is developed over friction faces extending inward to the shaft.

Stopping the drive is so often necessary to prevent damage or injury, again as in the crane example, that either kind of brake must generally be "failsafe"—that is, it must set and hold when electric power is interrupted for any reason (per Occupational Safety & Health Administration requirements appearing in the *Federal Register*, Vol. 30, No. 119, June 21, 1973). Hence, brakes are generally "spring set." Clamping action is generated by powerful coil springs. Electrical power releases the brake, usually working through a solenoid coil or electromagnet which compresses the springs, permitting the motor to rotate freely. Should the coil burn out, a wire break, or a fuse blow, the brake sets immediately.

Electrically-set brakes are less common, but may be used where safety is not involved, such as in printing press roll drives or other continuous web transport processes which must occasionally be stopped on command.

These and other options appear in Figure 8-23. All standard disc brakes are built for mounting directly to a flanged motor bearing bracket. (See Figure 8-24.) This automatically aligns the motor shaft with the brake mechanism, avoids any extra foundation structure, and permits removal and replacement of motor and brake as a unit.

Shoe brakes can be motor-mounted also. The motor is fitted with an L-shaped support or bracket arm to which the brake feet are bolted. Shoe brakes with wheels 30 or more inches in diameter have been motor-mounted, using either cast or fabricated brackets, for motor ratings as high as 3000 hp, 1200 rpm.

More often, however, shoe brakes are floor-mounted. This is cheaper, more rigid, and avoids the strain on the motor housing of the added brake weight. It is also possible, though uncommon, to floor-mount some disc brakes, by ordering them with foot attachments. (See Figure 8-25.)

Either a-c or d-c operating coils are available, single-phase a-c or (for some

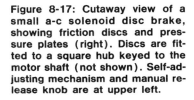

Figure 8-17: Cutaway view of a small a-c solenoid disc brake, showing friction discs and pressure plates (right). Discs are fitted to a square hub keyed to the motor shaft (not shown). Self-adjusting mechanism and manual release knob are at upper left.

Figure 8-18: Left: Operating mechanism of another solenoid type a-c brake, using a double-acting ball bearing cam linkage (right) to convert plunger travel into force on pressure plate. This model is available up to 15 lb-ft.

Figure 8-19: Right: A "direct-acting" a-c magnet brake with cover partially removed; manual release handle at left.

Figure 8-20: Typical shoe brake operated by a-c solenoid (left).

Figure 8-21: A large d-c shoe brake. Magnet coil is at the right between the two armatures.

More about Optional Accessories

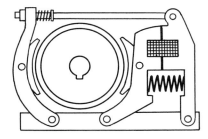

Figure 8-22: Mechanism of a double-armature d-c brake as shown in Figure 8-21 (previous page). The central coil (upper right) pulls both armatures together, compressing the spring to release the brake.

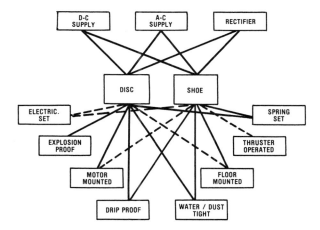

Figure 8-23: Right: The two basic types of motor brakes, with optional features for each. The dashed lines indicate versions not often encountered on industrial drives, though they are available.

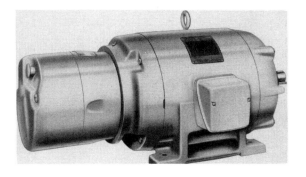

Figure 8-24: Left: Disc brake mounted on a drip proof motor.

Figure 8-25: A foot-mounted disc brake.

models) three-phase. In any event, the practical torque limit for a disc brake is about 1000 lb-ft. Most such brakes are in the 10 to 105 lb-ft range. For torques as high as 6000 to 9000 lb-ft the shoe brake is required.

The smaller disc brakes may be "direct acting." That is, the short travel of a magnet armature—1/16 to 3/16 inch—moves the discs enough to release the brake. No added leverage is needed. Higher brake ratings take more release power. So they use a solenoid with a fairly long plunger travel—¾ to 1½ inches—adding leverage to convert that long stroke into short disc or shoe movement with high force. (See Figures 8-26, 8-27, and 8-28.) In still larger sizes of d-c shoe brake, double armatures are used (Figures 8-21 and 8-22) to increase the releasing force still further.

For extremely high torque (10,000 or more lb-ft), thrustor brakes are available. Instead of a magnet or solenoid, these are worked by a hydraulic ram pressurized from a high speed 3-phase motor. Motor and ram are self-contained in a single cylinder called a thrustor. (See Figure 8-29, page 370.) This is a slow-acting device, like any hydraulic cylinder, so is unsuited to crane hoist use where quick stops are needed.

There are some other types of brakes or clutch-brake packages used in power transmission, especially for coupling two in-line shafts. Many are quite small, with torques measured in ounce-inches. But we will be concerned here only with the sizes of brakes normally encountered in industrial service, on motors—those in the torque ranges given by Table 8-I. Nor is there space here to comment on non-friction magnetic brakes such as the eddy-current type.

Brake selection

Choosing a brake requires a review of several items, described on this and the next five pages:

(1) Required torque:

For most applications, the minimum size brake should be the nearest standard torque rating above the full load rated torque of the motor. (See Table 8-

More about Optional Accessories

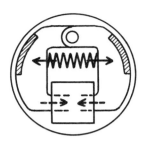

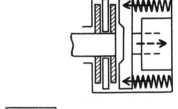

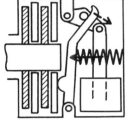

Figure 8-26: How some smaller brakes operate. The design in the middle is a direct-acting disc brake. The larger rating below it uses solenoid and leverage to develop more releasing force against stronger springs. The number of discs may vary from 1 to 5 depending on brake size.

The top version is actually an internal shoe-acting design, its housing resembling that of a disc type. It was once offered up to 15 lb-ft.

The solid arrows in each case show spring setting force; the dashed arrows, the releasing force.

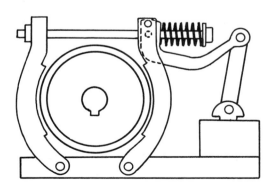

Figure 8-27: Large a-c shoe brake as pictured in Figure 8-20 (page 366), shown here is the released position with solenoid plunger fully seated in the coil.

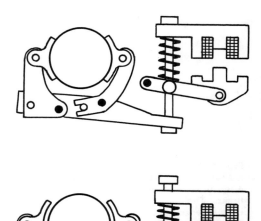

Figure 8-28: Use of levers to multiply magnet force in an a-c shoe brake, torque range 10 to 125 lb-ft continuous. Dark circles are fixed pivots for armature and shoes.

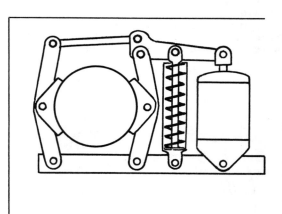

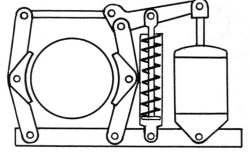

Figure 8-29: Thrustor brake operation. Above, set; below, released. The thrustor is on the right.

More about Optional Accessories 371

COMMON BRAKE TORQUE RATINGS IN POUND-FEET

a-c or d-c disc brakes	a-c shoe brakes	d-c shoe brakes
¾ — 1000	10 — 400[1]	25 — 9000
1½ — 575[2]	1½ — 2400[2]	65 — 9000[2]

(1) Up to 10,000 with thrustor.
(2) NEMA standard ratings per Publication ICS 2-220, 1978.

Table 8-I. Most industrial brakes will have torque ratings within these ranges.

II.) However, if the load is an "over-hauling" one, as for instance a hoist in which the descending heavy load may add considerably more torque than the drive motor rating itself, higher brake torque will be needed. The same is true if the load has high inertia, such as a large fan or bandsaw. Therefore, if a brakemotor application is changed—particularly the load inertia, or the required stopping time—a different brake will probably be needed.

BRAKE SELECTION GUIDE

Motor hp	RPM	Motor frame	"Normal" disc brake torque rating	"Service factor" brake torque rating
5	3455	182T	10	15
	1735	184T	15	25
	1150	215T	25	35
20	3515	254T	35	50
	1745	256T	70	105
	1175	286T	105	175
50	3555	324T	75	125
	1775	326T	175	330
	1185	365T	230	330

Table 8-II. Typical brake selection for some standard NEMA Design B drip-proof motors. The "service factor" selection, recommended by some brake manufacturers, provides margin for significant load inertia, which may not always be known. (Brake thermal capacity must be adequate for the duty cycle, and the torque rating desired must also be physically adaptable to the motor frame size.)

(2) Heating Capacity:

Like any friction device, a brake gets hot. Too much heat will destroy the friction elements, crack brake wheels, or cause other damage. Cycling, too, often can burn out the magnet coil. Brakes are usually rated thermally in terms of horsepower-seconds per minute, from which can be calculated the number of times per minute or per hour a given load can safely be stopped. Again, the use of a brakemotor in cycling duty different from that for which it was designed can lead to brake problems; when these appear, one of the first steps in diagnosis should be to find out if the load or the stopping frequency has changed. Brake coil duty may be continuous or intermittent and should be matched to the time rating of the motor itself.

(3) Mounting position:

Disc brakes may not work properly in both vertical and horizontal shaft positions. All standard brakes are designed for horizontal use. With the close clearances involved, when such a brake is mounted on a vertical shaft the weight of the free-running friction discs may cause them to drag on the pressure plates when the brake is released. Separator springs or other modifications will be needed to prevent that.

Equally important, the disc brake thermal capacity may decrease when the unit is mounted vertically, because heat dissipation is more difficult. So even if the brake works properly, it may still be unable to handle the duty cycle.

Some shoe brakes, in which gravity may cause magnet armature drag, are also sensitive to operating position. Also, these units are designed for specific right or left hand mounting (referring to the magnet position relative to the motor). If a motor with such a brake is relocated so that the brake must be reversed, it may then become impossible to replace or reline brake shoes with the brake in place.

(4) Electrical characteristics:

A-c versus d-c, the proper phase, and the correct voltage are obvious decisions. But voltage variation must also be taken into account. Tolerances on the pickup and dropout levels of the magnet coils will vary with the type of brake. This is especially true of d-c shoe brakes. Most a-c units are intended to work properly when voltage varies within plus or minus 10% to 15% of the rated brake value. This is similar to the variation permitted for standard a-c motors.

However, there is one big difference. Too high a brake voltage will overheat the coil, shortening its life, as will happen in motor windings when voltage is on the low side. But too low a voltage on the brake may cause its magnet armature to pull in only part way, a situation comparable to stalling a motor, and the brake coil may then burn out within a few moments. So although some 220-volt motors can safely be used on a 208-volt circuit, with some voltage drop besides, that could put the brake out of action in a hurry.

Many a-c brake coils are dual voltage. However, if a brake is rated 230 volts only, it can be matched to a 230/460-volt motor on either connection by using any of the arrangements shown in Figure 8-30 (next page).

Reduced-voltage starters are used with many a-c motors. The solid-state "soft starter" is becoming popular for many drives, with its advantage of stepless acceleration providing both reduced inrush current and reduced shock to the rotating parts. But any spring-set brake used on such a motor must be reconnected to get its power from a separate full voltage source, because the brake will not release at the reduced motor voltages applied by such starters. Hence the set brake will prevent the drive from starting at all.

(5) Enclosure:

Brakes may be dust or water tight, weather protected, or (for disc brakes

More about Optional Accessories   373

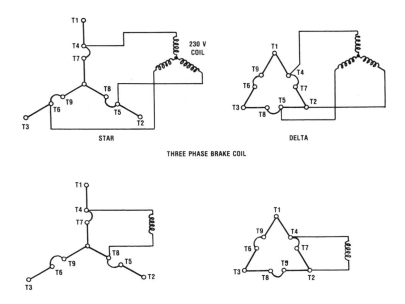

Figure 8-30: How 3-phase and single-phase a-c brake coils rated 230 volts may be connected to 460-volt star or delta connected motors.

only, because of their physical construction) explosion proof for various hazardous atmospheres. (See Figure 8-31.) Dust and dirt, especially metal particles, can wreak havoc in the brake magnet assembly and in the friction parts of a disc brake. Water and chemicals may cause corrosion or coil failure. So the brake enclosure must protect the working parts against the environment. Moving a brakemotor to an entirely different location, or applying it to a different machine or process, requires that the enclosures of both motor and brake be checked against the needs of the new environment.

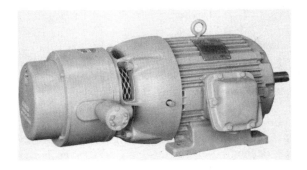

Figure 8-31: An explosion-proof brake mounted on a motor for hazardous area service.

Control methods

How are brakes controlled? Let's consider a-c brakes first. Figure 8-32 shows the simplest circuit. Whenever power is applied to the motor, the brake coil is automatically energized to release the brake. Should any accident interrupt that power, the coil drops out and the brake sets. If the motor circuit itself is not opened, but the brake coil fails, the brake will set, but not necessarily with enough torque to stall the running motor. Some brakes therefore contain thermal "overloads" of various types to give an alarm or shut off the motor starter when the brake overheats, particularly in hazardous atmospheres. (See Figure 8-33.)

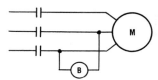

Figure 8-32: Simple a-c spring-set brake operating circuit for 3-phase motors. Brake coil is automatically energized to release brake whenever starter contacts close.

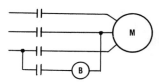

Figure 8-33: An extra starter contact is provided in the a-c spring-set brake operating circuit of Figure 8-32 so the brake sets more quickly when the motor is shut off.

Where combustible dust is present, Underwriters Laboratories requires that disc brake housing surfaces not exceed 165°C. At one time, some brakes for this service used a special friction disc material which disintegrated at about 200°C, rendering the brake inoperative before the exterior got too hot. But where braking was still needed for safety, this often unexpected disabling of the brake was not desirable. In other makes, a thermal/mechanical overload would release the brake, and keep it released, when the friction discs began to overheat.

Newer versions use self-contained thermal switches. For example, a lower temperature "alarm switch" trips first, to warn of the danger. If the brake stays in use and gets hotter, a second switch trips the motor off the line, automatically setting the brake and safely eliminating further operation. (See Figure 8-34.)

After any motor circuit is opened, trapped magnetic flux in the windings will continue to generate terminal voltage for periods ranging from a fraction of a second up to several seconds, depending on the design. Because the brake coil is connected directly across those terminals (usually inside the motor

More about Optional Accessories

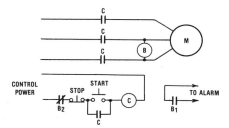

Figure 8-34: A-c disc brake circuit for explosion proof service. Brake thermal overload contact B_1 operates alarm circuit when internal brake temperature nears unsafe level. If brake is not allowed to cool, second overload contact B_2 opens to stop motor and set brake.

frame or terminal box), it will remain energized until that generated motor voltage dies out. This causes the brake to set sluggishly. Faster operation is possible with the circuit of Figure 8-33, which uses an additional starter contact to separate the brake coil from the motor when the contactor opens. Figure 8-35 shows the difference this can make in brake operating speed.

An a-c motor may use a d-c brake supplied from the a-c line through a rectifier control unit. One reason is the reduced noise of d-c magnets. Another is that it may be undesirable to add the peak brake inrush current, characteristic of a-c magnets, to the motor inrush occurring at the same time.

Typical rectifiers use a full-wave silicon bridge to produce 100 to 230 volts d-c from 115, 230, or 460 volts, single phase, 50 or 60 Hz power. (See Figure 8-36.) Note that all brake circuit switching is done on the d-c (output) side of the rectifier. If switching were done on the a-c side, the collapsing magnetic field in the de-energized brake could feed a short-time transient current through the rectifier circuit and brake coil, thus delaying release of the brake.

The d-c brake magnet is also more efficient and more compact. When used on a-c drives, however, the extra control components may offset those advantages. Each application must be decided on its own merits.

Figure 8-35: How the circuit of Figure 8-32 speeds up brake response for a typical drive (solid curve) compared to the simple circuit of Figure 8-33 (dashed curve).

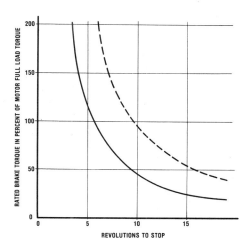

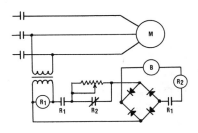

Figure 8-36: Rectifier power supply circuit for d-c brake used with a-c motor. Closing the motor contactor energizes a-c relay R_1, which completes the brake coil circuit to release the brake. In the process, relay R_2 operates to insert series resistance into the a-c side of the bridge, reducing the current load on the rectifier.

Shoe brake circuits with d-c motors contain some complications. For either shunt or compound wound motors, the shunt brake circuit is used. Brake voltage is relatively high, which means the coil must have many turns to operate on a fairly low current. Being energized continuously to keep the brake released while the motor runs, the coil would overheat if it drew a high current.

But a coil of many turns is so inductive that magnetic flux buildup, and operating force, is slow to appear when the coil is energized. This makes for sluggish brake release. To avoid that, a series resistor is provided, as shown in Figure 8-37a. This lowers coil voltage when the brake is energized, making it operate faster. Also, there may be a separate contact in the coil circuit between brake and motor, to prevent residual voltage in the motor from keeping the brake released after the motor contactor opens (just as in the a-c situation of Figure 8-33). Any change or defect in these resistor circuits can cause faulty brake action.

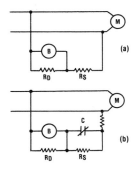

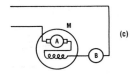

Figure 8-37: Brake control circuits for d-c motors. At (a), series resistor R_s is needed to control brake voltage and current properly with a shunt or compound-wound motor. At (b), part of R_s is shunted out by the contact C when the brake first operates. This may be a "forcing relay" with a short time delay before opening, or a contact on the brake mechanism itself. In the series motor circuit (c), the brake operates directly from motor armature current. No resistors are needed in this low voltage, low inductance brake circuit.

In each version, "B" may actually be two separate interconnected coils, for extra force. Shunt brake coils are connected in series; series brake coils are connected in parallel. Discharge resistor R_D dissipates the energy released by high coil inductance when the circuit is opened to set the brake.

Series wound d-c motors use an entirely different brake coil. This "series coil" has few turns of low resistance to carry full motor armature current. (See Figure 8-37c.) NEMA standards require such brakes to release when series coil current rises to 40 percent of rated motor current, and set when current falls to 10% of rated. (That is for ½ or 1 hour duty; for continuous duty the figures become 80% and 20% respectively.) On the other hand, shunt brakes should release when voltage rises to 80% of rated, and stay released down to 20% of rated voltage. Most d-c disc brakes are of the shunt coil type.

There are several advantages to series d-c brakes besides their simpler circuits because of lower coil inductance. Coils are more durable because of fewer turns and larger wire. Lower coil voltage is easier on insulation. Any opening of the motor's armature circuit, which might otherwise cause an overhauling load to drop or overshoot dangerously, will automatically cause a series brake to set and stop the drive. If the motor tries to overspeed, armature current drops to a very low value, permitting the brake to set and possibly preventing a damaging runaway. Similarly, however, when the motor load does vary quite widely, armature current may not always be high enough to keep a series brake released; in that case a shunt brake must be used.

Servicing brakes

Having seen how the common types of brakes work, we should next be concerned with their maintenance and repair. Many brakes, both shoe and disc type, are self-adjusting for normal lining wear. Make only those adjustments which the brake design provides for, and that you know how to make.

When attempting to replace brake shoes or friction discs, install new brake coils, or adjust brake torque (where that is possible), look for complete servicing information from the brake manufacturer. There are far too many design variations to include here. However, a few general principles can be outlined:

If the brake fails to release:

(1) Check for an open coil circuit. Is power reaching the coil from the motor winding or control system? Find out if the motor is being started, purposely or otherwise, on reduced voltage.

(2) Look for excessive voltage drop in brake leads or elsewhere in the circuit by connecting a voltmeter at the brake magnet terminals. Block the magnet travel with a piece of wood thick enough to hold the normal de-energized air gap. Then energize the brake circuit for a couple of seconds. Match the voltmeter reading with the brake coil rating. If the meter reads low, check out the power source and the wiring throughout the circuit.

(3) If the brake is a dual-coil d-c type, make sure the two coils are properly interconnected to "boost" rather than "buck."

(4) Look for shorted coil turns. Do this by carefully measuring coil resistance, checking it against manufacturer's data. If this is done when the brake

is hot after use, or in a warm location, measure the coil temperature, too, using Figure 8-38 to convert resistance to the usually "cold" coil data from the manufacturer. A hot coil with shorted turns may look like it has the right resistance, until you convert it to the proper reference temperature.

(5) If a d-c shunt brake coil is overheating or has burned out, make sure the correct value of series resistor is connected. If a forcing relay is used, see that it is working properly to reduce coil voltage once the brake has released.

(6) In some shoe brakes, badly worn linings may allow the magnet armature or solenoid plunger travel to increase so much that sluggish release results, or none at all. Check the lining. If rivets are showing, replacements are needed.

(7) Make sure all links, levers, pins, etc., are in place and workable. Look for binding. See that solenoid plungers are free from dirt which may prevent seating.

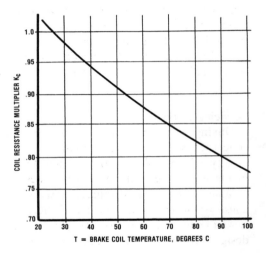

Figure 8-38: To convert brake coil resistance at actual temperature T to "cold" coil resistance R at 25°C, use K_c from this curve in this formula:

$R = (K_c)$ (coil ohms measured at temperature T)

If a brake fails to set:

(1) Look for broken springs, or binding in the mechanism.

(2) Observe the condition of linings or friction discs. Discs should be replaced if charred, cracked, or broken, if oily or greasy, or if worn to as little as half their original thickness. Some brakes have a lining "wear indicator" to show when replacement is advisable.

(3) Brakes without self-adjusting mechanisms will need periodic adjustment of magnet or solenoid travel as friction surfaces wear. This must be done in accordance with factory instructions.

(4) With disc brakes, watch for possible shifting of hubs on shaft which may throw discs out of engagement. Do the discs slide freely on the hub?

(5) If the brake has a manual release, see that it has not been engaged by mistake.

More about Optional Accessories 379

To check brake torque against specs:

(1) Set the brake *briefly* several times while maintaining power on the motor.

(2) Using a one- or two-foot torque arm with a suitable spring scale at the outer end, the inner end securely locked to the motor shaft, measure the scale force needed to rotate the motor shaft slowly with the brake set. Be sure to hold the scale at right angles to the arm while reading. Torque should be within plus or minus 10% of the brake rating. Listen for chattering or squealing during this test.

(3) For larger brakes, above about 125 pound-feet, this test becomes cumbersome. An approximate check can be made by finding out the inertia of the motor itself, bringing the uncoupled motor to full speed, then de-energizing it and setting the brake. Measure the stopping time carefully. The average brake torque will be approximately:

$$\text{Torque, lb-ft} = \frac{(\text{Inertia, lb-ft}^2)(\text{Full speed rpm})}{(308)(\text{Stopping time, sec.})}$$

Be sure to include the shoe brake wheel inertia in the formula. Typical brake inertia values appear in Figure 8-39. Note that shoe brake inertia tends

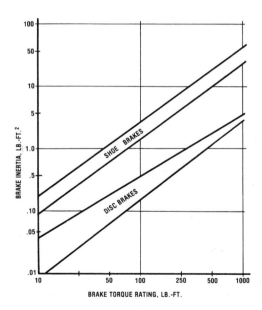

Figure 8-39: Inertia values for various makes of the two basic types of brakes.

to be about ten times as great as for a disc brake of the same torque rating. For brakes rated above 70 lb-ft, inertia of the disc type can generally be ignored because it is so small compared to the inertia of the motor rotor itself. Here are some figures for brakes rated to match 4-pole motor full load torque only:

Brake torque, lb-ft	Disc brake inertia as percent of motor inertia	Shoe brake inertia as percent of motor inertia
15	21	125
70	6.1	61
230	3	29

If an a-c brake is noisy in operation:

(1) Look for a broken or missing shading coil in the magnet pole face.

(2) Look for dirt or foreign matter interfering with solenoid plunger travel, or preventing armatures from seating.

(3) Look for solenoid assembly misalignment—shifted coil frame, or plunger guides out of line. Proper seating of the magnet is extremely important. Even when the travel is sufficient to operate the brake, partial seating will cause abnormally high coil current. This can overheat and burn out coils quickly, or destroy the rectifier supplying a d-c coil.

A brake checklist

Here are some other useful maintenance procedures:

Brake linkage pivot pins should be lightly oiled periodically. Be careful to avoid getting lubricant on brake shoes or friction discs. Some larger brake mechanisms employ grease-packed needle bearings, which should not need relubrication unless the linkage is dismantled. Follow the brake manufacturer's lubrication instructions.

Are all nuts tight and locked? Do pins show grooving, scoring, or noticeable wear? What is the condition of the brake wheel rim? Pieces broken out may not interfere with action of the brake, but they may unbalance the wheel enough to cause drive vibration. Look for wheel scoring, cracks, or discoloration indicating overheating. You may be able to clean up a bad surface by turning the wheel down, but ask the manufacturer first, because this will affect shoe travel and could require brake readjustment.

Cracked or discolored brake wheels, or disintegrated shoe linings, indicate too severe a duty cycle. Stopping frequency should be checked against the brake's design capacity.

Check the mounting of foot-type brakes, whether on the floor or on a motor bracket. The foot surface must be square with the shaft, locating the design wheel centerline within plus or minus 1/16 inch of the motor shaft centerline.

If one coil fails in a dual-coil brake, it is often possible to operate for a time with one coil only. The bad one should be either cut out or shorted depending on the interconnection between the two.

Coil circuit insulation can be tested by applying a-c over-potential as follows:

	Test volts to ground	
Brake rating	1 minute	1 second
0-50 v.	1100	1320
51-125 v.	1250	1500
126-240 v.	1480	1780
241-500 v.	2000	2400

A water- or dust-tight brake enclosure will have drain plugs. If these have not been periodically removed, water may have built up inside the brake housing, so check for that. Some explosion proof disc brake models come with a small internal heater, like a motor space heater, to prevent moisture condensation inside the brake housing. Make sure this is working, especially if examination shows rusting inside the brake.

All friction braking systems have these common features:

(1) Once the brake has been applied or "set," it exerts a constant stopping torque. With spring-set brakes, there is no way to significantly vary that torque. Stopping is relatively "jerky"—it subjects components to high mechanical stress or shock. This can be troublesome in such drives as conveyors carrying a fragile product.

(2) Friction brakes can be coupled only to motors that have an extra shaft extension for the brake wheel or discs. To mount the brake on the motor itself requires a special bracket. The brake mechanism takes up considerable space.

(3) Friction brakes have wearing parts which need periodic adjustment or replacement.

(4) In classified areas the risk of explosion requires expensive brake enclosures which may not be available for all ratings.

For all these reasons, engineers from the earliest days of induction motor development have sought ways to stop motor drives without an external mechanical brake. Half a dozen methods of "internal braking" have been patented. Approximately 14 different schemes (some of them with many different variations) have been tried experimentally, or on commercial machinery, with varying success. No one braking scheme, of course, can suit every drive.

How internal braking works

How does internal braking work? The three-phase induction motor rotates because the magnetic field produced by the individual phases combined is a rotating field. This drags the rotor around with it at a speed which changes only slightly with horsepower load on the shaft.

To produce braking action, one must "unbalance" that rotating field—destroy the symmetry between the three phases, or upset the normal relationship between the rotating field and the rotor itself. This can be done in many ways.

The simplest way is by plugging—reversing the direction of the rotating field without otherwise disturbing it. This can be done either at rated voltage or at reduced voltage. The advantages of plugging are that it works with motors that are standard both electrically and mechanically; it produces high stopping torque (Figure 8-40); the controls are relatively simple.

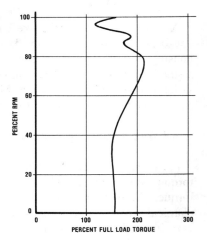

Figure 8-40: Typical full-voltage plugging torque for stopping a small induction motor.

But there are serious drawbacks. For example, plug-stopping can produce transient torques 10 to 20 times rated motor torque, severe enough to snap shafts, destroy couplings, or tear the drive from its foundation.

Whereas a friction brake stop adds no heat at all to the motor itself (all the energy loss of deceleration is dissipated in brake heating), plug-stopping a motor at full voltage heats it up as much as three starts do. This seriously limits the allowable stopping and restarting frequency. In larger sizes, manufacturers do not recommend plug-stopping at all because of its mechanical stress on windings, as well as the severe heating.

Another disadvantage of plugging is that the braking power must be shut down at just the right instant. Otherwise, the motor will simply "take off" in the reverse direction. This usually means putting a speed switch in the drive somewhere to sense a very low rpm at which to disconnect the motor. (See Figure 8-41.)

Both the motor heating and mechanical shock effects of plugging can be reduced by lowering the plugging voltage. However, that considerably complicates the control.

For all those reasons, plugging is seldom used as a normal stopping method for industrial machinery drives. It is common for emergency stopping of such loads as rubber mixers.

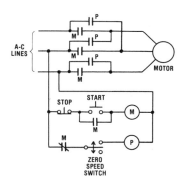

Figure 8-41: Plug-stop circuit showing speed switch to remove power when motor nears standstill so that it will not reverse. (For clarity, normal overload heaters and trips are omitted from these diagrams.)

Alternative schemes

A great many alternative a-c braking schemes are available, though not all of them can be found in drives now on the market. Some use single-phase power—normally available wherever three-phase power is in use. It's well known that simply "single phasing" a running three-phase motor won't stop it. But the connection of Figure 8-42 (one of several used in Europe) will generate a low braking torque, with zero torque at zero speed. To get reasonable deceleration with a low-friction load usually takes a double-cage rotor design.

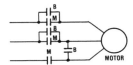

Figure 8-42: This simple single phase braking arrangement, developed years ago in Europe, is of little value for standard motors driving low-friction loads.

With special winding connections, however, the single-phase field may have typically six times the number of poles as the three-phase motoring connection. For example, one connection for a 4-pole motor, of 1800 synchronous rpm, produces a 24-pole braking field for which the synchronous rpm is only 300. When single-phase voltage is applied, then, the motor is slowed down to 300 rpm by "regenerative" or "overhauling" braking action—it is trying to rotate well above the synchronous speed corresponding to the magnetic field, so it is retarded by induction generator action until it reaches that speed.

At that point braking torque disappears. One patented scheme for overcoming that problem was introduced by a couple of motor manufacturers over 20 years ago. It divides the braking winding—a separate set of coils occupying about ⅓ of the stator slot space—into two sections. One of these, the "exciting" section, acts as just described. The other, or "short-circuited" section, wound with few turns of large wire and shorted upon itself, acts like a closed transformer secondary to absorb drive energy of rotation at still lower speeds, bringing the motor down to near zero speed where friction will normally stop it completely. Motors with such separate brake windings are oversize for the horsepower. The slot space required for braking coils is not available for useful motor output.

This basic brakemotor is still being marketed to manufacturers of machinery drives. A current version adds a "holding relay" for the formerly short-circuited portion of the brake winding. The high impedance relay coil responds to voltage induced in those coils. During motor operation that voltage is quite low, so the relay remains de-energized with its contacts open. Those contacts are connected in series with the "shorted" winding so that its circuit is actually open during motoring. During braking, the "exciting" portion of the brake winding induces a high enough voltage in the "shorted" coils to pick up the relay; its contacts close to complete the shorted circuit. Thus energy is lost in the shorted winding only when that is needed for braking.

In a second variation of the separate braking winding idea, the two sections (distributed in the slots at a 90 degree phase position like a two-phase winding) were alike. This was proposed on an auger drive for filling sugar bags. The object was to bring the auger to a stop and then have it back up slowly for a short time, thus preventing sugar from flowing by gravity down through the stopped auger after the bag had been filled. One of the two phases of the braking winding was weakened by a series impedance, thus providing a small amount of reversing torque.

Still another successfully tested idea was a three-phase auxiliary braking winding in two parts, the magnetic poles of one section arranged to oppose those in the other. Each of these two separate rotating fields act to slow the motor down at high rpm. At a lower speed, the torque of one field tries to reverse the motor, while the other merely tends to slow it to some lower synchronous speed (such as 600 rpm for a 1200-rpm motor). At zero speed, the two fields cancel each other entirely so the net torque becomes zero. (See Figure 8-43.)

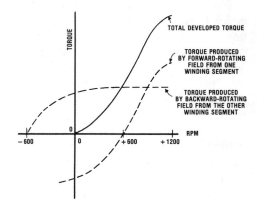

Figure 8-43: The two-section or "segmented" auxiliary braking winding, each section opposing the other, produces this kind of decelerating torque.

In a 4-pole motor, for example, each of the two segments of the brake winding produces four poles distributed over half the stator circumference. This has the same effect as an 8-pole complete winding, generating a rotating magnetic field at 900 rpm synchronous speed. The same is true for the other half of the braking winding, but in the opposite direction. The result might be termed "part-winding stopping."

Other methods

Capacitors, reactors, and tapped resistors have been used across or in series with separate braking windings to control the amount of braking torque. Other methods avoid the complexity of that extra winding by reconnecting the main motor winding in various ways to produce single-phase braking poles. One such method, patented in 1957, is illustrated in Figure 8-44. Here the contactor arrangement for switching from motoring to braking is fairly simple. There are dozens of variations in the connection, for using half, two-thirds, or all of the winding during braking, as well as for single voltage, dual voltage, or reversing motors. The motor may have as many as 18 leads, the contactors requiring from 5 to 15 poles. In one arrangement, four separate contactors are needed, two 4-pole plus two 3-pole.

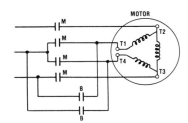

Figure 8-44: The simplest of a number of different ways to reconnect a three-phase motor winding for single-phase braking. Typically, the number of braking poles is three times the number of motor poles, and heating is low. (Control circuits for the motoring contactor M and the braking contactor B are not shown.)

The number of braking poles generated may be 2, 3, 4, 6, 9, or even 18 times the number of motor poles. Braking torque is relatively low—but so is current and heating. As a typical example: one 4-pole motor used 12 braking poles, and average braking torque was 45% of accelerating motor torque.

Some of these connections will work properly only with certain coil groupings, number of stator slots, or even coil pitch. Rewinding for simple voltage changes, which may cause no problem with standard motors, can render a winding incapable of braking properly.

Once disconnected from the a-c line, an induction motor can also be braked to a stop by connecting capacitors across the windings to excite the machine as an induction generator. The coasting load will then drive the "generator" until the rotational energy has been used up. The deceleration can be made to occur sooner by including energy-absorbing resistors in the capacitor circuit. Two versions of this appear in Figure 8-45. The resulting braking torque characteristics are shown in Figure 8-46.

There are several problems with such braking. One is the size (and cost) of the capacitors. To produce an initial peak braking torque of twice the motor's full load rated torque requires capacitance equal to about three times the no load or magnetizing kilovolt-amperes of the motor. This will usually produce

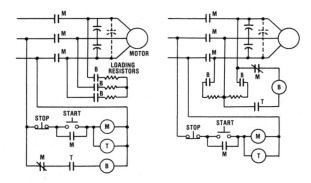

Figure 8-45: Two circuits for capacitor braking, used rarely with small motors. Each version uses resistors to absorb deceleration energy. Two capacitors are normal; a third, represented by dotted lines, is optional. The timing relay T operates to disconnect the braking circuit after a pre-set slowdown period.

Figure 8-46: Capacitor braking produces this kind of deceleration. Because braking torque disappears at a fairly high rpm, this scheme has limited usefulness.

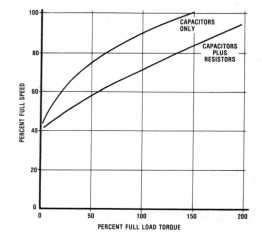

"over-excitation" of the motor—that is, the terminal voltage during braking may temporarily rise so high that it endangers insulation.

If capacitors sized for power factor correction were to be used for braking, the amount of braking torque would be quite low. However, it might be enough for high-friction loads needing little deceleration torque, especially at low speed.

Although often suited to a particular industrial application, most a-c braking methods have found little commercial acceptance. The main reason is the need for special motor windings, lead connections, or both. A second reason is that braking torque is relatively low, especially at low speed. A special high resistance (and low efficiency) rotor design may be required to compensate for that.

Other disadvantages are:

(1) Elaborate winding or control circuit connections are often needed.

(2) Those methods using a separate winding may exact considerable price penalty.

D-c dynamic braking

An internal braking system which avoids those drawbacks, one that has been well known for generations throughout the motor industry, is "d-c dynamic braking." It uses completely standard motors plus fairly simple control. The only disadvantage is the need for a separate source of d-c braking power. Until recent years, that limited the usefulness of dynamic braking. Now, however, as we shall see, solid-state electronics has made it an attractive alternative.

Here the normal relationship between polyphase rotating field and motor rotor is upset by changing that field from alternating to direct current supply. During braking, the three-phase leads are disconnected and a d-c braking voltage is applied to any two of them.

For comparison with the plugging torque of Figure 8-40, dynamic braking torque will vary as shown in Figure 8-47. The magnetic field generated in the motor is stationary, not rotating, with as many poles as the motor normally has for running.

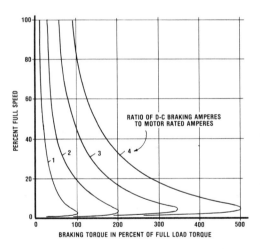

Figure 8-47: Speed vs. torque curves for d-c dynamic braking. The values of torque at a given rpm, or at a given current, are dependent on motor design, and particularly variable with rotor resistance.

There are various pros and cons to evaluate. A big advantage of d-c braking is that motor heating is only about ¾ that for starting. Thus, one d-c stop plus one start heats the motor only half as much as one full voltage plug stop plus a start.

The shape of the d-c braking speed-torque curve provides a couple of benefits: (1) braking torque rises rapidly at low speed, so that the motor initially slows down rather gently, the more abrupt deceleration occurring only after it has reached a low speed; (2) the high torque at low speed tends to allow drives with high friction to coast quickly to a stop once past that torque peak, and prevents "windmilling" as with some types of fans.

The connections shown in Figure 8-48 include contactor poles to open the circuit between motor winding and rectifier when the motor is running. If this were not done, a-c line voltage would damage the d-c supply. That extra contact can be eliminated by applying the d-c voltage between two points within the winding that are normally at the same potential when the motor is running. Since no voltage appears between or across those two points, the rectifier can remain in the circuit without damage when brake power is off. The braking circuit feeds the neutrals in the two parallel winding paths. Figure 8-49 illustrates the effect.

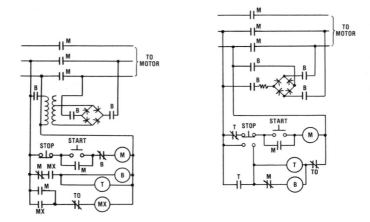

Figure 8-48: Two versions of d-c braking control using conventional rectifier power. Both include the timing relay T, which cuts off braking power (via the time delay contact TO) after the motor has had time to stop. Leaving braking power on longer would simply waste energy in winding heating. The circuit on the left uses an auxiliary contactor, MX, plus a tapped transformer to vary braking voltage—a purpose served in the circuit at right by the resistor shown.

Mechanical interlocking is usually provided to prevent both B and M contactors being energized simultaneously. Relay contacts prevent motor restarting during the braking cycle.

It is claimed that this kind of circuit also uses less d-c power than a conventional scheme. But there is another benefit of greater value. Usually a winding will develop as many d-c braking poles as it does motor poles. With the hookup of Figure 8-49 (left), of which there are at least a dozen variations, the number of braking poles is typically only half the number of motor poles. Because of what that does to rotor resistance, the braking torque curve changes its shape considerably, as seen in Figure 8-49 (right). This can be helpful for applications in which braking torque at the higher speeds would otherwise be inadequate. In addition, less d-c current is needed for a given braking magnetic field strength if the number of braking poles is reduced. This saves energy, may reduce contactor/rectifier capacity, and permits more stops per hour.

More about Optional Accessories   389

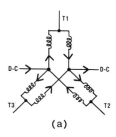 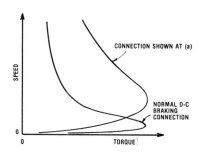

Figure 8-49: With the winding connection shown at left, d-c braking voltage is applied between two neutral points across which no voltage appears during motor operation. A rectifier power source can therefore remain connected whether the motor is on or off. The number of braking poles is less than the number of motor poles, which results in higher braking torque for the higher rpms—as shown in the curves in the graph at right.

We have seen that single-phase a-c braking has high torque at high rpm, whereas d-c braking torque peaks at low rpm. It therefore occurred to one inventor to combine those two effects in a single braking circuit using both a-c and d-c. The result was a 1956 patent under which some motors were successfully built. One circuit appears in Figure 8-50, having a braking torque characteristic per Figure 8-51. A variable resistance for torque adjustment can be included in the d-c part of the circuit.

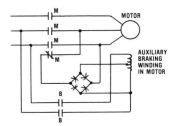

Figure 8-50: A circuit used for combined a-c and d-c braking. The main braking torque comes from the auxiliary single-phase winding; initially high, it drops to zero when the motor nears standstill. A d-c supply is tapped into part of that winding, giving a low d-c torque that peaks near zero rpm. Auxiliary braking windings are usually placed in the tops of the stator slots, above the motor coils.

A disadvantage of dynamic braking is that the average braking torque sufficient to stop the drive in the required time brings with it a "peak torque" (the maximum torque point on the curves of Figure 8-47, page 387) that is sometimes dangerously high for drive components. This needs careful checking in any drive.

Here we must consider a somewhat controversial point about dynamic braking, which also arises with other forms of internal braking: Does this system provide a "holding torque" to anchor the drive at zero rpm?

According to the curves of Figure 8-47, it does not. There is zero torque produced by the d-c field when the motor is at standstill. Authorities have disagreed on this, however. One writes that "while this statement appears plausible from an academic standpoint, it is, as a matter of fact, not true."

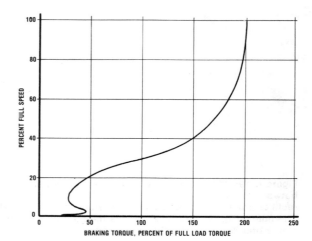

Figure 8-51: Combined a-c and d-c braking torque curve for the circuit of Figure 8-50.

Others make flat statements like "dynamic braking torque drops to zero at standstill," or that such braking "does not provide holding power," or that "no holding torque is exerted at exactly zero speed."

What really happens is this. The d-c braking torque does become zero at zero motor rpm. But if the stop occurs quickly enough, transient effects may give rise to some unpredictable torque at zero speed—a sort of overshoot effect. This has led some observers to claim that holding torque can generally be expected. But for normal stopping cycles that will not be true.

Also, the torque peak at very low rpm means that a slight rotation of the shaft will abruptly encounter large restraining torque. This too has led some people to consider this as a holding torque. And it is often quite useful, as in preventing windmilling of cooling tower fans. But it is not a true "holding torque at standstill." In any drive that must be held, then, such as a crane hoist which could otherwise drop its load, a friction brake must also be provided.

Note that even the existence of a sizeable torque at zero speed doesn't mean the load will be held. The plugging torque of Figure 8-40 demonstrates that fact.

Electronic d-c brakes

The flexibility of modern solid-state power supplies gives today's d-c braking a high degree of simplicity and flexibility for standard motors. A typical "electronic d-c brake," as these are sometimes called, appears in Figure 8-52. This particular line is available for motors from 1 through 700 hp. At least one specialty control firm has offered a similar package for years up through 200 hp. These braking systems have been marketed under various trade names. Features available include adjustable deceleration rate and jogging.

More about Optional Accessories

Figure 8-52: A modern electronic brake controller, for a 150-hp motor. Shipping weight is 90 pounds. Units of this kind are readily available for drives through 700 hp, 460 volts.

Figure 8-53 shows how the cost of this type of braking compares with that for friction brakes (disc type—shoe brakes will generally cost more, but for the largest motor ratings are the only choice). If positive holding action is not needed, it is a considerable advantage to be able to use standard motors without special mounting brackets or shaft extensions. Where mechanical shock is a problem, the modern d-c circuit can often provide a far smoother, controlled deceleration.

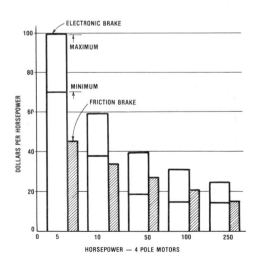

Figure 8-53: For standard 4-pole motors with braking torque equal to 100 percent of rated motor full load torque, prices for electronic d-c braking vs. friction (disc) braking are about as shown here. Disc brake cost includes motor mounting, shaft extension, and wiring as well as the brake itself, and will vary with brake enclosure (values shown are for drip proof brakes). Shoe brakes, required for the larger horsepowers, are much more expensive. A wide variety of options causes great variation in the electronic brake data.

Normally, dynamic braking uses a voltage source rated at no more than 5% to 10% of rated motor voltage. Figure 8-54 shows how much braking current is needed for a given torque. Assuming current of 150% of motor full load amperes (adequate for many drives), here are typical braking parameters for standard 4-pole, 460-volt motors:

Rated motor hp	D-c braking amps	D-c volts (percent of motor rating)
5	10	38 (8.3)
10	22	37 (8.0)
25	50	23 (5.0)
50	95	14 (3.0)

This brings out another advantage of the modern electronic braking circuit. Its voltage variation capability permits one unit to suit several horsepower ratings. The older rectifier systems required a tapped transformer or other less efficient means of adjusting output volts.

Users may realize how plug-stopping severely stresses a motor. But they sometimes aren't aware that d-c braking (like any other type of internal braking) imposes its own stresses. Stopping a motor too often can burn it out.

Energy loss in a motor during braking includes two components. There is a loss W_r in the rotor, which is present during any deceleration of a rotating system. When a friction brake does the work, that heat loss will appear at the brake shoes and wheel. The energy involved depends upon the rotating inertia, the total change in rpm, and the amount of retarding torque supplied by the load during slowdown.

Commenting on this, one supplier has said that "all the load energy goes into heating the rotor of the motor...." By "load energy" he means the W_r component.

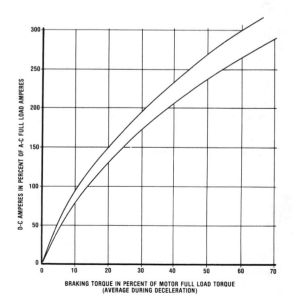

Figure 8-54: To produce a desired "average" braking torque throughout the slowdown period needs d-c braking amperes as shown here. The width of the band allows for variation in motor design; "high torque" (high rotor resistance) designs generate more torque at a given current than standard motors do.

But there is also stator heating, W_s, caused by the flow of braking current through the winding. This cannot be neglected. It depends on the d-c amperes, the winding resistance, and how long the current flows.

Because braking time is shorter when braking current is higher, W_s tends to be fairly constant for various slowdown times provided the current is cut off as soon as the drive stops. The W_r component is not a function of time. However, a high inertia load is more difficult to slow down, so will raise both the stopping time and the rotor heating. On the other hand, a high friction load is easier to slow down, so will reduce both stopping time and rotor heating. When W_r and W_s are added up for a typical drive, the total usually won't exceed the total motor loss for one start, and will often be much less.

In applying d-c dynamic braking, then, the user should check with the motor manufacturer to be sure that the motor can safely handle the heat input of at least twice the number of starts being considered. This will be conservative.

The calculation methods in handbooks or catalogs for figuring starting time, motor heating, and braking power all assume the use of "pure" d-c for braking. But the output of a battery, d-c generator, or full-wave rectifier, is quite different in waveform from the output of an SCR circuit using adjustable firing angle to vary braking voltage. So the calculations may not be completely accurate. Most electronic brakes feature adjustment of braking time and power by the operator.

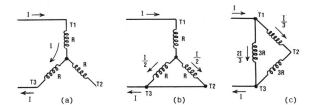

Figure 8-55: The three basic connections for applying d-c braking power to three-phase motor windings. Resistance R_T between terminals is 2R in case (a); 1.5R in case (b); and 2R in case (c).

Does it make a difference how the braking power is supplied to the stator? These are several possible winding connections, as Figure 8-55 shows.

For case (a), the total braking power P_{dc} in the winding, which is what produces braking torque, is:

$$I^2(R) + I^2(R) = I^2(2R) = I^2(R_T)$$

For case (b), the power is:

$$I^2(R) + (I/2)^2(R) + (I/2)^2(R) = I^2(1.5R)$$

For each connection to use the same power, and develop the same braking torque, the current for case (b) must be $\sqrt{2/1.5}$, or 1.15 times as great as the current for case (a). Because of the difference in line-to-line resistance, the applied d-c voltage for (b) can then be 15% less than for (a).

In case (c), the motor being equivalent in performance to the wye-connected version (a), we find that:

$$P_{dc} = (\tfrac{2}{3}I)^2(3R) + (I/3)^2(3R) + (I/3)^2(3R) = I^2(2R)$$

$$R_T = (\tfrac{2}{3})(3R) = 2R$$

$$\text{so } P_{dc} = I^2(R_T) \text{ as in (a)}$$

Summing up: Allowing for the inability of any non-friction braking method to hold a drive at standstill, d-c dynamic braking appears to be the best system for general industrial use. It is the most common. Its advantages are:

(1) It uses standard off-the-shelf motors; no special windings or leads.
(2) It requires a minimum number of contactor/control components.
(3) It is suited to all motor environments.
(4) The circuitry permits simple adjustments of braking torque, current, and time.

Keeping motors dry

Like braking, the next motor auxiliary function of interest may be provided either by self-contained accessory devices built into the motor, or by a special external circuit producing the desired effect within the motor winding. This is "space heating." As we shall see, it may not be "space" heating at all, but winding heating. Unlike braking, this function is needed for many other types of electrical equipment besides motors—switchgear, for example.

The object of electrical apparatus space heating is not merely to keep the equipment warm, but to keep it dry. Air always contains moisture, as expressed by the relative humidity. When air temperature falls low enough to reach the "dew point," that moisture will condense out as liquid on the surface of windings, terminal boards, etc., exposed to the air. Figure 8-56 shows that this can occur at fairly high temperatures if the air is humid enough.

Condensation is prevented either by warming the air itself within the equipment enclosure (indirect heating), or by warming the electrical parts themselves (direct heating). In switchgear or control cubicles, the wiring and various devices are so distributed throughout the enclosure interior that the entire space must be heated to make sure everything is warmed at least 5°C above the surrounding air temperature.

In motors, either the entire motor interior may be heated, or heat may be applied only to the windings themselves which are a fairly concentrated, regular shape rather than being a group of widely scattered parts.

There are three questions to consider when planning space heating:

(1) If heat is needed at all, when will it be needed?
(2) How much heat is required?
(3) Where and of what type should the heating elements be?

More about Optional Accessories

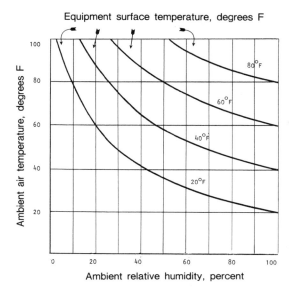

Figure 8-56: These curves can help predict possible condensation problems. Find the intersection point of a vertical line through the relative humidity you have, with the horizontal line through the ambient air temperature. If that intersection lies above the surface temperature curve for whatever you want to keep dry, condensation can occur. But note that if the equipment or winding surface temperature remains always above ambient temperature, condensation cannot take place.

Atmospheric conditions

Figure 8-56 can help determine if the atmospheric conditions themselves dictate space heating—whether the equipment is indoors or out, and seasonal variations in temperature and humidity. Another factor is the nature of the devices to be protected: Are there exposed terminals, or windings which are not waterproofed or encapsulated? Are materials subject to corrosion?

Consider also the nature of the environment. Are there dusts which become highly conductive when damp? What about vapors which can form corrosive liquids in the presence of condensed moisture?

Finally, consider whether or not the apparatus will generate enough heat of its own. Will it normally be energized, carrying load, at all times? If so, extra heating may not be required. On the other hand, if equipment is sometimes idle for long periods of time—days, weeks, or even months—or if it contains many accessories far away from the main heat source, it is a good idea to provide space heating. If heaters later appear unnecessary, it is simpler to leave them unused than to have to install some when no provision has been made.

Usually, heat will be needed only part of the time. To conserve heating energy, there are two basic methods of heater control:

(1) Manual or automatic switching to energize heaters only when the equipment is idle, common practice with motors. (See Figure 8-57.)

(2) Thermostatic controls to turn on heaters only when ambient temperature falls below a certain point. This is infrequently used with motors.

Thermostats present some problems. One is that it can be difficult to coordinate the thermostat setting with the surface temperature of the parts involved as well as the temperature of the surrounding air. As Figure 8-56 shows, both

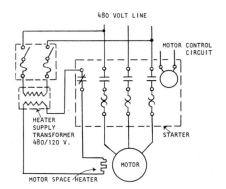

Figure 8-57: Typical scheme for automatically energizing motor space heater when the motor is idle.

are important. With outdoor equipment, it may be hard to vary thermostat settings properly with seasonal changes, especially with humidity.

How much heat to supply can be estimated from a few simple rules. Remember that too much heating is preferable to too little, and that there is no substitute for experience, or field testing, in deciding what amount is best suited to the application.

For large cubicles, such as typical switchgear or dry-type transformer housings, a useful rule of thumb calls for two watts of heating power per square foot of exposed surface. Thus, a cubicle 2 × 3 × 6 feet, with 66 square feet, would take about 130 watts of heating power. For convenience, switchgear manufacturers may provide heaters in readily available increments of 250 watts. Most large cubicles have some ventilation openings, for "free convection" air circulation, to carry away heat generated in operation of the equipment. If there are no such openings, space heating power can be reduced somewhat.

The '2DL' rule

Several rules have been suggested for sizing motor heaters. One is: watts needed = 30 times D times L, in which D is the overall frame diameter in feet, and L is the length over the end bells in feet. This rule can work fairly well for small motors up to about the 250 diameter frame, but above that it doesn't give enough watts for adequate heating.

Hence, the "2DL" rule is more useful: Supply watts about equal to twice D times L, in which D and L are stator core O.D. and stack length respectively, in inches. Heating power will typically range from 100 watts for a 10-hp, 1800-rpm motor up to 2200 watts for a 3000-hp, 2-pole machine. Figure 8-58 shows some actual test results on large machines.

This rule presumes indirect heating. The air itself inside the motor enclosure is heated, which in turn keeps the windings warm. If the heat is instead applied directly to the windings themselves, the amount of power needed is drastically reduced, as we shall see.

More about Optional Accessories

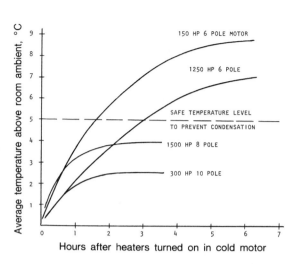

Figure 8-58: Actual space heater tests in various motors showing validity of the "2DL rule" for heater size. The two lower curves were produced by heating watts only 25 to 35 percent of 2DL; the upper curves, by wattages about equal to 2DL. Note that the lower curves do not reach high enough temperatures to ensure that all parts of the winding will always be free of condensation. (The time required for temperature to reach final value is not important, because when a motor is shut down after a long run and the heaters have energized, the motor will retain its operating heat for hours.)

This brings up the final question, as to the type and location of heating elements. Figure 8-59 shows some of the many different heaters available. Other types come in sheet or strip form, rigid or flexible. Some are self-sticking, like an adhesive label, to be placed directly on the inside wall of an equipment enclosure.

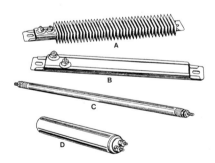

Figure 8-59: Several common types of space heaters for electrical equipment. A is a finned strip (low surface temperature); B, a strip heater; C, a tubular heater; and D, a cartridge.

The most commonly used elements are encased in a rigid metal sheath, usually of tubular or strip form. The exact choice depends on the room available for the wattage needed, as well as on the heater surface temperature that can be tolerated. Most commercially available metal-sheathed elements are designed to heat solids, such as press platens, die blocks, metal melting pots, and the like. They concentrate a lot of heat in a small space, counting on close contact with large masses to draw the heat away quickly. Watts per square inch of heater surface is 25 to 35. If mounted in free air, such elements will get red hot—from 750°F to 1500°F. Inside electrical apparatus, a flat metal surface with such elements clamped directly to it is likely to be dangerously overheated. Standoff insulators or brackets of some kind must be used.

For obvious reasons, such heaters aren't permissible in explosive atmospheres. Elements of much greater surface area, with watt densities from 20 down to 7, will provide the same total wattage at safe temperature—typically 200°F to 300°F. These are usually finned (as in Figure 8-60) in order to get increased area without too much increase in installation room.

Figure 8-60: Typical large motor with finned strip heater below stator, accessible through removable cover plate. (Photo courtesy Louis Allis.)

Tubular heating elements can be bent to adapt them to the space, such as around the end of a cylindrical motor housing. Follow these rules if you are bending such heaters in the field:

(1) Bend only when the heater is cold.

(2) Use a minimum bending radius of at least twice the tube diameter.

(3) Don't bend within a couple of inches of the end of the heating element inside the tube. With steel sheathed heaters, apply power and observe the end of the heated zone. If the sheath is non-magnetic, use a magnet to find the end of the internal steel "cold conductor" that extends from the end of the tube to the end of the heating coil itself.

(4) Clamp the heater down to a flat surface while bending, to prevent its buckling or bowing.

(5) Check heater continuity and insulation resistance afterwards.

Locate heating elements with due regard for the expected circulation of warmed air. Because heated air rises, heaters should always be near the bottom of the enclosed space. Avoid mounting them directly under interior structures or devices that may block vertical air flow. Sometimes it may be wise to split the heaters up, putting elements in several different places. This is especially true, for example, in large vertical motors, when heaters in the bottom alone may be too far removed from the top. To get windings several feet above the heaters up to 10°F above ambient, tests have shown that it may be necessary to raise the temperature of parts near the heating elements as much as 70°F.

When the enclosure has ventilating openings, don't place heaters so the warmed air will be wasted directly to the outside. The apparatus, not the surroundings, needs the heat.

Will the heater fail?

Are space heaters reliable? Experience of both heater manufacturers and the builders of electrical equipment using such heaters confirm that heater failures are rare. But they do occur. The metal-sheathed element is not vapor-tight. It can absorb moisture from the air, which has caused some heaters to fail when first turned on. Similarly, water splashing or dripping into a motor enclosure has been known to short-circuit heater terminals.

However, bear in mind that the normal function of space heaters is to keep idle equipment dry. This is particularly true for motors. If heaters are energized when the motor is not running, especially during periods of long storage prior to motor installation, there will be no opportunity for de-energized heaters to absorb moisture.

Motors being readied for initial startup should have heater insulation resistance checked. It may be low if heaters have not been in use. If so, they should be thoroughly dried out by operating for several hours at half voltage.

Unduly concerned about lengthening heater life, some users specify that motor space heaters be operated at voltages below their normal rating—anywhere from 80% down to 50% of the design value. The motor, unfortunately, needs the same amount of warming whatever the heater voltage. Since total heater watt output varies as the square of the voltage, working the heaters at half voltage cuts the power to one-fourth, requiring four times as many heating elements to be installed for the same total watts. In many machines, it becomes quite difficult to find the room for that many. Such needless overdesign should be avoided.

Users may have trouble with heaters in extremely corrosive environments. Heaters and terminals, though using alloys or platings resistant to corrosion, obviously cannot be coated with epoxy or protective varnishes as a motor winding can be. One remedy may be to enclose heating elements within junction boxes, as has been done in some pipeline motors to further reduce the exposed surface temperature. But enclosing the heating elements takes up a great deal of space and also requires "warm-up time" when heaters are energized.

Another solution is shown in Figure 8-61. The principle here is the same as in the electric hair dryer or "heat gun." A small 275 cfm motor-driven blower is attached to a box containing an open resistance-wire heater of the type used in the load boards seen in many shops. Heated air is blown directly into the motor housing. The total cost is between $100 and $150. The installation must be designed so the air doesn't blow right out again without some circulation through the motor interior. Sometimes motors that were never equipped with heaters are moved to different surroundings where heat is needed, and this is one way of supplying it if the auxiliary power is available.

Figure 8-61: Two-speed vertical motor with separate forced-air heater, developed by service shop, attached to air opening in frame.

Neither strip nor tubular heaters come in sizes suited to small NEMA frame motors, which will normally use one of two other types: the "cartridge" heater (Figure 8-59), or the "wrap-around" flexible heater tied right onto the stator end coils.

The wrap-around heater, available in various lengths and wattages, is either a glass-insulated tape or ribbon, or a silicone rubber molded strip. Because it can be tied directly around the winding, it applies heat where it is most needed, using only a small amount of power—typically 25 to 150 watts, or less than half the amount needed for other types of heaters. Surface temperature is no problem. Wrap-around heating elements have been successful through the 680-frame diameter, but the largest frame is usually the 440 or 500 size. With large formed-coil stators, the heater doesn't contact enough of the total winding surface to keep it all warm.

If the motor is a multispeed two-winding machine, wrap-around heaters may be unsuitable, unless the speeds are such that there is room to attach a separate set of heaters to each winding. Similarly, it may not be feasible to keep the rotor winding warm in a slip-ring motor by heating only the stator coils.

For machines up to about 250 hp, particularly 600 volts and below, separate space heaters can be eliminated. Applying low voltage single-phase power to the idle motor will develop sufficient heat directly within its windings, with far less power consumption. A typical scheme for doing this appears in Figure 8-62. The required kVA does go up rapidly with motor size (Figure 8-63), which tends to restrict the usefulness of this method. But it has worked even with a

More about Optional Accessories 401

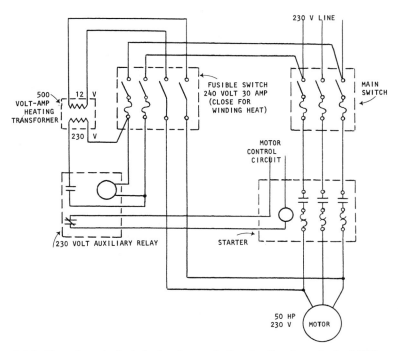

Figure 8-62: Manual control of single-phase winding heating for a typical 50-hp motor. Note that the starter is locked out when heater power is on.

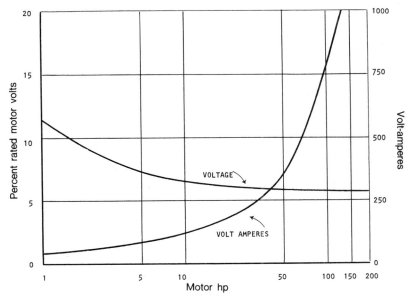

Figure 8-63: Single-phase heating voltage and volt-ampere levels needed to warm standard motor windings. Sharp rise in volt-amperes above about 75 hp shows why larger motors are seldom heated this way.

1500-hp, 28-pole, 4160-volt synchronous machine, although it's important to avoid opening the field discharge circuit with heater power on, because rotor induced voltages can endanger personnel. Such heating took 55 stator amps at 208 volts, or about 11.4 kVA. Although the 2DL rule would have required only about 4 kVA for separate space heaters, the enclosure for such large diameter machines of short length, usually of open pedestal-bearing construction, doesn't lend itself to uniform heating in that manner.

Direct winding heating

Direct winding heating of high voltage motors always involves a close look at local code compliance and personnel safety. A typical scheme that has been successfully used is shown in Figure 8-64.

Because actual heating watts generated in the single-phase heating method are only about 1/6 of the a-c kVA supplied, it may be economical to use rectified d-c power instead. The availability of solid-state devices has made this more popular.

Another advantage of low voltage winding heating is that all the heating connections can be made at the motor starter, saving the cost of extra trenching and conduit runs to the motor itself. A refinement of the method, eliminating the need for a separate low voltage transformer or rectifier, is shown in Figure 8-65. The electronic "black box" producing low voltage directly from the motor power supply lines via an SCR circuit is only about 3 by 3 by 5 inches. Similar circuitry is now commercially available.

One supplier has offered a combination "ground detector" and solid-state winding heater package for motors up through 4160 volts (Figure 8-66) that has been used with a number of chemical plant drives. This device contains a monitoring circuit to "track" the insulation resistance of the idle motor by applying 200 volts d-c to the winding, like a low voltage megohmmeter. Figure 8-67 shows how insulation resistance may lessen with time in an idle machine. When it falls below a pre-set level, the winding heating power is turned on, while at the same time the starter is locked out so the motor cannot be started until windings are thoroughly dried.

For motors that cycle on and off in response to pressure, temperature, or flow controls in a process line, the combination ground detector and solid-state winding heater may not be suitable. It may endanger the process for a motor to unexpectedly fail to start even with reduced winding "integrity." So such devices must be carefully matched to the job.

Single-phase a-c or d-c heating of a multispeed single winding motor can generally be handled the same way as for a single speed machine, except that if heater control is automatic it is advisable to wire contacts from both high and low speed starters into the heater circuit, so that power cannot be connected to the winding at either speed if the heating circuit is energized. Some increase in heater voltage may be necessary, because motor rated current (from which heating kVA is normally estimated per Figure 8-63) will be lower

More about Optional Accessories

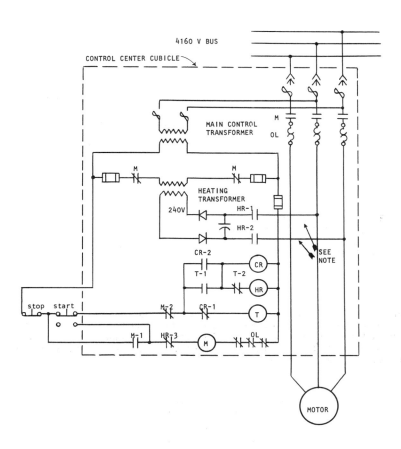

Figure 8-64: Circuit schematic for single-phase heating of high voltage motor. All components except start-stop buttons are inside control cubicle. Rectifier circuit supplies d-c heating power to motor, permitting low rating for contactor HR. (Note: the two d-c wires must be insulated for 5 kV.)

A note about this circuit's operating sequence: Upon closing of control transformer switch, thermal time-delay relay T is energized through start-stop button contacts. This relay opens contact T-2, then closes contact T-1, energizing control relay CR. Contact CR-1 opens, dropping out relay T while relay CR seals in through contact CR-2. Thermal relay T then cools and restores T-1 and T-2 to their original conditions. When T-2 recloses, heater relay HR is picked up through CR-2. This opens contact HR-3 to lock out the motor starter, at the same time applying heating power through contacts HR-1 and HR-2.

Pressing the motor start button drops out both HR and CR, and the closing of contact HR-3 permits the motor starter to operate. Starter contact M-2 then locks out the heater circuit. Any power interruptions during this operating cycle do not prevent cycle from resuming when power is restored.

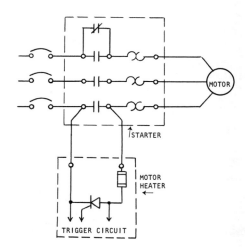

Figure 8-65: A recently patented method for using a thyristor or triac to provide winding heating in an idle motor with no auxiliary transformer.

Figure 8-66: A combination insulation resistance monitor and solid-state d-c heater package for 460- through 4160-volt motors. Low insulation resistance locks out the motor starter while energizing the heating supply. Pilot light shows when heater is in operation. Cost of this package is between $150 and $400, depending on voltage rating and degree of enclosure.

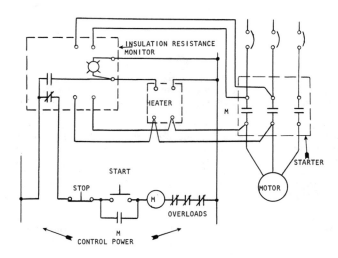

More about Optional Accessories 405

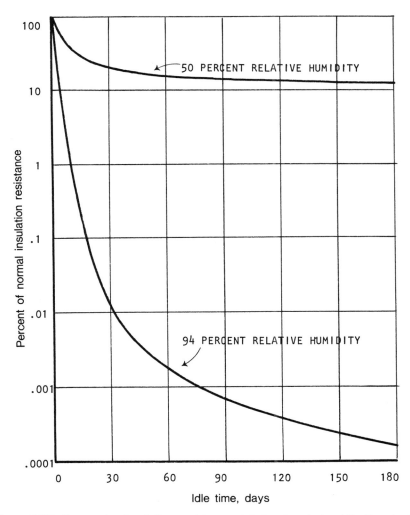

Figure 8-67: How motor insulation resistance deteriorates during idle time in humid atmospheres.

for multispeed ratings than for a single speed rating in the same frame.
In a two-winding multispeed motor, single-phase heating should be ap-

proached with caution. Parallel circuits or a closed delta in one of the two windings may act as a short-circuited transformer secondary to alternating heating current flowing in the other winding. This can greatly increase the heating kVA and even damage the windings. Shop trials, starting at very low voltages with careful measurement of actual winding temperature and current, should precede any field installation.

When heating power is applied automatically, local practice may require that a warning sign be placed at each motor, calling for the heating circuit to be de-activated before any work is done on the motor. De-activation might be unnecessary if a low-capacity, low voltage power supply is used for heating. But if a large supply is used, to serve a number of motors, or if a circuit like that of Figure 8-65 is employed such that line voltage might appear at the terminals of an idle motor, safety must be given greater consideration.

Heating power can also be applied manually. Some users simply provide a toggle switch for the heater circuit. One electrician, after repeated motor failures because of low insulation resistance on startup in a very wet and dirty plant, provided heating power by reconnecting some spare 4160/240-volt transformers with 440 volts applied to the primaries. The 26-volt secondary output is ample for single-phase heating of motors up to 200 hp. Low voltage cables were strung throughout the motor control center cubicles, having bared sections covered by plastic hose at each starter. To heat an idle motor, the hose is slid away, allowing lengths of No. 8 flexible cord with alligator clips to be connected from the bared sections of heating cable to the load side starter terminals. Starters for the larger motors, with separate disconnect compartments, carry warning signs and special locks to prevent motor starting with heating power connected.

Providing adequate space heating, whatever the method used, is cheap insurance. If equipment is expected to endure dampness, outdoor climatic changes, or long periods of shutdown and yet be ready for quick startup, keeping it dry inside is well worth what it costs.

Terminal box sizing

Although it may not seem like an "accessory," the motor terminal box often contains some of the optional devices mentioned in earlier chapters, such as current transformers and surge capacitors. It always contains the interconnections between motor and power system without which the drive cannot function. It is often required to be "oversize" even on motors of only 100 or 200 hp. Standards such as NEMA MG1 dictate minimum box dimensions—but when users bring in extremely large power cables for fairly low amperage, the minimum box size may be much too small.

Why is too small a box such a frequent complaint in the industry? Because the motor manufacturer has no way of knowing how much internal space is needed. He cannot predict the size or type of incoming power cable, nor the size of the connectors that will be used in the field.

Generally, the feeder cable size will be larger than the motor lead cable size. Because motor leads are grouped loosely together, are not enclosed within conduit, and are cooled by the motor's ventilating air, they can often carry more amperes for their size than cables in an external circuit. (See Table 8-III.)

Cable size, Gauge	Allowable Amperes					
	A	B	C	D	E	F
16	13	15	7.3	18	24	10
14	26	22	11.4	25	35	20
12	35	29	18.1	30	40	30
10	49	47	28.9	40	55	40
8	62	75	45.6	55	80	60
6	85	105	75	75	105	85
4	112	150	117	95	140	120
2	146	205	185	130	190	170
1	170	235	233	150	220	190
0	200	275	292	170	260	210
00	225	315	367	195	300	240

Column A: Cable manufacturer
Column B: Jobber catalog
Column C: Electrical consultant: Class B windings
Column D: National Electrical Code Table 310-16: 3 conductor: 90 degree
Column E: Ditto except single conductor, Table 310-17
Column F: Motor manufacturer's standard

Table 8-III. Half a dozen different "ampacity tables" for low voltage Class B lead cable. Note the wide variations for both small and large sizes. No one of these is "standard" throughout the motor industry. For No. 8 or larger, note how Column F figures exceed those in Column D.

The NEC ampacity tables aren't much help. First, the motor designer does not know whether copper or aluminum will be used. Second, he does not know whether conduit, tray, raceway, or direct burial will be involved. Third, he does not know the cable temperature rating.

Finally, in many large plants today cable size is not determined by ampacity. Two other practices govern the user's choice. One is voltage drop. Power system design practice commonly allows up to 2% voltage drop in a motor feeder circuit. That presents no problem when the motor is running. But on starting, the drop may reach 15%—causing accelerating torque to diminish by one-third. Many motors, especially above 250 hp, would then overheat or fail to start entirely. Hence, system designers have tended to increase their cable sizes.

More recently, still further increases have been made to permit the cable to withstand a fault on the circuit. Unless it remains safely cool until the fault can be cleared, the cable might burn out, perhaps in an underground duct where replacement would be quite costly.

In a typical 1974 utility project, for instance, a 4000-volt motor with full load current of 160 amperes was supplied by 350 MCM cable. Yet the motor lead wire size was only No. 2. The NEC might have dictated a 1/0 feeder cable size. Such extreme discrepancies cannot be predicted by a motor designer.

Users attempt to deal with this problem in ways that are not always effective. For example, take the common spec requirement for a box "one trade size larger than standard," or for a box that is "oversize." Just what is a "trade size"? And how large is "oversize"? The standard box normally used on a motor of the next larger horsepower, or the next larger frame diameter, may not be any bigger.

Or the specification may read "incoming cables will be sized per the NEC." Figure 8-68, for copper cable, based solely on ampacity, shows why that information does not help in deciding how much cable room to provide in the box.

Even giving the proposed incoming cable size doesn't suffice. Consider the spec reading "conduit box to be sized ... to allow for makeup of 2 —300 MCM conductors per phase." The motor manufacturer does not know:

(1) What size lugs will be on the incoming cables. There are no standards.

(2) The cable outside diameter, which determines cable bending radius. The box must provide room to "train" or position the cables as they pass smoothly from a bundled group at the conduit entrance to their individual connections to the motor leads. Big cables need a longer bending radius than small ones.

(3) For higher voltage motors, the stress cone space needed.

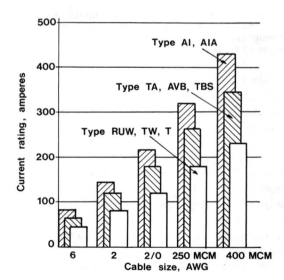

Figure 8-68: How copper feeder cable ampacity varies with cable type; taken from NEC Table 310-16 as it appeared in the 1981 Code.

Stress cones are required for termination of shielded incoming cable. Such cable is normal for any circuit at 4 kV and above, and is also used for some 2300-volt feeders. Motor specs often call for motor terminal boxes to "allow for" stress cones, without any clue as to the make or type of cone to be used. Figure 8-69 shows the wide variation possible. The designer can estimate average values of L_1 and L_2. But L_3 depends entirely on the incoming cable type and size.

Why not make all boxes big enough for the "worst case"? When the motor is relatively small, overly large terminal boxes can seriously interfere with the motor foundation, with the usually-expected 90-degree steps of box rotation to suit different conduit positions, with access to motor base dowels or to space heaters, and especially with ventilating air openings in the motor frame.

The solution is for the user to apply his knowledge of the incoming cable system, as well as the practices of the installer and maintenance crew, to decide what box configuration is needed—then specify those dimensions.

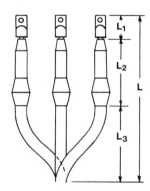

Figure 8-69: How much terminal box space is needed for 4000-volt stress cones? Dimension L_1 may range from 2 to 6 inches, depending on terminal length. The stress cones themselves cause L_2 to range between 5 and 17 inches, depending on type and make. Cable type and size results in L_3 from 10 to 20 inches. The overall dimension L is typically 20 to 24 inches, but can be as low as 18 or as high as 43.

Regardless of box dimensions themselves, uncertainty about incoming cable also makes it difficult for the motor manufacturer to pick the right size conduit entrance. He could use NEC ampacities for a commonly used cable type to estimate average conduit sizes for "standard" boxes. Table 8-IV illustrates such a practice. But that won't take care of cable selection based on fault withstand capability, or voltage drop. So this, too, is something the user should specify.

Maximum continuous motor current (amp.)	Assumed user cable[1]		Conduit entrance provided[2]	
	No. per phase	Size	No.	Size, in.
201-300	2	4/0	1	3
301-400	2	350 MCM	1	3½
401-500	2	500 MCM	1	4
501-625	2	750 MCM	1	5
626-725	3	600 MCM	1	5
726-750	4	400 MCM	2	3½
751-800	4	500 MCM	2	4
801-900	4	600 MCM	2	4
901-1000	4	700 MCM	2	5
1001-1100	4	750 MCM	2	5

[1] Based on NEC (1984) Table 310-16 using Type THHN copper cable, derated per Note 8 to table, and for 40°C ambient.

[2] Based on NEC Table 3A, Chapter 9.

Table 8-IV. How one motor manufacturer estimates size of required conduit entrance for standard large motor terminal boxes.

Concern about physical damage to boxes, particularly during installation, has recently been expressed in such requirements as "Box must be able to sustain a force of 300 pounds applied in any direction without deformation." How does one comply with that? No recognized test methods exist. The terms are vague ("without deformation" is impracticably broad). Is the force a steady load, or an impact? Is it to be applied over a square foot of box surface, or at a single point? So this is not a useful basis for achieving terminal box strength or rigidity. The common request for "⅛ inch minimum material thickness," or simply "heavy" box construction, won't suffice.

What about corrosion? All motor terminal boxes are normally made from either cast iron (up to a maximum dimension of about 18 inches) or steel. Aluminum or even plastic may seem attractive in certain environments, but both are generally ruled out by high cost, lack of strength or stiffness, difficulty of fabrication, or unsuitability for proper grounding.

Steel boxes may be given a variety of corrosion-resistant coatings. However, avoid specifying boxes of stainless or galvanized steel because of their high cost and limited availability. Besides, if the box is so subject to corrosion, what about the iron and steel parts of the motor frame itself?

Code rules

After box sizing and material, the next most frequent cause of concern among users is the presence in some motor terminal boxes of different circuits at different operating voltages. The possible safety hazard of mixing "high" and "low" voltages within the same working space has long been recognized. However, what constitutes "high" or "low"?

The National Electrical Code seems to make it clear. Articles 300 and 310 show the dividing line is at 600 volts, in such wording as "Conductors of over 600 volts shall not occupy the same equipment wiring enclosure ... with conductors of 600 volts or less." However, Article 300-1(b) states that "The provisions of this article are not intended to apply to the conductors which form an integral part of equipment, such as motors...." Article 300-32 repeats the point in these words: "Conductors of high-voltage and low-voltage systems shall not occupy the same wiring enclosure or pull and junction boxes. *Exception No. 1: in motors....*" High voltage is here defined again as "over 600 volts, nominal."

It is standard practice among motor manufacturers to use separate terminal boxes for space heater connections, temperature detectors, and the like on any motor rated 2300 volts or above. But the NEC clearly does not require that. Moreover, the interpretation by some users that "low voltage" means 50 volts or less, and "high voltage" includes 115 or 230, does not seem justified. If local codes do require that, of course, the motor manufacturer can comply, but he should not be expected to assume such a requirement.

Figure 4-22 in Chapter 4 (page 166) showed how one terminal box assembly was arranged to provide some degree of separation between low and high voltage circuits, at user request. The current transformer secondary wiring, including a secondary protector plus resistor loading network for a compressor load control circuit, was run directly from the main terminal box (necessarily containing the CT itself on the motor's high voltage leads) into a separate small box alongside, through a conduit nipple. Such arrangements can be worked out for almost any condition. But they require prior consultation between specifier and designer.

Just as the NEC makes no attempt to impose its design rules on wiring within motor terminal boxes, other national or industry standards also have little say about how such boxes are made. Especially on large motors, such boxes become complex structures—sometimes as large as walk-in refrigerators. (See Figures 8-70 & 8-71.)

Such enclosures resemble switchgear cubicles yet differ from them in several ways. They lack the outdoor protection of switchgear, with its double housing and controlled ventilation. Motor terminal boxes are attached directly to a strong heat source, whereas switchgear produces little internal heat and has a very large internal volume to absorb it. Switchgear doesn't see constant drive vibration. Finally, incoming cable systems are designed, tested, and rated by numerous nationally recognized standards. Motor winding insulation is rated by IEEE/NEMA for temperature stability, moisture resistance, and voltage

Figure 8-70: Left: Typical large motor terminal box during assembly, containing current transformers and surge protection equipment.

Figure 8-71: Right: This 800-hp motor is almost dwarfed by its 5-foot high terminal box, which contains several hundred pounds of accessory devices.

endurance. But what happens when these two standardized equipments are joined together in the motor terminal box? No general standards at all govern the assembly, insulation, or test of the electrical connections or of the box itself.

"Windings may fail, but terminal boxes themselves don't blow up." Don't you believe it! A short circuit at the connections will shut down the drive as surely as a wiped bearing or grounded coil. Moreover, winding failures in large rotating machines may be costly, but they seldom threaten the safety of plant personnel. A circuit failure in the terminal box is another story.

A 1975 IEEE conference speaker showed pictures of a motor in which the leads had been damaged inside the box during installation, causing a fault when the machine was energized. The resulting high energy arc caused the box to explode. Its flying parts killed a nearby worker. In another plant, an employee who lived to tell about it described a similar failure: "When we started the motor, the first thing I knew was this big bang and flash, and I could hear the conduit box cover bolts sing right past my ears on both sides."

The reason such failures seldom occur in the United States is because the high voltage fuses common in motor circuits act fast enough to hold fault current below an explosive level. When that doesn't happen, internal pressure reaching several hundred pounds per square inch is built up by heated gases resulting from arc vaporization of conductors and insulation. The arc may also destroy parts of the box structure itself. Even large boxes are a tightly enclosed space which cannot sustain this rapid pressure rise without breaking apart.

Standards development

Years ago in England, numerous such explosions took place, leading to the 1963 development of box standards intended to eliminate the safety hazard. Terminals are rigidly supported, boxes are built to stand high internal pressure, arc chutes isolate and help extinguish arc flames, and explosion vents relieve pressure above the design limit. Sample boxes are tested under actual high-power faults (typically a fault current of 75,000 amps at 3300 volts).

British engineers concede that one reason they had more trouble than engineers in the U.S. was the American use of circuit breakers rather than fuses in most motor circuits above 600 volts. Despite that difference, however, as power system capacity increases, operating margins are shaved, and emphasis on personnel safety grows, we may expect more demand in this country for what might be called "damage limited" terminal boxes for large motors. Some users are already asking for them. (See Figure 8-72.)

So far there are no U.S. standards for design or test of fault-resistant boxes. Until there are, and until products are on the market, installers and servicemen must do their best to prevent damaging faults. It is as important to safeguard terminal insulation, keep connections tight, and reduce conductor movement inside the terminal box as it is within the winding itself—perhaps more so, because within that box "extra dips and bakes" or VPI treatments don't help.

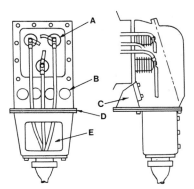

Figure 8-72: Typical 3300-volt motor terminal box developed around 1960 by British Electrical Research Association. Connection support insulators are at "A"; blow-out plugs "B" relieve internal pressure during fault; "C" is pressure vent; "D" is horizontal joint allowing removal of motor from its base without disturbing incoming cables; sealed cable entrance compound chamber is at "E." One problem with use of similar boxes in the U.S. is the common presence of bulky surge protection devices, almost never used in Britain.

Why boxes fail

Loose connections are a major cause of failure. In small motors, terminal lugs, connecting bolts, and the cables themselves tend to be considerably oversize for the current. In large machines, many factors contribute to a greater risk of trouble.

To begin with, users, contractors, and consultants often install their own compression lugs on the motor lead cables. Because such cables always use extra flexible stranding, the diameter over the copper itself is not the same as for the so-called "code" or "commercial stranding" cable used in the power distribution system. It's quite likely, then, that the lugs chosen will not be an ideal fit. Users are not prepared to judge lug tightness, nor to apply proper pullout tests in the field. Heating and cooling cycles after motor startup make loose lugs get looser, resulting in local overheating, then failure. (See Figure 8-73.) The solution is to use only lugs—and crimping tools—which the motor manufacturer recommends for the cables he furnishes.

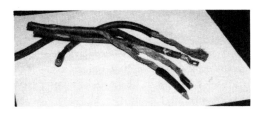

Figure 8-73: Motor lead cables ruined by overheating at loose terminal lugs.

A second problem is incorrect tightening of bolted connections. As Figure 8-70 shows, some large terminal boxes contain as much buswork as a piece of switchgear. The terminations will have adequate thermal capacity only if joint pressure is correct, which means the right number, size, and tightness of the joint bolts.

A November 1977 editorial in an electrical trade magazine was titled, "Why not torque all terminations?" Asked the writer, "Because field connections are so vitally important to system operation, why not make them as precisely correct as possible? A vast background of industry experience supports the conclusion that terminations should always be torqued...."

What torque values are proper? Use these figures if no other information is available:

Bolt diameter inches	Tightening torque, pound-feet
1/4	8
5/16	14
3/8	20
1/2	40

Whenever a large motor is down for cleaning, insulation resistance testing, or other de-energized maintenance, all terminal connections should be checked for tightness, cleanliness, and evidence of overheating. Many operators and maintenance firms are also checking connection temperature with the motor running as an "early warning" of looseness, which may not be apparent when the machine is shut down and cool. This is easily done with a hand-held infrared scanner.

Recognizing the size and complexity of connections in a box like that in Figure 8-70, designers have looked for new ways to lower the installation cost. In some large drives, it may take an electrician and his helper a week of steady work to fully insulate all the connections by hand taping. Particular skill is needed to build up stress cone terminations for the shielded incoming cable usually required above 2500 volts.

Therefore, two alternatives have now arisen. The first is to eliminate all joint insulation by leaving the termination bare. For power plant auxiliary motors, a 1977 American National Standard defines the terminal box construction of Figure 8-74—which is designed for bare connections. Although a stress cone is still needed on the incoming cable, there is no taping otherwise.

Clearance

There must be ample clearance through air between live parts, as well as enough creepage or tracking distance along surfaces. The porcelain insulators shown in Figure 8-74 provide that. In servicing such a unit, make sure insulator supports are intact and tight. Keep the insulators clean. There is a tendency to imagine the box interior as a clean, dry, indoor environment. But this is often untrue. Even a tightly gasketed box will "breathe" air in and out, perhaps loaded with dirt and moisture. In petrochemical plants, condensation of corrosive chemicals may drip into the box from within conduit entering the

More about Optional Accessories

Figure 8-74: Terminal box per ANSI power plant motor standard for 6600 volts. Leads in background are motor neutrals, brought out only for testing. Note ground pad and cable at lower right, used for grounding the incoming cable shields.

box from overhead. Any cables used inside such a box should include "drip loops" to prevent moisture from running back into the motor itself.

The designer, in choosing his clearances, can't allow for all possible cable connections that might be used in the field. If extra large lugs are used, or if they are not attached properly, clearance may be lost without anyone realizing it. (See Figure 8-75.)

How much clearance should there be? A very old rule of thumb was to provide a minimum striking distance of ¼ inch plus an added ¼ inch per thousand volts, and twice those amounts for creepage along surfaces. That would provide the distances in Table 8-V. Such figures may have been usable generations ago in clean, dry locations. But as more has been learned about apparatus life, and as motors have been more widely used outdoors or in chemically contaminated areas, such clearances have proven inadequate.

In Safety Standard MG-2, NEMA calls for the minimum clearances of Table 8-VI, which for comparison also shows the much more liberal figures specified for apparatus in general by the National Electrical Code. More usual design practice, however, is indicated by Table 8-VII, which provides plenty of safety margin for manufacturing variations, or wet, dirty conditions.

The second alternative to joint taping is the use of high voltage heat-shrinkable tubing in various forms instead of tape. If original installation cost were the only factor, taping might still be acceptable despite its expense. But two other problems have arisen. One is the need for greater reliability. People trained to do a taping job that will retain its insulation integrity for many years are getting hard to find. Besides, even the best hand-taping job cannot compete in quality with modern winding insulation systems. The other problem is that connections are no longer left undisturbed for years. Periodic

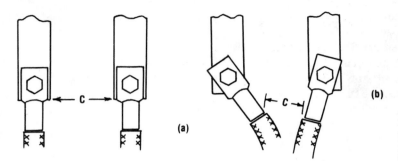

Figure 8-75: Adequate clearance C, provided for by design at (a), may be lost by improper assembly as at (b).

Rated voltage	Clearance, inches	Creepage, inches
600	0.5	1.0
2500	0.88	1.75
5000	1.5	3.0
7000	2.0	4.0

Table 8-V. Minimum electrical clearances suggested about 1910.

Rated voltage	Minimum clearance, in.	Creepage, in.	NEC clearance, in.
2300	1	2	5
4000	2½ line-grd., 3¼ line-line	3 line-grd., 4 line-line	6
6600	3 line-grd., 4 line-line	3½ line-grd., 5 line-line	7

Table 8-VI. Clearances proposed by NEMA MG-2, as compared with those in the National Electrical Code. Strictly speaking, the Code (Article 710-33) doesn't apply to spaces "within apparatus."

Rated voltage	Minimum clearance, in.; Indoors / outdoors	Creepage, inches; Indoors / outdoors
2300	2½ / 4	3 / 5
4000	3½ / 5	4 / 6
6600	4 / 6	Use ANSI standard insulators (7.5 kV class)

Table 8-VII. Typical terminal box design practice of large motor manufacturer.

insulation resistance tests require disconnecting the winding from the feeder circuit. So the joints must be taken apart—as quickly as possible—then reinsulated again after test. This may be done every year.

Neither the motor manufacturer nor the service industry can encourage such maintenance testing if it is costly and leads to reduced drive reliability. A solution can be the use of single-layer insulation coatings which can be applied or removed in a few minutes, and that give a high quality, compact, moisture-sealed and corona-resistant termination with little effort.

The motor manufacturer may use such material either as straight tubing or as molded caps or boots over exposed studs or bars. After final connection, the user may apply other tubing items, sometimes using layers of semi-conducting tubing as stress cones.

It's easy to recognize when this material has been used. But it may not be so easy to replace it with the proper item after repair or testing. All heat-shrinkable tubing is not alike. Much of the tubing on the market is designed only for 600-volt circuits. It will not withstand corona. It may not hold up in a damp or dirty atmosphere. Find out from the motor manufacturer what the right material is. The tubing supplier can tell you how to apply it. Most common for high voltages is "EPR" (Ethylene Propylene Rubber) which is corona/ozone resistant, usually with a 2 to 1 shrinkage ratio, meaning that after heating it shrinks to about half its original diameter. Round tubing will conform to other shapes, such as rectangular bus bar, or bolted joints, provided the perimeter around the surface is at least half the inside circumference of the unshrunk tubing. Of course, the larger the object, the thinner the wall of a given size tubing after shrinking in place, and the more tendency there is for the tubing to be highly stressed over sharp projections or corners. Over lugs bolted to projecting terminal studs, for instance, it's usually best to build up a reasonably smooth contour with electrical grade putty (in tape form) before applying the tubing.

An alternative to the supported bare connection is some form of separable connector. Motors up to 20,000 hp at 13,200 volts, or down to 100 hp at 2300 volts, are now being disconnected from the line for servicing by simply "pulling the plug." It's almost as easy as disconnecting a small appliance from a home wall socket.

The plugs and sockets themselves, of course, are much larger and more complex. Because they can be damaged by improper manipulation, they must be handled carefully.

Why are these new connectors being used? First, because regardless of motor size, the terminal box connections at 2300 volts or higher are normally fully insulated. Preventive maintenance testing (such as insulation resistance checks) recommended for motor windings, once a year or more often, requires removing all that insulation, breaking the connections for test, then fully reinsulating. Whereas most users can supply technician time and training for the test itself, the qualified electrician's time for reinsulation is more difficult to come by today.

Consider a typical 6000-hp, 4-kV motor having six cables per phase from the winding, each to be joined to an incoming shielded power cable, plus an additional six neutral cables per phase terminated on a common shorting bar. Also in the terminal box are three current transformers for differential protection, a surge capacitor, and three lightning arresters. To bolt up all those joints, then hand-tape all connections, took an electrician and his helper five working days. That included stress cone makeup on the incoming cables, which wouldn't need reconstruction each time the connections were opened up for a winding test. But the other work would have to be redone.

In another example (a 25,000-hp, 13.2-kV motor using strap leads rather than cable, again with current transformers plus surge protection) connecting the supply circuit in the field took one man eight days. Shortly after the motor went in service, trouble developed on the circuit and the motor had to be disconnected. Just opening all the joints took 30 hours.

Figure 8-76: Connectors similar to those described in Figure 8-88 (page 428) are used here for 400-hp (background) and 2000-hp (top foreground) motors, each 2300 volts.

Separable connectors

A second reason for the growing popularity of quick disconnects is the development and standardization of "separable connectors" during the past decade for the utility industry. The most common type is shown in Figure 8-77. Such devices were originally developed for underground power distribution circuits. Transformer and oil switch vaults below ground level are subject to flooding, yet they must often be temporarily disconnected for some service work on the circuit. It isn't always easy to be sure the circuit is dead before doing that. Furthermore, a common reason for working on the circuit is a fault somewhere on it, so that the serviceman may inadvertently close or open the connection while the fault exists.

So the connector system was developed—and nationally standardized—with these features:

(1) Minimum voltage rating 15 kV line-to-line.

(2) "Loadbreak" construction—connector separable under full load current which is limited to 200 amperes.

(3) Connector parts will withstand a certain number of "fault close" cycles without failure.

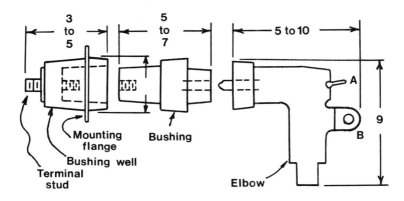

Figure 8-77: In this loadbreak connector assembly, the bushing well is mounted in the apparatus, and the replaceable bushing screws into that. The incoming cable is terminated in the elbow, which snaps over the bushing to complete the connection. The dimensions shown (in inches) are typical for a 200-ampere system. Elbow pulling eye is at "A"; removable cap over voltage test point is at "B."

(4) Special rubber compounds molded into the connecting surfaces provide a conducting ground path to act as part of the cable shielding system, so that the safety shield is automatically extended over the joint when the connectors are fitted together.

(5) Elastic, matching interfaces between connector parts provide a tight seal for safe water immersion under load.

A number of manufacturers now offer these connectors, all using national standard interfaces between mating parts for interchangeability.

The 200-ampere loadbreak connector assembly is in three parts—the elbow, the bushing or "plug" which the elbow fits, and a "bushing well" (sometimes called a "flower pot" because of its shape) into which the bushing is screwed. (See Figure 8-77.) The plug is a separate, replaceable part because it is subject to damage by a fairly low number of operating cycles involving fault current; a damaged bushing can easily be unscrewed from the well and replaced with a new one.

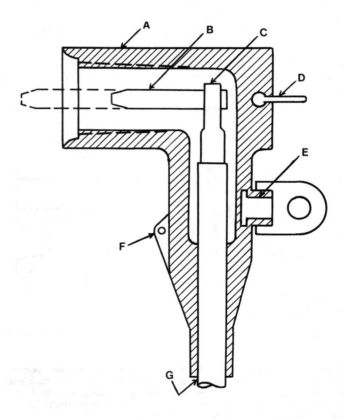

Figure 8-78a: A molded rubber "elbow" or "terminator." The basic shape is the same for both loadbreak and non-loadbreak versions, but the dashed line shows the "arc follower" contact tip used for the loadbreak type. This special material helps extinguish the high-power arc formed when the male contact snaps free from the bushing receptacle as the elbow is pulled off under load. The loadbreak elbow also includes a molded-in ring that snaps over a groove at the end of the bushing for positive locking.

A is the body with molded shielding. B is the contact pin (male contact). C is the cable lug. D is the pulling eye. E is the test point. F is the grounding eye. G is the incoming cable.

More about Optional Accessories   *421*

Later, "non-loadbreak" connectors of a similar nature were made available. This assembly uses no well. Instead, the single-piece bushing is permanently mounted in the equipment (the motor's terminal box, for example). Some 200-ampere bushings have been made of skirted porcelain. Lead cable connections from the motor winding to these bushings can be left uninsulated. Most bushings, however, are epoxy, with fairly short shanks. Though designed for 15-kV systems, 8.6 kV to ground, they aren't warranted for use even at 4000 volts in contaminated air. The original intent was for the apparatus end of the bushing to be under oil, as in a transformer tank. In air, dampness and dirt can collect on the epoxy surface, eventually resulting in tracking from the bushing mounting (ground) to the stud where the lead cable is attached.

Both 200- and 600-ampere non-loadbreak parts are in use. In the lower current design (Figure 8-78), the elbow omits the "arc follower" because opening a de-energized circuit produces no arc. Also, there is no need for an abrupt "snap" in opening or closing the circuit to minimize arcing, so there is no locking groove in the bushing or ring in the elbow to engage that groove. Instead, the tapered surfaces of the two parts simply slide together. They are then held in place by a wire yoke or "bail" secured to the bushing mounting and locked over the elbow by a screw. (See Figure 8-79.) The 600-ampere elbow and bushing are screwed together as shown in Figure 8-80.

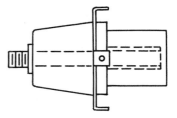

Figure 8-78b: A typical 200-ampere non-loadbreak bushing. For motor use, the mounting flange at center is bolted or welded to a plate inside the terminal box. A motor lead cable is attached to the threaded stud at left; the elbow slides over the right-hand portion of the bushing. The lugs on the mounting flange are for attaching the bail (Figure 8-79) that secures the assembly.

Figure 8-79: Terminal box of a small power plant motor. These are 200-ampere non-loadbreak connectors, with the "bails" in place to keep the elbows from loosening on the bushings. (Photo courtesy Elastimold Div., Amerace Corp.)

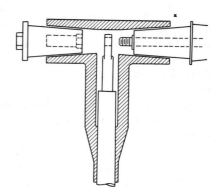

Figure 8-80: In the 600-ampere non-load-break connection, a bushing stud (right) passes through the cable terminal lug, and a removable cap (left) is screwed over the stud to make up the complete joint. The projecting test point, at far left, is not connected to the molded-in nut within the cap, but is coupled to it only by capacitance. The complete assembly looks like those in Figure 8-81.

Although there is no reason to interrupt motor terminal box connections under load, some users may want loadbreak bushings furnished with the motor because matching elbows for installation in the field are readily available for them from many sources.

Vibration

Some motor manufacturers have discouraged use of 200-ampere connectors because the loadbreak type is held together only by friction and the bushing groove, and the wire bail is the only securing mechanism for the non-loadbreak version. Unlike a transformer or switchgear application, motor service does involve vibration that, though slight, is continuous. The 600-ampere connector, held together by ⅝-inch bolts torqued to 50 or 60 lb-ft, is usually considered more reliable for rotating machinery use. In a small motor, however—below about 1000 hp—they waste a good deal of terminal box space. They are better suited to the larger sizes and higher voltages.

These parts also offer a modification in which a separate 200-ampere connector can be tapped off the 600-ampere elbow to an auxiliary circuit. In some motors this is used for leads to surge protection capacitor terminals. Thus, the capacitor can be unplugged from each phase for winding tests, just as easily as the line leads themselves can be separated.

All types of elbows include a "test point." The serviceman can use a capacitive voltmeter to see whether the circuit is "live" or not. This is of little value in motor use because it will be obvious whether or not the motor is energized.

A second feature for the 200-ampere size is a "pulling eye." Because these connectors were devised for use in live circuits, they are made for opening and closing with a hot-line tool or "shotgun stick"—a long insulated pole with a hook or claw on the end to grasp the pulling eye and move the elbow on or off the bushing. The 600-ampere size, used only in non-loadbreak service, can be worked by hand. These connections are entirely "dead front"—that is, anyone who has occasion to open the terminal box of an energized motor will not be

More about Optional Accessories

Figure 8-81: A large motor terminal box with its cover removed to show assembled separable 600-ampere elbow and bushing connectors. Each bushing's mounting flange has been spotwelded to a steel plate, which in turn is bolted to support bars welded into the box; this provides a continuous grounding path. (Photo courtesy Elastimold Div., Amerace Corp.)

exposed to any live parts on the elbow or line side of the bushing. (See Figure 8-81.)

It may seem wasteful to use a 15-kV device on a 2300- or 4000-volt motor. However, parts developed for another purpose entirely are being adapted to motor use. Demand for motor connectors does not yet justify making parts for 2.3 or 4.0 kV only.

Although the higher voltage rating makes the parts larger than they could otherwise be, the 200-ampere connectors have been successfully used with 2300-volt motors down to the 100- to 200-hp range. Figure 8-82 shows one

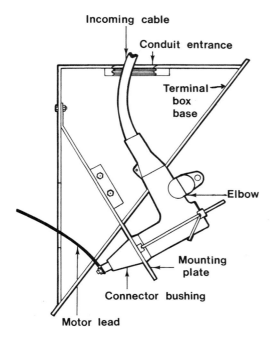

Figure 8-82: Small 2300-volt motors have used compact terminal box arrangements like this for 200-ampere separable connectors.

fairly compact terminal box installation. The box may not have to be much larger than for conventional lead connections, because there is no need for working room to tape the joints. When other accessories are present, however, some extra space may be required. (See Figure 8-83.)

Because of the elbow size, the terminal box will normally be designed so a large plate to which the conduit is attached can be unbolted from the box. After the elbows are pulled off the bushings, the motor with the remainder of the box attached can be removed from its base, leaving that plate on the conduit. It's possible to take the elbow apart to remove the incoming cable

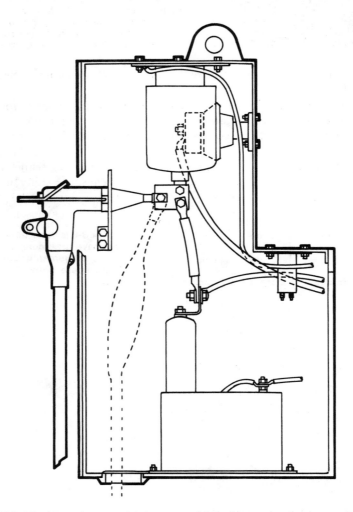

Figure 8-83: In a larger box, containing current transformers and surge protection, the connector system takes up more space. Note here how the box would have to be deeper (extended to the left) to accommodate separable connectors.

from it, but that is a time-consuming job and can lead to stripped connector threads or other damage.

When working with separable connectors on any motor, observe these two important precautions. First, the elbow must be removed with direct pull, in line with the bushing axis, to avoid elbow damage. Unless the surfaces have been properly lubricated before assembly, disassembly can be difficult. Some lubricants apparently "freeze" with age. Therefore, never fit an elbow to a bushing without coating the bushing surface with silicone grease. This lubricant is always supplied with the elbows, so it should be available on the user's premises.

As one manufacturer described separation of a loadbreak connection: "The lineman ... must exert considerable force in order to break loose the vacuum and the lock ring.... When the elbow finally lets go, there is sufficient energy built up in the elastic eye that the elbow snaps off and provides correct contact speed to guarantee satisfactory loadbreaking performance." This may take 400 pounds pull. Although non-loadbreak connectors don't work quite that way, they will need to be worked loose by slight twisting while pulling. Figure 8-84 shows what has often happened when maintenance people carelessly pulled off an elbow, forgetting that it is designed to be removed by the pulling eye.

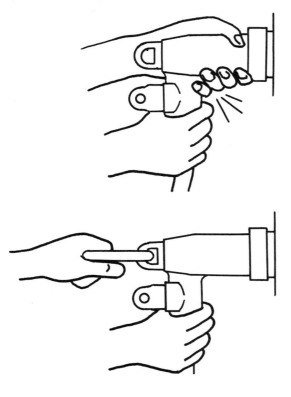

Figure 8-84: Careless maintenance. Pulling the elbow off a bushing by the cable end can crack the elastic material. One user calls this damage "pistol grip rip." It's avoided by pulling carefully in line with the bushing axis, as in the lower view, while rotating the elbow on the bushing through a slight arc to break the tapered interfaces free.

The second precaution: For maintenance testing, it's usually impractical to disconnect the winding leads from the bushings or wells in the terminal box. Hence, such testing can't distinguish wet, dirty, or defective insulation at those connection points from an actual winding defect. Before assuming a bad winding, examine those terminations carefully.

Sometimes the motor lead connection to the bushing assembly is left bare, as with the porcelain type mentioned earlier. Some bushings have longer shanks that provide adequate creepage distance. But when insulation is required, it may consist either of special tapes or a shaped plastic "boot" which is heat-shrunk over the area. (See Figure 8-85.)

Figure 8-85: Motor lead cable connection to a 600-ampere bushing, insulated with a heat shrinkable boot. A sample like this passed a high voltage water immersion test after six months of outdoor exposure in an industrial environment.

Disassembling

If it is ever necessary to take apart or reinsulate such a termination (to rewind the stator, for example), consult the motor manufacturer for methods and materials. The joint has to be sealed against dirt and moisture. Here is a typical taping procedure:

(1) Clean all surfaces. Clean and tighten the nuts that hold the cable lug to the bushing stud.

(2) Cover the entire stud/lug area with insulating putty, building the surface up to a smooth contour to fill all voids. Let this material set until fairly firm.

(3) Depending on the rated motor voltage, tape over the puttied area with multiple layers of half-lapped mica or silicone rubber tape. This taping must lap tightly over the edge of the putty onto the bushing surface to form a seal.

(4) A layer of shrink film tape may be applied and then shrunk with a heat gun (typically for 30 minutes at 300°F) to tighten the layers. Remove the shrink tape. (Above 4000 volts, such tightening may be called for at an intermediate stage of the taping process.)

If made properly, such an insulated connection will be watertight.

The heat-shrinkable boot has these advantages over taping:

(1) It is much quicker to apply.

(2) It is one piece, not susceptible to inter-layer voids.

More about Optional Accessories 427

(3) It is lined with a thermoplastic adhesive, which oozes out around the edges during heat-shrinking for visual evidence of a seal.

(4) If the boot must ever be removed, cutting it off with a knife takes only a moment.

Remember that some outer surfaces of the elbow are conductive. Unless such areas remain in good contact with well-grounded parts of the motor terminal box or frame, the incoming cable shield will not be adequately grounded. This is a safety hazard. So the elbow "grounding eye" must be securely wired to ground. Such ground connections may have to be taken apart when a motor is removed for servicing, so be sure they are properly replaced.

Also, some bushings have conducting collars as part of the overall shielding system. (See Figure 8-86.) These also require tight contact with ground.

Standard separable connector parts can handle only one cable per elbow, whatever its size. However, adapter straps or special lugs can easily be made to attach two or more motor lead cables per phase to the bushing. (See Figure 8-87.) Insulating the connection may be a little more difficult. If an additional incoming cable per phase is needed, a second elbow and bushing must be added.

Figure 8-86: A non-loadbreak epoxy bushing for a 200-ampere connector assembly. This uses no metal flange for apparatus mounting, but instead has a molded-in ring of conducting material over which a clamp is bolted. For proper shielding of the bushing, that clamp must keep the ring in good contact with ground.

Another type of connector is the "in-line" coupler, typified by the mine service assembly shown in Figure 8-88. This groups all three phases in a single set of parts, though allowing only a single cable per phase. It is available in large sizes for 5- to 15-kV cables used in excavating or mining service. The parts are not so easy to adapt to motor terminal boxes. Usually, the socket portion is mounted in the box with the plug portion outside, as illustrated in Figure 8-89. Such connectors normally include a ground circuit on a separate pin. This maintains cable shield continuity. The socket end of that circuit must be properly grounded at the motor.

Other connectors using only one conductor per plug and socket are made for motor use at lower voltages but high current. They are especially popular for d-c machines in oil drilling service. (See Figure 8-90.)

Figure 8-87: Inside the terminal box of a 1750-hp, 4000-volt motor, each of the 600-ampere bushings is fitted with a copper bar adapter for connecting two motor lead cables to each bushing stud (in background).

Figure 8-88: A typical "mine cable coupler" rated 200 amperes, 6600 volts. Different socket configurations suit various types of motor terminal box mounting. The plug is held tightly in place by a threaded ring that screws onto the socket.

Neutral and ground connections also deserve attention. Figure 8-91 shows a neutral shorting bar mounted on insulators to which the neutral side of each motor phase is wired. This bar is usually left bare. Nevertheless, it must be separated from ground—the steel box structure—fully as well as the main line terminals. The power system in the user's plant may have a neutral ground of some sort, but grounding will probably be through a carefully selected impedance to control ground fault current. Shorting out that impedance by permitting a direct neutral ground at the motor can wreak havoc. So don't take shortcuts on neutral insulation any more than you would for the line side.

More about Optional Accessories

Figure 8-89: Closeup of a 1000-hp, 2300-volt motor terminal box, showing only the socket portion of a mine coupler type connector. (Photo courtesy Joy Mfg. Co.)

Figure 8-90: Single conductor, high current connectors like this are common for d-c motors on oil drilling rigs. They are rated up to 1000 volts, 900 amperes. Weather-proof seals are included. Other makes include three conductors, up to 625 amperes, 600 volts.

Figure 8-91: Neutral shorting bar, supported on two insulators, can be seen at the top of this 1500-hp motor terminal box.

Surge protective devices—capacitors and lightning arresters—are effective only when properly grounded. The case of each device is usually cabled directly to a ground post somewhere in the terminal box (as recommended by an IEEE standard). This makes sure the ground path does not depend on conduction through bolted joints where the case itself attaches to the box. So make sure these cable joints, too, remain tight.

Recently some customers have requested that the surge protective equipment be "floating"—that is, mounted on insulators, with the only ground path through an easily removable cable. This permits cutting those accessories out of circuit easily when hipotting the motor winding itself. Here it is even more important to make sure ground cables are tightly reconnected after the test. Otherwise, surge protection of the winding may be lost entirely.

When reassembling any terminal box connection, replace all hardware—nuts, bolts, and washers—using the same material in the same location. Many joint loosenings are caused by unequal expansion and contraction of hardware, as motor load varies and the terminations heat and cool. Though not common with copper connections, Belleville spring washers may be used in the joint. Make sure they are not left out at reassembly.

Should you build your own terminal box? Some users of large motors make a practice of throwing away the original equipment and substituting their own. There are no standards to guide you. But it may help you to know what design practices manufacturers use, besides those already discussed here.

Be sure to support and protect the motor leads. They should be clamped at or near the point where they enter the box. (See Chapter 4, Figure 4-22, page 166). Even a sheet neoprene "lead separator gasket" is helpful. This limits movement of cables under starting currents. Extra sleeving around leads in

Figure 8-92: Field modification of this 350-hp motor terminal box resulted in a far from ideal job.

this area eliminates chafing or cutting of cable jackets. Holes in steel plates through which leads must pass ought to be chamfered, or at least deburred. In any event, you want to avoid the kind of installation shown in Figure 8-92.

Use low current densities in joints, whether insulated or not. Tin or silver plating of copper parts is sometimes helpful, though no longer standard even in switchgear. Try to stay under 500 amperes per square inch of joint surface (continuous current) for brazed connections, or 300 for bolted joints.

What kind of insulators are best for bus supports? The user may have his own rules. Either porcelain or epoxy may be available. Each has advantages, but above 2500 volts the skirted porcelain insulator is hard to beat. Whatever you use, make sure it is fully rated under ANSI standards for the voltage class. Use the kind with two or more mounting holes at each end, so that it cannot rotate should the mounting bolts loosen.

Big, heavy boxes need extra support from motor frame or base. This is most likely with any box over 18-20 inches high, especially if it contains heavy accessories. When you observe large drives in operation, watch the light reflected from flat terminal box surfaces. You may see a "shimmer" indicating box vibration of fairly high amplitude, even when motor frame vibration is quite low. Such shaking may eventually crack the box, break bolts, or damage the leads. When in doubt, brace it. Don't let the terminal box become the weak link in the chain of drive reliability.

9 Installing Motors Properly

"WE JUST PUT IT on our asphalt floor and started it up."

That's how one user described the installation of his new motor-driven compressor. He might not have made that boast if he had known how often such casual methods generate serious electric motor problems.

Said an experienced engineer several years ago, "Statistics show that in oil refineries and chemical plants, 75% of electric motor failures are mechanical rather than electrical, when primary cause is considered." Plant surveys by the IEEE confirm that for other industries, too. And the "primary cause" is often a poor job of installation—especially a supporting structure that's too weak, or too flexible. (See Figure 9-1.)

Added another consultant: "Foundation design and machinery installation require more care than is usually practiced.... Many facilities must be renovated or the equipment regrouted long before the life of the equipment is exhausted."

Those called upon to diagnose drive problems or rebuild a motor seldom have the chance to design the foundation ahead of time. Knowing how such problems develop, however, can help them find trouble faster and avoid making it worse. The accompanying checklist (see box) is a helpful summary.

A machine operator, if plagued by unexplained bearing failures, high vibration, or shaft breakage, is likely to assume that the fault lies with poor motor design or construction. He seldom suspects the true source of his difficulty—a poorly designed or badly made motor base, or one to which the motor has not been properly mounted.

The 'coplanar' base

Consider the essentials of a proper base. Motors generally have four points of attachment to the supporting structure. All must be "coplanar"—that is, they must all lie in a single flat plane (normally horizontal). High school geometry teaches us that any three points automatically determine a plane. But some care is necessary to ensure that the fourth point will lie in that same plane. When that doesn't happen, tightening down the motor's four hold-down bolts, one at each corner, will always "spring" or distort one or more feet, or bend the base structure. The result will be overstress or misalignment.

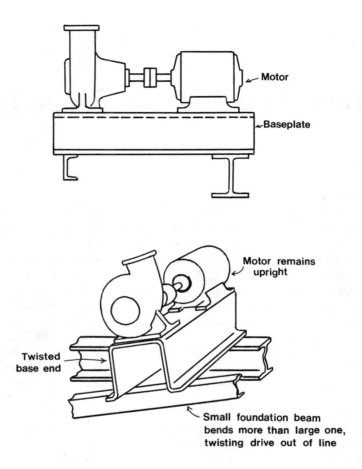

Figure 9-1: A commonly encountered drive support problem. The baseplate itself is strong and stiff enough but the underlying structure is not, so twisting and misalignment result.

That is why motor vibration problems can frequently be quickly diagnosed by simply loosening one foot bolt at a time, then observing the effect on drive vibration. If the problem disappears, the mounting wasn't right.

So the first essential for a proper base is: *all mounting surfaces must be in the same flat plane.* Two conditions may prevent that:

(1) Use of several separate plates or rails instead of a single baseplate. It's hard to line up all the surfaces.

(2) Poor grouting, or welding on a machined structure, so that the base twists out of shape after it's completed.

What's the purpose of grouting? First, this thin cement which is poured and tamped in and around the motor base during installation, between steelwork and the concrete foundation block, provides full support for drive weight. A large drive base resting on a concrete pad at only a few support points can sag, warp, or vibrate between those points. But when properly grouted in, the entire base is fully supported by rigid concrete.

Second, grout anchors base or sole plates so they cannot shift. That isn't necessary when the entire supporting structure is steel. Then, bolts and dowel pins can connect all the separate parts so no "working" or movement occurs during drive operation. But when the steelwork ultimately rests on a concrete pad, foundation bolts alone don't ensure stability.

Although maintenance personnel seldom put grout in place, they should realize its importance. Whenever you check out a drive, especially when investigating vibration, study the grouting. Are there cracks, places where the grout is missing, gaps beneath the motor base, or evidence of powdering and crumbling of the material? All these can lead to excessive vibration.

Equally important in achieving the coplanar four-point support is proper shimming. Shims are space-fillers, which raise or lower motor feet to match unavoidable irregularities in a baseplate surface. They also serve, of course, to raise or lower the entire motor for correct shaft alignment with the driven machine.

Either way, they must match the mating surfaces, and they must be solid.

A Motor Foundation Checklist

☐ Vibration means trouble. Often the drive base is at fault. Examine it and the shaft alignment first; the motor itself last.

☐ Check for dowels at all separable joints in the assembly between base and drive components, right down to the underlying foundation.

☐ Watch for proper placement of shims—no gaps left unshimmed—and for good shim condition.

☐ Be sure all hold-down bolts are in place and tight.

☐ Remember Murphy's First Law of Foundations: "Adjustable braces are usually out of adjustment."

☐ Don't assume the floor itself is solid. A solidly-assembled drive on a shaky floor will be troublesome.

☐ Watch for distortion caused by overhung loads.

☐ Expect twisting or distortion if base welding is intermittent rather than continuous.

☐ Look for signs of heavy welding done after installation on machined bases; it has probably caused warpage.

☐ Although box sections or tubes can be much stiffer than simple beams or channels, their stiffness is greatly reduced if the section is "split" or open along one side.

☐ A drive base made in one piece is always better than one made in several sections; deep bases are superior to shallow ones.

☐ Look for stiffeners or gussets near mounting bolt holes. (See Figure 9-12, page 445.)

☐ For maximum stiffness, support braces should be directly beneath the motor.

Poor shimming is so common that one heavy equipment engineer said this about it: "It is my experience that the only reasonable assurance of proper shimpacks is to physically remove and inspect the shims at every machine support just prior to final alignment. Obvious problems with shims include rust, improperly cut shims, folds and wrinkles, burrs, hammer marks, and dirt. I once found a turbine installed on shimpacks wrapped in masking tape."

Shims should be anchored so they cannot "work their way out." Use a few thick shims in preference to a great many thin ones. (Ideally, there would be only one precision-ground shim at each foot, but this is seldom possible.) If the baseplate and motor foot surfaces cannot be made parallel, a tapered shim is recommended. Otherwise, when the foot is tightly bolted down, it will bend to fit the tilted pad surface. Trouble inevitably results.

Stable motor base

A good motor base must also be *stable*. That is, once correctly installed to furnish good support, it must remain that way. Expect trouble whenever you see a steel drive base which has been modified by welding on new pieces. Even adding stiffeners, to increase base rigidity, can have a bad effect. The extreme localized heat of welding causes differential expansion in the parts which will distort the structure. It will add new stresses, perhaps relieving old ones. Avoid this unless it's possible to remove the entire base, stress-relieve it, then remachine the mounting surfaces as necessary.

As an example of what can happen: A 1250-hp pump-motor combination operated successfully for several years. Changing conditions required replacement of the motor by a 1000-hp unit having different shaft height. Rather than install a costly adapter to raise the foundation, the operators welded "stilts" onto each corner of the existing structure. These stilts were strong enough to support the new motor's weight, but they weren't stiff enough to absorb drive vibration. (See Figures 9-2 and 9-3.) Only after a motor bearing failed was the problem corrected.

Figure 9-4 illustrates another sort of thermal problem. This motor drives a fan handling hot furnace gas. As the fan heats up it expands, causing the fan

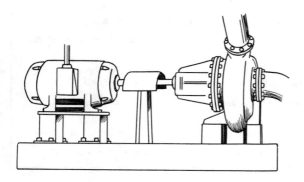

Figure 9-2: Mounted on slender stilts to line up with the pump shaft, this 50-hp motor can easily be thrown out of alignment, leading to excessive vibration.

Installing Motors Properly 437

Figure 9-3: A horizontal pump drive mounted on slender pipe supports. The pipes can allow the assembly to vibrate in two directions, especially when forces are transmitted from connected piping to the pump.

Figure 9-4: Separate, high concrete foundation for this fan drive motor means that the fan shaft, heated by process gas, will "grow" vertically much faster than the motor shaft.

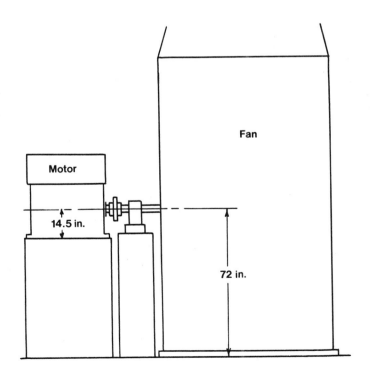

shaft centerline to rise well above its "cold" position. The motor, however, remains much cooler. And even if it did get hot, its shaft centerline height can increase vertically only one-fifth as much as that of the fan shaft. Thermal expansion at any given temperature is a fixed amount for each inch of initial length, and the initial fan shaft height is five times as great as the motor's.

The result was extreme misalignment between the two halves of the gear coupling. Torque lockup in the coupling transmitted much of this to the motor shaft, which broke twice before investigation by the motor manufacturer revealed the true cause.

There was a vibration pickup on the drive which could have given warning of impending trouble. But it wasn't located at either motor or fan bearing. Instead, it was at the intermediate pillow block bearing, mounted on its own concrete pad, which was fairly insensitive to the disturbance.

A third essential for any good motor base is *rigidity*. This isn't the same as strength. A wire rope has little rigidity or stiffness. It can be wound around the wrist yet is strong enough to lift an automobile. On the other hand, a concrete slab may have great stiffness, but little ability to withstand a direct pull. Unless a motor mounting is stiff, forces such as belt tension can pull it out of alignment, or cause it to vibrate dangerously. (See Figures 9-5 and 9-6.) Figure 9-1 also shows the importance of stiffness—resistance to twist or bending.

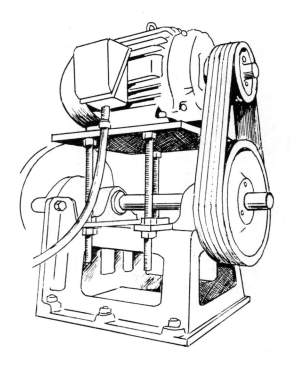

Figure 9-5: Supported by four long, slender studs, this motor is free to shake in all directions. Restraint comes only from the downward belt pull, which would not help if resonance developed. Mountings of this type are common.

Installing Motors Properly 439

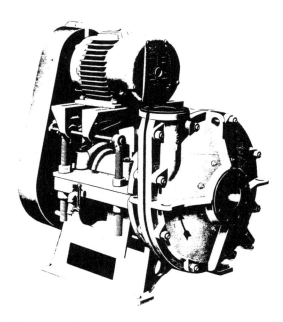

Figure 9-6: Another type of belted pump drive that can easily develop vibrations in several directions.

Proper tuning

Closely related to stiffness is the fourth characteristic of a good drive foundation: *proper tuning*. This means that the natural vibration frequency of the entire drive assembly, with all components in place, must not coincide with operating frequencies—especially the normal drive rpm. Figure 9-7 illustrates that coincidence, which is called resonance.

Most of us are familiar with this condition. A simple swimming pool diving board becomes resonant when the swimmer jumps up and down on the end of it at just the right frequency. The board's natural vibration frequency, if coincident with the jumps, leads to very large oscillations that tend to reinforce themselves. The same thing happens on a trampoline.

Changing the mass or weight of the assembly will change its natural frequency. Hence, it isn't enough simply to strike a motor baseplate with a hammer and measure its natural frequency with a vibrometer. That frequency will be entirely different once the motor is bolted down to the base.

We usually think of resonance as something to be avoided at motor full load rpm. There are, however, two kinds of drive in which resonance may occur at other frequencies. One is that type of machine which produces periodic impact loads—particularly torsional shocks. For example, a hammermill or a log chipper undergoes a torque impact or shock each time a hammer or a chipper knife passes some fixed point. These vibratory impulse loads may occur many times during each revolution. They can therefore put a sudden extra twisting load on the drive shafting at a frequency many times the motor rpm. If this matches a natural response frequency in the drive, severe and damaging torsional oscillations may result.

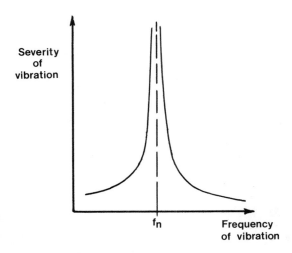

Figure 9-7: All structures have at least one "natural" vibration frequency to which they will strongly respond if acted upon by any external force at that frequency. This condition, called "resonance," is illustrated here, f_n being the natural or resonant frequency.

Reciprocating compressors can generate similar high frequency vibratory forces depending on the number of cylinders. These forces are fairly low frequency, and can reach high amplitude, to transmit damaging vibrations through foundation structures to remote parts of a building or even outside the plant.

The second type of drive involving many different resonances is the energy-saving variable frequency motor. Its popularity is growing rapidly. Because such motors may operate for long periods at not one, but many different speeds, there may be a wide range of possible drive rpms—several of which may lead to resonance.

Critical speed

The diving board, a motor shaft, and many other structures can vibrate naturally in several different "modes" at different frequencies, which are normally multiples of one another. That's why a motor will have more than one "critical speed" (coincidence between natural shaft vibration frequency and rpm). The lowest frequency is called the "first critical," and usually it is the only one of concern. A second, much higher natural frequency is called the "second critical." There may be a third, and a fourth, etc. Since a normal motor has only one running speed, it is sufficient to design the machine so its first critical is well separated (usually on the high side) from running rpm. Even a two-speed motor can be readily designed to avoid the first critical coinciding with either rpm.

But when that operating rpm can span a wide range, avoiding all the criticals can become extremely difficult. A foundation in such a drive can be especially vulnerable to resonance. Anyone installing this type of unit should be aware of this. Even when the foundation does avoid resonances, trouble may develop later when changes in drive operating conditions change the

Installing Motors Properly

frequency range. Look for such a change whenever a variable frequency drive develops vibration or bearing trouble.

When checking a drive for resonance, it's necessary not only that all components be firmly in place, but also that all normal working loads be applied. For example, all belts should be in place and properly tensioned. The overall stiffness of motor mountings like those in Figure 9-5 and 9-8 vary considerably with the amount of belt tension. That is one drawback to such arrangements.

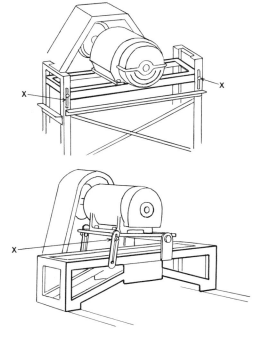

Figure 9-8: Two 150-hp crusher drives using 1200-rpm motors. Upper diagram: under full downward belt pull, foundation crossbeams beneath motor are near resonance at motor speed. Lower diagram: heavy angle bracing supports center of beams, giving much more rigid support with resonant frequency well above 1200 cycles per minute. Belt tension adjustment is at Point X in both diagrams.

In Figure 9-8, we see two versions of belted drive, in similar service, one subject to resonance at running speed because of a lack of stiffening support directly under the motor. The belt pull is downward, such that variation in belt tension can generate vertical "bounce" unrestrained by the base. The other version illustrated avoids that, because bracing is located to oppose the belt pull directly.

Resonance in vertical pump/motor assemblies is common. For one thing, the motor is held in place at one end only, unlike the usual horizontal motor mounting. This leaves the other end free to shake.

More important, no matter how the motor is attached, the mounting structure or "pump head" may be extremely limber. Note the typical constructions in Figure 9-9. These heads are often of thin material, containing large access openings for making up shaft couplings, so that there is little sidewise stiffness in one or more directions. Such a structure resembles the diving board, stood on end. It also resembles a cattail growing in a swamp, with a large weight

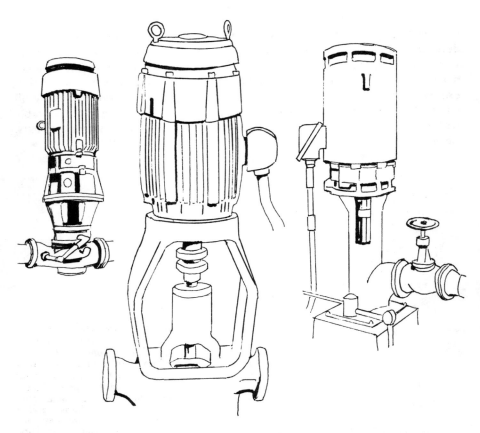

Figure 9-9: Several typical vertical motor pump head mountings. Shown are both their large height-to-width ratio, which leads to low natural reed frequency, and the cutouts that take away much of the desired stiffness.

(the motor) fluttering on the end of a thin reed swaying in the wind, so that such movement is known in the motor industry as "reed frequency vibration." With some large machines it can reach 1/16-inch.

Altering natural frequency

At least one pump manufacturer has begun writing motor specifications that include provisions for adding weight to the top of the motor as necessary to alter the natural frequency of the assembly. In emergencies, this has been done by piling on sandbags.

Adding weight is a crude means of lowering the natural frequency. Another method is to insert spring washers on the motor hold-down bolts. It may also be possible to raise the natural frequency beyond the danger point, by stiffening the mounting structure. As the curve of Figure 9-7 shows, it is only necessary to separate natural and exciting frequencies to avoid resonance; the separation can be in either direction. But those methods are poor substitutes for designing resonance out of the assembly to begin with.

With pump drives, even a stiff support may not prevent vibration or misalignment introduced through connected piping. Consider the in-line pump of Figure 9-10. Even minor vibration—surges in the fluid system—may set this entire assembly bouncing uncontrollably. Although strong enough to support the motor weight, the piping cannot provide the needed stiffness.

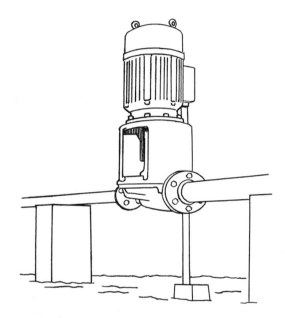

Figure 9-10: "In line" pump and motor, supported only by connected piping resting on concrete blocks. Blocks won't damp out vibration originating in either drive or piping.

Even a horizontal pump/motor drive, properly mounted on a firm foundation, can be twisted or shaken by forces imposed through the attachment of heavy pipe. In one example, identical pump motors side by side exhibited entirely different vibration levels—one perfectly safe, the other dangerously high. The serious vibration disappeared when individual flange bolts were loosened in that unit's discharge pipe coupling.

One way to avoid the effects of piping forces is to support the piping from its own foundation close by, as in Figure 9-11. This can prevent the tail from wagging the dog, so to speak.

Piping may transmit not only physical shock, but damaging thermal growth as well. One service shop reported that hot air blowing over piping caused 30 mils of coupling misalignment in a fuel oil pump drive.

'Free baseplate'

A new development in pump drives is the "free baseplate," supporting both

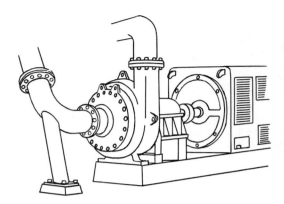

Figure 9-11: One type of support used to minimize drive vibration transmitted to a pump from the external piping.

pump and motor. It's been described this way: "... [A free baseplate is] not bolted or otherwise fastened down to its supporting concrete or structural steelwork.... It is free to move during operation in the horizontal plane, in response to changing horizontal piping strains.... It is restrained in this plane solely by the connecting pipework, and by friction between the baseplate and its support."

This may seem to contradict what has been said here about foundation requirements. But a free baseplate is no ordinary structure. It must be stiff enough to resist all twisting and bending—permitting only horizontal, not vertical, movement—so that no shaft misalignment occurs. Careful design is essential. It then offers the advantage of eliminating foundation bolts and grouting.

For more conventional bases, the final important step in saving a motor from destructive vibration is to make sure it remains bolted in the right place. Every maintenance check on a drive should include foundation bolt tightness. How tight should the bolts be? If anchored in concrete, they can pull loose if the concrete is overstressed by too much bolt tension. To avoid that, nuts should not be torqued above the limits given in Table 9-I.

Make sure also that dowels are in place. Bolts alone will not keep parts from shifting. Sometimes motor dowelling is deliberately omitted. "We have to shift motor position, to realign the unit, every time we start up after a long shutdown," argued one plant operator. With his well-run maintenance and inspec-

Bolt size, inch	Maximum nut torque, lb-ft
½	15
⅝	30
¾	50
⅞	80
1	125
1⅛	180
1¼	250
1½	420

Table 9-I. Suggested maximum tightening torque for foundation bolts anchored in concrete.

tion program, he seemed able to get away with it. But that's rare. Here is a report from an industrial insurance firm called in to assess the damage in a less successful application: "... The feet of the motor were located on unmachined steel plates and were only half shimmed and not even dowelled. During operation, the strong vibrations loosened all screwed fasteners and the shims could become displaced. Finally, the rotor rubbed against the stator and severe damage to the motor resulted."

Such troubles aren't limited to small, inexpensive machines. Consider the 6000-hp cement mill motors that weren't dowelled to their bases because "it wasn't necessary." One stator shifted after startup, allowing a rotor rub which burned out the winding. A few hours of dowelling time was saved, but repairing the motor took two months. Similar stator-rotor rubs in several 2500-hp, 720-rpm fan motors at a Midwestern power plant were in part caused by lack of dowels between motor and base, despite instruction books and drawings specifying them.

A reminder about both dowels and shims when they are used: Whenever a motor must either be taken off its base for servicing or shifted because load changes require realignment, be sure dowels and shims are carefully replaced.

Clutch selection

After a motor is properly fastened down to a suitable foundation, the next step is coupling it to the driven machinery. (Only "direct coupled" drives will be considered here, because the selection and installation of belts, chains, and gears would fill an entire volume.)

Even the in-line coupling connecting two shafts on a common centerline comes in many forms. The type needing the most thorough understanding, and careful selection, does not "lock" the driving and driven shafts together at all times, but instead can permit the motor to reach its full speed before the load does.

We have seen in earlier chapters how starting heavy loads, especially at reduced voltage which is common for larger drives, may overheat induction

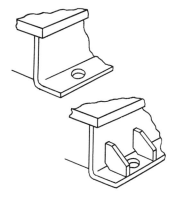

Figure 9-12: Mounting foot may bend or vibrate (upper diagram) unless adequately stiffened by gussets or braces (lower diagram). Watch for this in baseplate design.

motors. A motor big enough to do the job may be too expensive, may have poor performance, or may not be available at all.

An economical solution is likely to be "unloading" the motor during starting by using a controlled acceleration coupling. Technically, such devices may be either "couplings" or "clutches," but the overall term "unloading coupling" seems fitting.

As long as the drive need not run at different speeds, such couplings are simpler and less expensive than electromagnetic drives. Depending upon motor size and voltage, they're often preferred over variable frequency "soft" starting—which has severe limitations for high inertia or high torque loads.

Figure 9-13: A typical centrifugal clutch, shown assembled (at left), as well as disassembled (center and right).

The centrifugal clutch (Figure 9-13) is one simple device enabling standard motors to accelerate such loads. It's compact, as you can see in Figure 9-16 (page 448). Available in sizes up to 6000 hp at 1200 rpm, it includes an input hub driving a set of friction shoes, and an output drum. Rotation of the input member forces the shoes centrifugally outward against the drum. As speed increases, so does the centrifugal force, resulting in a clutch speed-torque characteristic as shown in Figure 9-14.

When the motor first starts, the shoes slip readily against the drum, allowing the motor to reach almost full rpm before the driven load gets moving. Figures 9-14 and 9-15 show what happens. In the region from Point A to Point B, the motor is accelerating rapidly with the clutch shoes slipping and the load not rotating. When the motor reaches Point B, the shoes continue to slip but transmit to the load the high torque exerted by the motor at that speed. The friction between shoes and drum is "sliding friction."

Once the clutch output drum (and its connected load) reach the motor speed at Point B, the shoes cease to slip. Friction between shoes and drum

Installing Motors Properly

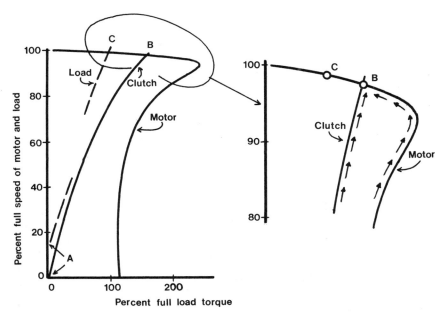

Figure 9-14: How a centrifugal clutch works. At Point A, the motor starts to rotate and sees only the counter-torque imposed by the clutch curve. The motor accelerates from A to B as the clutch shoes slip; the load may or may not begin to rotate. At B, motor and clutch curves intersect so the motor cannot go any further. It continues to supply the torque at that point until, through clutch friction, the load reaches that same speed. Clutch and motor are then locked together, and the motor torque is well above what the load requires so the drive accelerates up to full speed at Point C. A delayed engagement clutch (see accompanying text) will not begin to slip at zero speed, so the Point A will shift upward in rpm as indicated.

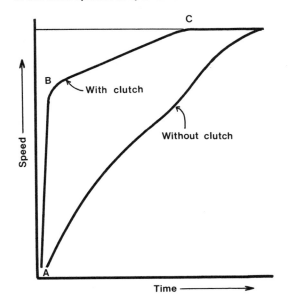

Figure 9-15: A comparison of load acceleration time vs. speed with and without a clutch.

becomes "static friction"—which is much greater than sliding friction. The two members then become "locked" together, the drive running at full speed throughout with no power loss in the clutch, and quickly reaching Point C.

Several different clutch design types span the range from ½ hp up to several thousand. One is the "delayed engagement" type, in which each shoe is restrained by a spring. The shoes cannot move out against the drum until input member speed reaches a high value, because the springs resist centrifugal force. Such behavior is useful if the load may sometimes be driven at some other rpm by a second prime mover, and a slipping clutch would then add unwanted "drag" or friction loss in the system. (See Figure 9-16.)

Figure 9-16: Where a delayed engagement clutch is needed. When the normal electric motor drive (left) operates the pump at center, the emergency drive (right) must be disconnected. So engine and pump are connected through the clutch shown (note the small amount of space it takes). Internal springs prevent the clutch from engaging until the engine is running above its idling speed. The engine can therefore be tested occasionally without shutting down the pump or interfering with the motor drive.

Even if the driver speeds are the same, as in a twin motor drive, with only one motor energized the shoes in a "free engagement" (no springs) clutch would rest through gravity against the bottom of the output drum. If that drum is spinning because the load is being driven by the other motor, friction would prematurely wear out the shoes. Springs that hold the shoes away from the drum at low input member speed will prevent that contact.

Another type of clutch is the "dry fluid" design, which uses no friction shoes. Instead, a "flow charge" of steel shot contained in the clutch housing is forced outward by centrifugal force as rotation begins. The charge packs around a rotating impeller, which then transmits torque to the output shaft.

Most clutches are best suited to fairly high speeds—such as 1800 rpm—because at low rpm the transmitted torque becomes large, requiring the clutch diameter to be large also. "Pulley" or sheave clutch configurations are made for belt drives, but they should always be on the higher speed end of the belting.

At least one "safety starting clutch" on the market does not use the centrifugal engagement principle. Its friction element is continuously interposed between driving and driven members, so that constant slippage is present at all motor speeds during acceleration. A spring adjustment varies the amount of friction to suit changes in motor or load characteristics. (See Figure 9-17.)

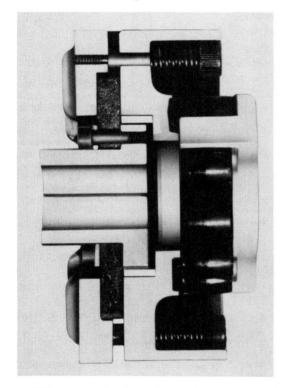

Figure 9-17: A "soft starting clutch" that does not use centrifugal force to engage shoes as input shaft rpm increases. The clutch facing slips continuously during an acceleration, transmitted torque being fixed by the spring-loaded adjustment screws (top, bottom).

Using any clutch to best advantage requires a proper match between the speed-torque curves of clutch and motor, as in curve (a) of Figure 9-18. Here the two curves intersect at a torque well above the full-load rated value. Motor torque much higher than normal is therefore available for load acceleration, reducing the starting time and thus minimizing heat losses in both clutch and motor.

What the clutch does is to "take over" those heat losses itself. Temperature rise in the shoe/drum assembly, generated by the sliding friction of clutch slippage while the load comes up to speed, replaces the temperature rise in the motor.

Curve (c) represents a poorer match. Accelerating torque delivered to the load is even lower than the motor would furnish if direct coupled without the clutch. Acceleration time will be much longer. The motor may not overheat,

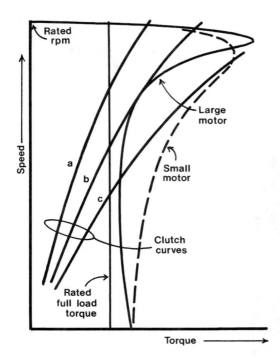

Figure 9-18: Many different clutch characteristics may be available for one application. That shown at (a) is best for a large motor (the solid speed-torque curve). Faster acceleration and best use of the clutch results from curve (c) if the motor has a curve like the dashed line, more typical of a small machine. Know what the motor will do before you select a clutch.

because the load does reach a fairly high rpm before full clutch engagement, but the clutch itself is more likely to exceed its thermal limits.

Curve (b) in Figure 9-18 is another condition to avoid. First, the torque available to start the load is again fairly low. Second, acceleration may be inconsistent from one start to the next because motor and clutch curves are essentially tangent to one another over a wide range of rpm—no precise intersection point exists. Choosing the proper protective relays or overloads is difficult when starting time is unpredictable. Clutch heating will also be variable.

Finally, a severe mechanical overload may not break some clutches free. A motor may have a service factor of 1.15, or even 1.25 in some instances. But the clutch "breakdown torque," at which static friction is overcome, may be 140% or more of the lock-in or engagement torque. Hence, the clutch may not unload the motor when drive output exceeds the motor's overload capacity. If the lock-in point (Point C, Figure 9-14) is close to motor breakdown torque, a jam in the driven machine is likely to stall the motor. That is why, according to one clutch manufacturer, "a clutch too big for the job does more harm than good."

A centrifugal clutch, like any friction device, has definite thermal limits. Acceleration heat must not exceed what the clutch can safely dissipate. Sometimes heat sensors can be furnished to shut down the driving motor if prolonged slippage overheats the clutch. But sensor selection is tricky, because they must not trip during a normal load acceleration.

Installing Motors Properly 451

If no overheating results from a single start, but the clutch is intended to allow many starts within a short time, review the application carefully with the clutch supplier. Otherwise you could simply trade motor burnout for clutch burnout.

The fluid coupling

What's the alternative when starting requirements exceed friction clutch capacity? Consider a fluid coupling. Because heat developed is through fluid friction rather than between solid members, heat exchangers can readily carry it away. In contrast to the centrifugal clutch, a fluid coupling will have slip when operating at full speed (typically 3% to 4%). This will add some energy loss to the system.

One type of fluid coupling (Figure 9-19) avoids that inefficiency by including a centrifugal friction clutch to give a "one to one lockup" connecting input and output shafts at running rpm. If an overload or jam forces a reduction in output shaft speed, that clutch disengages to protect the drive. Another version

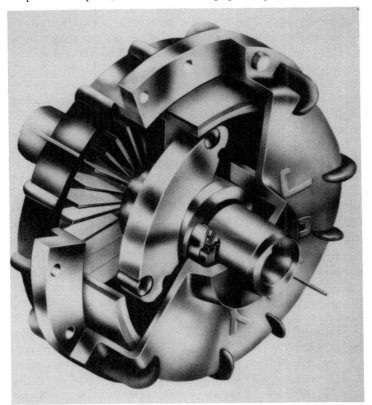

Figure 9-19: Cutaway view of a combination fluid coupling and centrifugal clutch. Fluid impeller vanes are visible at left; the clutch is visible at center.

contains a planetary gear set with individual fluid impellers that eliminate relative motion between input and output shafts at full speed. Hence there is no internal slip loss. This so-called "lock-in" coupling has been used up to several hundred horsepower.

From ½ hp up to about 1000 hp, most fluid couplings are self-contained or "constant fill." Impeller housings hold a fixed volume of fluid. That fluid charge, and the way it circulates between impellers, determine the coupling speed-torque characteristic. Therefore, the only way to change that characteristic is to baffle or block part of the fluid circulation path, or to remove part of the charge from the unit.

Larger couplings, up to 15,000 horsepower, have heat exchangers to cool the fluid. Also included is a means of varying the speed-torque curve, such as the "scoop tube." This tube is moveable in or out of a stationary impeller housing, its position determining the internal fluid level. The further in it moves, the higher that level; as it is withdrawn, fluid drains out through the tube to a lower housing level.

With a low coupling fluid level, the motor can start more fully unloaded. Only after the motor is up to speed is the scoop tube moved back in to charge the coupling for load acceleration. This may be done rapidly or slowly, and the total load acceleration time may be calculated based on the coupling's fluid circuit behavior. Hence, that time isn't simply a function of inertia and system torques, as is true for either a rigid coupling or centrifugal clutch.

Another fluid coupling variation features a separate storage chamber for the fluid charge. Under centrifugal force, fluid escapes into the impeller housing through an orifice, gradually engaging the coupling as with scoop tube control.

A fluid coupling, like most clutches, will not transmit stalling torque back to the motor if properly applied. This builds automatic mechanical overload protection into the drive. However, at the stalling point—when the driven load is stopped—the coupling may impose on the motor an overload torque between 125% and 170% of rated torque.

Matching the fluid coupling to both motor and load follows the same principles as for a centrifugal clutch. Figure 9-20 shows a poor match, involving a 450-hp ball mill drive. The motor accelerated unevenly, usually with considerable delay in bringing the mill up to speed. The curves show why. At breakaway (zero speed), load torque exceeded motor torque. Therefore, the coupling slipped 100%—yielding no rotation of the output shaft—until motor speed reached a point at which motor torque did exceed what the load required. From there, the motor accelerated on up to about 60% rated rpm, with the coupling slipping less than 100% and the load beginning to roll. There the motor and coupling curves intersected, bringing about full coupling engagement once the load rpm reached that point. But no further acceleration could occur, because just above that speed the load torque again exceeded what the motor could provide. So the coupling began slipping again, letting the motor go on up to about 90% rated rpm, where the coupling could once more engage to complete the load acceleration.

Such a combination of characteristics precludes any calculation of acceleration time, motor heating, or motor protective relay settings. Had a somewhat smaller coupling been chosen instead, one with the "alternate" curve in the illustration below, those difficulties could have been avoided.

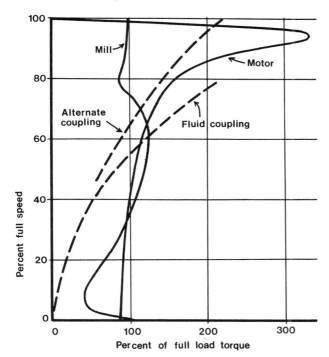

Figure 9-20: An unwise choice of fluid coupling characteristics caused trouble in this ore mill application.

No energy saving

Because any "unloading coupling" permits the drive motor to accelerate faster, without risk of overheating, it's perhaps only natural to assume that the coupling acts as an energy saver. That's an extra benefit meaning a great deal to users in this time of emphasis on machine efficiency.

High motor starting current, said one engineer, "can be directly translated into dollars of energy consumption. The amount of current draw required to start . . . can be drastically reduced by using a fluid coupling." Another engineer contended that centrifugal clutches "will fight high energy costs." Still another has hailed such clutches as "the clutches that save energy," arguing that energy saving is "one major advantage" of the device. Because, he said, the motor "can reach nearly full speed within fractions of a second, the time spent drawing high current is greatly reduced." (See Figure 9-21.)

Unfortunately, although the theory is valid, the benefit is seldom significant. Let's consider how an unloading coupling can save energy. Doesn't getting a

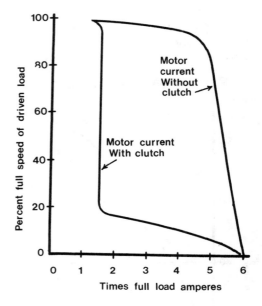

Figure 9-21: These curves are often used to emphasize energy saving obtained by starting through an unloading coupling. Lower current does mean lower energy cost, but in some applications a lower starting current may last for a longer time. In any event, the saving may never show up on the utility bill.

drive up to speed take the same amount of energy no matter how you do it? True, a certain minimum amount of work must be done. The only source of that work is the motor, which can get the necessary energy only from the power line; the coupling provides none of its own.

But that amount of energy is a *minimum*. Without going into all the mathematics involved (see Chapter 3), the energy in watt-seconds needed to accelerate any drive can be stated as:

Energy E = 2.31 (total inertia of system, lb-ft^2)(final rpm/100)^2

For example, suppose the inertia is 500 lb-ft^2 (load plus motor combined), and full speed is 1785 rpm. Total energy to bring the system from rest up to that speed is:
E = 2.31 (500)(17.85)^2, or 368,000 watt-seconds (same as 368 kilowatt-seconds)

If the load is purely inertia, the motor will have to supply energy E whether there is an unloading coupling or not.

The laws of physics will show that an equal amount of energy (E) simultaneously shows up as heat in the accelerating motor, to raise rotor temperatures according to the amount of copper, aluminum, etc., in the parts. The stator winding is also heated. If we start through an unloading coupling, the same E will again appear as heat—but now most of it will appear in the coupling, raising the temperature of shoes, drums, or fluid.

However, many loads—such as conveyor belts—do not possess only inertia. They also exert considerable counter-torque, holding back the motor as the system accelerates. (See Figure 9-20, previous page.) Applying physical laws again reveals what is often overlooked in the simple "cookbook" formulae often used to calculate what happens in an accelerating motor. At every point on the speed-torque curves of motor and load, during acceleration, the energy E that is appearing as heat must be multiplied by the ratio:

$$\frac{T_M}{T_M - T_L}$$

in which T_M is the motor torque at that speed, and T_L is the load torque at the same speed. If T_L is almost as great as T_M, the difference between the two becomes quite small—and the multiplier on heating gets very large, as in Figure 9-22. (In the most extreme situation, illustrated by Figure 9-23, at some speed T_L is greater than T_M and the drive cannot start at all without an unloading coupling.)

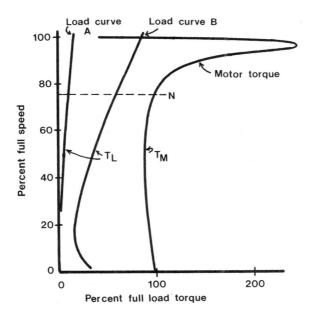

Figure 9-22: Two loads of the same inertia but different accelerating torque demand will take entirely different amounts of energy to reach full speed. In the process, the driving motor will see a corresponding difference in its heat losses. Those basic losses derived only from drive inertia must be increased by the torque ratio (see accompanying text) throughout the acceleration period. At the speed N, in this example, here's how the two compare:

	T_M	T_L	$\frac{T_M}{T_M - T_L}$
Load A	1.0	0.1	1.0/0.9 = 1.11
Load B	1.0	0.6	1.0/0.4 = 2.5

Thus, a motor accelerating Load B will have to absorb more than twice as much heat as the same motor accelerating Load A. A clutch can "save the motor" by absorbing that extra heat within itself.

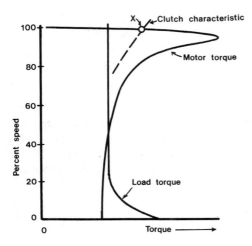

Figure 9-23: When T_L becomes greater than T_M, a direct-coupled motor cannot start the load. But an unloading coupling makes the start possible, bypassing the lower rpm range where motor torque is inadequate and causing the motor to apply a higher torque—at Point X—to the load through the starting period.

The coupling can prevent that great increase in system heating by exerting on the load a torque much greater than the T_M it would see without the coupling. Figure 9-23 makes that clear.

What the coupling does, then, rather than reduce the minimum energy needed for acceleration, is avoid a major increase in that minimum which would otherwise take place. Therefore, if you want to take advantage of energy savings by using an unloading coupling, make sure the load torque/inertia values will permit that cost advantage. High torque loads, such as conveyors, pulverizers, or ball mills, are likely to offer greater savings than a purely high inertia load, such as a fan.

But don't expect the difference to pay off the mortgage. Remember what the power bill covers. You pay for both kilowatthours of energy used and for maximum demand. Kilowatts of power lost during starting isn't important at all. What's important is kilowatts multiplied by time. How long does the start last, with or without a coupling? And how does the cost of starting power, kilowatts multiplied by time, compare to the cost of running the motor between starts?

Here's an example. Suppose a typical 50-hp, 4-pole open motor draws 600% inrush current and takes 15 seconds to start a load. Its full load running power loss totals 3.7 kilowatts. Starting loss will be very nearly in proportion to the square of the starting current, which isn't constant throughout the start, but let's assume that it is. The starting loss in the motor will then be 3.7 times $(6)^2$, or 133 kilowatts. That lasts for 15 seconds. The kilowatthours you pay for will then be 133 times 15 divided by 3600, or 0.4. If power cost is 5 cents per kilowatthour, the start will cost just 2 cents. Probably an unloading coupling can reduce that by a third, or by half—but the savings will not exactly be startling.

Suppose the motor, once started, runs at full load for an hour. The kilowatts used will include the 3.7 loss *plus* the 50-hp output which is equivalent to

37.4—a total of 41.1 kilowatts. In an hour that runs up a power bill of just over $2. In comparison, the 2 cents for one start is insignificant.

Of course, in some applications a 50-hp motor might be started and stopped more frequently. But it would take ten starts per hour before the accelerating power cost rises to 10% of the running cost in this example. Figure 9-24 gives similar data for 100- and 1000-hp machines.

What about the demand charge? This is normally based on the "integrated" or summed-up total demand over a 15-minute period. Whether or not any motor is undergoing 5-second versus 30-second starts during such a period will never be noticed by the usual thermal element integrating demand meter.

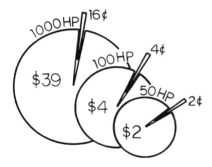

Figure 9-24: How the cost of a single start (the small wedges—15 seconds long for the smaller motors, 10 seconds for the largest) compares to the cost of running at full load for one hour (the rest of the pie), with power at 5 cents per kilowatthour. Clearly, cutting the starting cost in half, or doubling it, will have little effect on total operating cost, even if several starts occur each hour.

Size versus efficiency

There is another way, however, in which unloading couplings are said to be energy savers. There is indeed a potential reduction in operating cost if the coupling allows the motor to be a smaller, more efficient machine.

If a motor must start a heavy load unaided, two design solutions are possible, singly or in combination. First, the motor can be larger. Increased physical size provides greater thermal or "heat sink" capacity to absorb the energy loss produced during starting.

What does that mean to efficiency? It depends entirely on how much bigger the motor is, and on how those larger parts are used. If you look into oversizing of motors, you'll see statements like this: "... Because of partial load operation, the motor will be less efficient, resulting in higher operating costs." Don't let that lead you astray. One of the persistent myths about induction motor efficiency is that an underloaded motor is always "less efficient." On the contrary, it is quite often true that:

(1) A motor is more efficient at ¾ load, even at half load, than it is at full load.

(2) An oversize motor, running at partial load, is more efficient than a smaller machine running at full load.

Here are some examples:

(1) 5-hp running load:

10-hp motor running at 5 hp — efficiency .901
 5-hp motor running at 5 hp — efficiency .855

On an energy cost comparison only, this clearly favors the larger machine.
(2) 7½-hp running load, using motors from another manufacturer:
 10-hp motor running at 7½ hp — efficiency .89
 7½-hp motor running at 7½ hp — efficiency .88

Again, the larger unit will cost less to operate.

A second approach to severe starting duty is to change the rotor cage—increase bar resistance, change bar (and slot) size, and so forth. This may be possible without any change in motor size. It tends to increase the full load rotor loss and therefore increase operating cost. But again, no easy formula exists to estimate how much the change may be.

So do consider unloading couplings for severe starting duty. But don't install them for the wrong reasons. They will permit motors to start more often; reduce motor size, weight, and cost; and sometimes avoid a special motor design with long delivery time and perhaps lower reliability. In general, though, don't expect the coupling to be a big energy saver.

Drive acceleration time can be important in properly setting up time-current protection for motors. When a centrifugal clutch is used, the time will be:

$$T = \frac{(\text{total inertia})(\text{rpm})^2}{(16.1)(100{,}000)(\text{hp})} \text{ in seconds}$$

All the items in the formula are considered at the intersection point of the motor and clutch speed-torque curves.

This becomes more complex when a fluid coupling is used, especially with scoop tube control. If the motor manufacturer is asked for the acceleration time, he can only quote on the basis of a direct-coupled start—which will be considerably in error.

Whatever the type of unloading coupling used, the basic application requirements are: Know the motor behavior, know what the coupling will do, then make sure the two are properly matched. This will ensure the most economical, reliable, and long-lived drive.

The importance of alignment

Careful alignment of the driving and driven shafts is essential for a trouble-free assembly. This is an exacting art about which a great deal has been written. In a volume such as this, space does not permit a complete discussion of the subject, but here are a few essentials:

First, keep in mind that the greatest care in foundation design can be nullified by lack of attention to alignment. According to a maintenance management consultant with decades of experience, "45% to 80% of all bearing and seal failures ... can be traced to misalignment."

Flexible couplings, designed to "accommodate" some amount of shaft misalignment, are no substitute for a proper job of installation. Here is what some motor manufacturers have to say about that:

Manufacturer A: "The flexible coupling may, by readily permitting connection of two shafts which are not aligned, be responsible for vibration ... and internal wear of the coupling itself, which will tend to cause breakdowns. ... For these reasons the shafts in all coupled applications should be lined up with the same high degree of accuracy, regardless of the type of coupling applied."

Manufacturer B: "Flexible couplings are not intended to permit permanent misalignment. Even slight misalignment will reduce bearing life and cause other problems even with good flexible couplings. Flexible couplings permit some temporary slight change in alignment or end play to allow for thermal expansion during start up or during some unusual momentary load conditions."

Manufacturer C: "Efficient, quiet, troublefree service can be obtained ... only if the motor/load combination is properly aligned and the components are set on a proper, firm base that will permit alignment to be maintained. Permanent misalignment will create a number of problems—most notably, a reduction in bearing life." (We saw an example earlier in this chapter.)

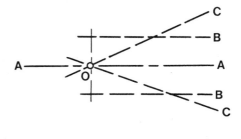

Figure 9-25: Driving and driven shaft centerlines in perfect coincidence, or "zero alignment," are shown at A. Each passes through the "coupling point" O, where the faces of the two coupling halves come together. Shafts at B are parallel to A, but "offset" so they do not pass through O. This is offset misalignment. Condition is evaluated by method of Figure 9-27a (page 461). Shaft centerlines C pass through point O, but are not parallel to A. This is angular misalignment, measured by method of Figure 9-27b. Often shafts will exhibit both types of misalignment, being tilted like C but not passing through O.

What do we mean by "alignment"? Here is a simple definition: "Co-linearity of shaft centerlines at the coupling point." (See Figure 9-25.) From simple illustrations found in many handbooks or instruction manuals, plus what was said earlier about shimming, one might assume that alignment involves only raising or lowering the motor shaft. The problem is far more complex. It's true that the alignment process normally involves moving only the motor (the driven machine is first carefully placed, then secured, in its final position, and external stresses that could affect alignment are removed). But as Figure 9-26 shows, the motor can then be moved in many different directions, any or all of which may be necessary to correct a mixture of offset and angular misalign-

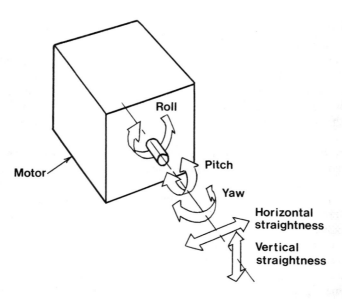

Figure 9-26: Motor misalignment results from not just one or two possible incorrect shaft positions but from six different kinds of variation, or "degrees of freedom." Five of them are illustrated here. The sixth is endwise shift or "end play" (see accompanying text). "Roll" is of little concern in motor alignment, but the other modes of movement are all important.

ment. For example, the shafts may perfectly coincide vertically, but be out of line sideways, or vice versa, or anything in between.

Checking alignment

The basic checks for misalignment involve dial indicator readings taken as shown in Figure 9-27. In doing this, observe these precautions:

(1) Make sure the indicator support is rigid and solidly secured to the shaft. Before taking readings, check the mounting by sliding a feeler gauge under the indicator button, then seeing that the dial reading returns to its initial value when the feeler is withdrawn.

The indicator mounting bracket or arm will deflect of its own weight. Hence, even with perfect shaft and coupling concentricity, readings taken on top of the coupling will differ from those beneath. Determine the amount of this deflection carefully before taking final alignment readings.

One way to do this in the shop is to machine a piece of bar stock the same size as the shaft on which the indicator bracket will be mounted. Make it long enough to also mount a post or disk at the indicator button location. With this "dummy shaft" in the lathe, set up the indicator bracket and post with the indicator at the top of the shaft. Zero the indicator. Then rotate everything 180 degrees so the indicator is at the bottom. The reading there will give the

Figure 9-27: (Right and below) The two basic alignment checks using a single dial indicator. Readings are taken every quarter turn as both shafts are rotated together. Rim or hub reading at (a) establishes offset misalignment. Coupling face reading, (b), determines angular misalignment. Why rotate both shafts together? Rotating only one shaft—the one with the indicator—will usually reveal misalignment, but can also indicate the apparent presence of misalignment when there is none. The actual trouble might be a bent shaft, or a coupling hub not concentric with the shaft. (See Figure 9-28.)

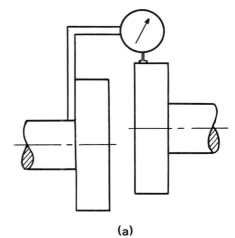

(a)

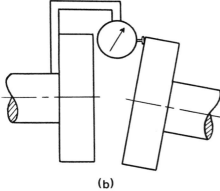

(b)

Figure 9-28: This simple example shows why rotating only one shaft while reading coupling hub runout can give misleading results. Eccentricity of the right-hand coupling rim indicates offset misalignment even though none actually exists—shaft centerlines do coincide.

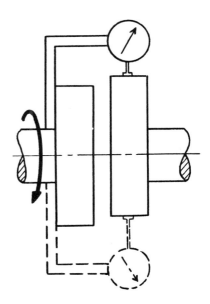

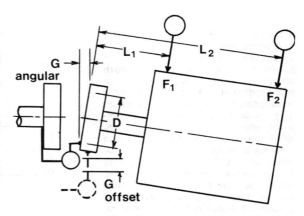

To correct angular misalignment, all movements at location F_1 should be:

$$G \text{ times } L_1/D$$

and at F_2:

$$G \text{ times } L_2/D$$

Suppose that $L_1 = 10$ inches and $L_2 = 55$ inches, and the coupling readings are:

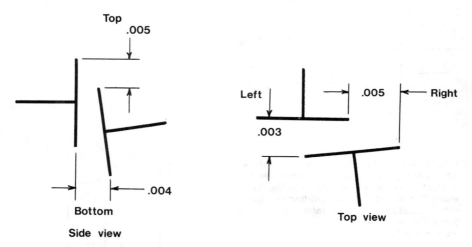

Then the vertical corrective movements should be:

	At F_1		At F_2	
	Mils	Direction	Mils	Direction
To correct offset	5	Up	5	Up
To correct angular	4	Down	22	Down
Total	1	Up	17	Down

And the horizontal movements should be:

Offset	8	Left	8	Left
Angular	3	Left	16.5	Left
Total	11	Left	24.5	Left

total "sag" to be added to or subtracted from whatever alignment readings are taken on the drive itself. In many cases it will be negligible, but this should not be assumed.

(2) Because a motor can shift sidewise while being raised or lowered during alignment, two more dial indicators should be placed at either side of the machine, alongside the feet (Figure 9-29), to reveal any such movement. They will also measure any intentional sidewise movement at either end of the motor that may be needed to correct misalignment in the horizontal plane.

(3) To properly coordinate the "moves" for correcting both types of misalignment in both horizontal and vertical planes, record all readings on sketches or in tables with consistent directions. While doing so it's convenient to visualize angular correction as a swinging of the motor about an imaginary pivot point on the face of the motor coupling. Such charting is illustrated by Figure 9-29. Without such a systematic approach, one can easily lose track of the various adjustments.

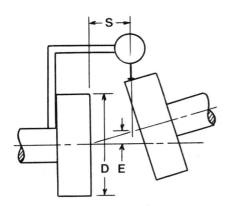

Figure 9-30: As the distance S increases relative to coupling diameter D, the indicator will appear to show an increasing amount of offset misalignment, even when the actual misalignment is entirely angular. The amount of this "error" appears as E in the figure.

(4) As Figure 9-30 shows, angular misalignment alone can sometimes produce indicator readings showing apparent shaft offset where none exists. If the distance S is more than about one-tenth of the diameter D, this effect should be allowed for in calculating corrective moves. That is one reason for correcting angular misalignment first, then checking for offset. Ignoring this refinement will result in some over-correction to be eliminated in a second trial.

The methods of Figures 9-27 and 9-29 are best suited to initial installation when it is convenient to separate the coupling hubs. For in-service alignment checks, when coupling separation is not convenient, many other methods of alignment make use of the same principles in different ways. A set of two

---

Figure 9-29: (Opposite page) An example of how to calculate motor foot movement at both ends of motor, up-down and sidewise, for correction of both angular and offset misalignment. Note additional indicators at feet, F_1 and F_2, to monitor sidewise movement. The up and down movements are made by inserting or removing shims; left-right adjustment is done with jackscrews while observing the indicators at the feet.

indicators (or even three) permit simultaneous checking of both angular and offset alignment, as in Figure 9-31.

Sometimes the coupling can be separated but it is impractical to rotate both shafts simultaneously—for example, when a high speed motor is coupled to a high ratio gearbox. Detailed recommendations for such situations must come from the machinery builders.

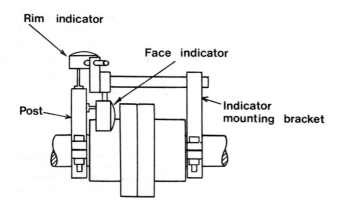

Figure 9-31: The "two indicator" method of checking alignment when it is not convenient to break the coupling.

How "perfect" should alignment be? The most widely accepted rule is that angular shaft misalignment should not exceed a total indicator reading of .001 inch per inch of coupling radius to the point of measurement. Total indicator reading for offset misalignment should not exceed .002 inch regardless of diameter. Remember that these are suggested maximum figures. Aim for the lowest readings possible, rather than being content with these limits simply because they are considered tolerable.

Allow for varying operating conditions. A notable example of this, similar to the fan mentioned earlier, is the pump handling extremely hot or cold liquid. Motor and pump enclosures (and shafts) will be at about the same temperature after the drive has been idle for some time. Upon startup, the pump temperature will begin either to rise or to fall with respect to the motor temperature, making the two assemblies shift position relative to each other. One shaft will rise in height more than the other, causing increasing misalignment. Often this must be allowed for by "building in" some initial misalignment which the changing temperatures will automatically correct. Know what temperature/growth differential to expect, and be prepared to deal with it.

For drives subject to such differentials, how can the compromise between hot and cold conditions best be made? There is no easy or precise method. The following procedure has been suggested:

(1) Bring the shafts into "zero alignment" with both machines cold.

(2) Operate the drive long enough to bring everything up to normal full load operating temperature. Shut the motor down. Immediately realign the shafts under hot conditions.

(3) Restart the drive; check vibration and noise. Stop and start alternately, adding or removing shims each time until the equipment is running as smoothly and quietly as possible.

(4) Stop the drive; let it cool. Read the offset and angular displacement of the shafts at this point, and record them for future reference. This particular degree of cold misalignment will be "normal." Future maintenance checks should assure that this condition is maintained.

Once properly made, whatever the machinery operating conditions, alignment should not be neglected. Plan to check it regularly, just as you would check bearing oil level or winding insulation resistance. With the passage of time, foundations may settle, motor mounts warp, bearings and shafts wear. All these variations can result in damaging misalignment. Remember: "Misalignment of shafts causes more machine failures than any other agency except unbalance."

Dealing with end play

As Figure 9-26 indicated, there is another kind of misalignment with which the motor installer and operator must be concerned. The drive coupling process must also take proper account of end-to-end adjustment—which involves motor end play or "end float." What is this, and why is it present?

End play is the axial free movement possible for the rotor-shaft assembly when nothing is coupled to the motor shaft. It exists for three main reasons.

One is that machining variations on bearing brackets, shafts, etc., require some allowance for tolerance buildup. Hence, bearing position can vary somewhat from the nominal design value.

Second, any bearing has some internal looseness or "play." If this is unduly restricted, bearing parts can "bind up" and overheat.

Finally, the rotor-shaft assembly heats up under load. As we have seen, severe starts can produce high transient temperatures. In a small motor, the rotor may reach 50°C to 100°C, especially if the machine is totally enclosed with a Class F temperature rating. So the shaft will expand endwise, requiring additional end play of about 5 mils per 60°C per foot of shaft length between bearings.

Rotor movement cannot be unlimited, of course. In a running motor, magnetic forces alone (though relatively slight) tend to hold the rotor in the stator. But no such forces exist in a de-energized motor coasting to a stop. Furthermore, those forces may vary in direction during acceleration, so an unrestrained rotor could oscillate back and forth far enough to cause damage. Another necessary limitation on rotor movement is the need to "close up" bearing assemblies so lubricant will not escape.

End play in standard horizontal ball bearing motors may vary from $\frac{1}{32}$ to $\frac{1}{8}$ inch depending upon bearing size. Larger machines often include some kind of bearing locking arrangement limiting the play to much smaller amounts. Sleeve bearing play is greater—normally ¼ to ½ inch. Such bearings exert endwise restraint on shaft movement through "thrust faces" as shown in Figure 9-32.

Such faces are incapable of sustaining continuous thrust, only such momentary "bumping" as occurs during acceleration. Tolerance accumulation, plus variations in magnetic force, requires that end play be large enough to prevent thrust faces from ever riding continuously against a shaft journal. Some motor manufacturers fit sleeve bearings into their housings with adjustable endwise stops so that the assembly can compensate for manufacturing variations.

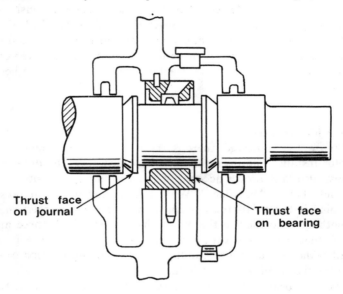

Figure 9-32: Typical sleeve bearing assembly illustrating how endwise movement of shaft is limited.

Many special anti-friction bearing arrangements depart from the pattern just described. One is the grinder drive, requiring the shaft to have no endwise movement at all. A locked bearing assembly is used, at the drive end, having the bearing inner race clamped to the shaft, the outer race clamped in the housing. The bearing itself is of special construction, such as the preloaded double-row type, so that it contributes no looseness of its own. Tolerance buildup and thermal expansion in the rotor-shaft assembly remain inevitable, of course, but are allowed for by a freely-sliding fit of the other motor bearing within its housing.

A less severe limitation is the high-pressure hydraulic pump load. Its tight internal clearances will tolerate some limited shaft movement. The motor will still use a locked drive-end bearing, but of a standard type having some internal looseness to permit a small end play. (See Figure 9-33.)

Another special case is the vertical turbine pump. Not only does it have close internal clearance between pump impellers and bowls; it also imposes

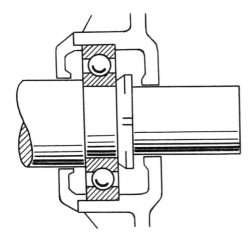

Figure 9-33: Restrained ball bearing assembly that limits motor end play to the amount of internal looseness in the bearing itself. Inner race is clamped by locknut on shaft; outer race is clamped in bearing housing by an inner cap.

high thrust loads—sometimes both upward and downward—on motor bearings. Normally the motor's upper bearing takes the high downward thrust. The restrained lower bearing must resist upthrust while usually restricting end play between .005 and .025 inch. Special assembly procedures prescribed for such motors must be closely followed whenever bearings are replaced.

Vertical motors are free from one complication of horizontal design. Because the shaft is vertical, its weight acts endwise to hold the rotating parts against the lower limit of the end play except during those intervals when pump upthrust exists.

Rotor centering

But the standard horizontal machine is not held in any particular position by anything except magnetic "centering force" (assuming the load is properly prevented from applying any endwise thrust of its own to the motor shaft). Where will such a rotor-shaft normally position itself under the influence of those forces? How great are they? Three definitions are involved in the answers:

(1) Magnetic center. This is the position the rotor will try to take when running as long as it remains undisturbed by any forces outside the motor. The term "center" leads to confusion, as we shall see, but simply means that if the rotor is moved to the left of that point, it will pull back to the right when released, and vice versa. It's important to note that under load the magnetic center *may* shift slightly compared to its no-load position.

(2) Mechanical center. The location of the rotor-shaft assembly exactly halfway between the physical limits of its possible movement.

(3) Centering thrust. The amount of magnetic endwise force exerted on the rotor when any outside influence moves it away from its magnetic center while the motor is running.

Many users expect magnetic and mechanical centers to coincide. In some machines this happens, because (as with the adjustable stops mentioned earlier), bearing position—which fixes the mechanical center—may be adjustable at assembly to suit the magnetic center location.

No one is concerned about this in a ball bearing motor. Not magnetic forces, but the internal bearing constraints, will then determine the rotor's running position.

The sleeve bearing machine, though, will normally be so constructed that magnetic center can lie anyplace within the mechanical limits set by the bearing thrust faces. There is no inherent reason for magnetic center to be at mechanical center. Left to itself, the rotor will "float" so that bearing thrust faces and shaft shoulders will be out of contact with each other. How far out of contact is unimportant.

The magnitude of the centering thrust is often of undue concern. This force will vary with motor voltage and current (hence with load), as well as with overall rotor length and the amount of off-center displacement. No simple formula exists to calculate this complex variation. Nor is there a simple test.

Theoretically, it should be simple enough to push a running shaft towards either end of its mechanical end play, measuring the force needed to do so. However, it can be dangerous to contact the spinning shaft with a spring scale or other hand-held instrument. What tests have been reported show that total force rises rapidly with initial movement off magnetic center, then falls again as displacement is increased towards the mechanical limit. Values have ranged from 30 to 200 pounds. Such low numbers show that it is far more important to protect motor bearings from driven machine thrust than to safeguard the driven machine against thrust imposed by the motor's centering force.

Driven machinery will also include some built-in restraint on its end play. Compressors, fans, pumps, and many other loads usually contain thrust bearings limiting end play to a few mils. Often such a load is connected to a motor with a sliding gear-type coupling (the flexible type, designed to take some shaft misalignment, that was mentioned earlier in this chapter). Such a coupling will have more end play between its hubs than that allowed by the motor bearings.

If misalignment is present, uneven forces between hubs and the outer coupling sleeve may cause driving and driven members to slide apart. In one direction, this "hub separation" is limited only by the driven machine's thrust bearing; in the other direction, only by contact between motor bearing thrust face and shaft shoulder. That can impose continuous thrust load on that motor bearing face, possibly damaging even if small.

Misalignment is not always accidental (recall the hot or cold pump application mentioned earlier). And even when alignment is perfect, "coupling lockup" can still transmit thrust to motor bearings. We have all observed the tendency of a sleeve bearing machine, uncoupled, to "bump" back and forth during starting. The same often occurs during coastdown. When a coupled motor stops, then, the shaft may have drifted over against the bearing thrust face farthest from the load. Upon restarting, torque transmission through coupling gear teeth creates high friction that resists any endwise movement between coupling hub and sleeve. The motor shaft is then forced to stay where it was—with shaft contacting thrust face.

One gear coupling supplier claims this friction force won't exceed 7% of tooth pressure caused by torque. Another gives 10% as the limit. Even those small fractions, however, may represent several hundred pounds. In a worn or poorly lubricated coupling, the force can be much greater. Pin-and-bushing type couplings, or spring grid units, can create the same effect.

As the motor and driven machine warm up under continued load, both shafts expand thermally. The result is thrust applied to bearings in machines up to the limit of the coupling friction. Early bearing failure sometimes follows.

Mechanical end play limits in sleeve bearing motors may be at either end of the machine depending on size and manufacturer. The occurrence of bearing failures at the end furthest from the coupling does not mean the coupling is not responsible. One service shop repaired several outboard bearing failures on a 2000-hp motor. Investigation finally revealed the user had installed a coupling that did not control end float, and was unaware of the damaging consequences of his choice.

The solution to such problems is the "limited end float" coupling. In the common sliding version, "thrust buttons" or spacers are fitted to the end of each shaft, permitting the shafts to pull apart to the limits of the coupling sleeve but to come together no closer than allowed by contact between the two buttons.

Theoretically, the same result could be achieved by "over-pressing" the coupling hubs so that the shaft ends themselves (rather than buttons) would properly limit movement in one direction. However, that could cause scraping, allowing metal chips to get into the coupling gear teeth, unless the two shaft ends were perfectly square and flat. Flat spacer washers, if used instead of buttons, wouldn't allow for any angular misalignment between shafts.

For other types of coupling, similar arrangements are available, such as neoprene or metal "rung spacers" to limit the end float in spring grid couplings. It may also be possible to obtain motors having end play greater than normal. In the largest drive sizes, however, unlimited coupling end float will exceed ½ inch, and motors with end play greater than that may not be available. One large ore mill designer once asked for some 1500-hp motors having ¾ inch end play. This could not be done, because shaft oil flingers at each end of the journal would not have remained within the bearing housing for all possible shaft journal positions.

The proper installation of a limited end float coupling is shown in Figures 9-34 and 9-35. In Figure 9-34a, a locked thrust bearing in the load is assumed, resulting in little or no end play of the driven shaft.

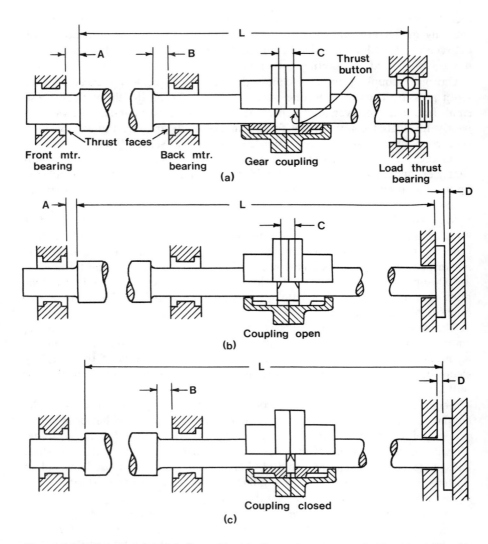

Figure 9-34: Coupling and shaft positioning for various types of drive. At (a), with locked thrust bearing in driven machine, coupling position allows enough clearance in motor bearing at A for expected shaft expansion plus an extra 1/16 inch margin. Dimension B need only be a minimum of 1/16 inch, because shaft expansion tends to increase it. The driven shaft itself can produce no pull on the motor shaft through coupling friction or otherwise.

When the driven machine uses a Kingsbury type thrust bearing, as at (b), bearing play D may be as much as 1/32 inch. With maximum coupling hub separation C, and the load thrust bearing movement closed up in the forward direction, dimension A must be adjusted to allow for expansion over the length L.

When the coupling is closed, as in (c), and the load thrust bearing is fully shifted in the opposite direction, B must be at least 1/16 inch.

If both directions of end play are limited in the motor at the same bearing, as in Figure 9-35, then A and B will still be determined in the same way.

Installing Motors Properly

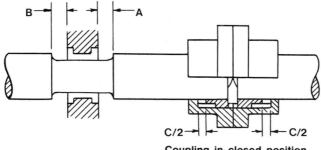

Coupling in closed position

Nominal Motor End Play E	Limited Coupling End Float C	A	B	C/2
½	3/16	11/32	5/32	3/32
¼	3/32	11/64	5/64	3/64

Recommended motor end play, from NEMA MG1-14.37B, 1978:

Motor Horsepower	Synchronous rpm	Minimum Motor End play	Maximum Coupling End Float
125 through 250	3600 and 3000	0.25	0.09
300 through 500	3600 and 3000	0.5	0.19
125 through 500	1800 and below	0.25	0.09

For larger motors, NEMA MG1-20.81, 1981, adds this line:

| 600 and higher | All speeds | 0.5 | 0.19 |

as well as this note about coupling end float: "Couplings with elastic axial centering forces are usually satisfactory without these precautions." (Such couplings are little used for smaller machines.)

Figure 9-35: Proper installation of limited end float couplings should follow these rules. All dimensions are in inches.

Why ground?

Having planned for a proper installation "from the ground up," including coupling and alignment, the motor user may consider his work complete. Electrical connection to the power system is equally important, of course, and has been touched on in earlier chapters. Because the main power leads carry motor operating current, designers and installers alike are well aware of the importance of clean, tight, properly-sized connections. The ground connection, so necessary for both apparatus and personnel safety, is often much less well understood. Hence no discussion of motor installation can be complete without a review of grounding practice.

If electrical systems were haunted by evil spirits, the top candidates for the job would be transient surges. But second place must go to the sometimes equally unpredictable behavior of equipment grounding circuits—which sometimes work so well when not wanted, yet at other times are useless when most needed.

How, and why, do we maintain the integrity of the first two links of the protective chain in Figure 9-36? The major objective of such grounding is to limit the potential between non-current-carrying parts of the equipment, and between those parts and earth, to a safe value under all operating conditions both normal and abnormal. A second objective is to furnish a low-impedance return path for ground fault currents. A high impedance can permit dangerous voltages at equipment surfaces during a fault. Also, says one authority, "high impedance at joints and connections . . . may cause arcing and heating of sufficient magnitude to ignite nearby combustible materials or explosive gases."

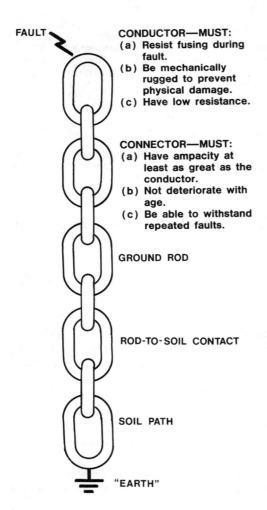

FAULT

CONDUCTOR—MUST:
(a) Resist fusing during fault.
(b) Be mechanically rugged to prevent physical damage.
(c) Have low resistance.

CONNECTOR—MUST:
(a) Have ampacity at least as great as the conductor.
(b) Not deteriorate with age.
(c) Be able to withstand repeated faults.

GROUND ROD

ROD-TO-SOIL CONTACT

SOIL PATH

"EARTH"

Figure 9-36: The grounding chain for equipment. Each link is important. If the conductor/connector arrangement at the top itself includes several "sublinks" of uncertain quality (such as bolted joints between steel parts), can the entire chain be considered reliable?

Unfortunately, as the same author pointed out, equipment grounding "is often provided as an afterthought and, consequently, may or may not be adequate for the purpose intended.... A well-planned equipment grounding system must be provided whether the system neutral is grounded or not. Ungrounded neutral systems often operate for extended periods with a single phase faulted to ground. During such periods, a contact between another phase conductor and a metallic enclosure raises the enclosure to full line potential above ground."

Grounding in general is important enough to rate an entire IEEE standard to itself: the "Green Book," IEEE 142. Isn't it likely, then, that adequate "frame grounds" for motors would be so standardized as to leave no uncertainties? No—that isn't the case.

Then how are motor grounding provisions being specified today by motor users? According to some current specifications:

(1)(Gas processor) "... furnish one copper lead connected directly to the stator core for grounding." Question: What size?

(2)(Public utility) "Provide grounding pad." Question: How large? Drilled for what sort of attachment?

(3)(Auto manufacturer) "The conduit box shall be of ample size.... Means shall be furnished to terminate a grounding conductor of suitable size." Question: What size is "suitable"?

(4)(Consultant) "All motors shall have provision for frame grounding." Question: What kind and size?

(5)(Oil company) "Motors shall be provided with a tapped hole in the frame foot ... to facilitate attaching the grounding conductor lug." Question: What size threads, and how large is the lug?

NEMA drilling

Certain requests tend to recur. One is for ground pads having "NEMA drilling." Actually, no NEMA motor standard mentions ground pad drilling or any other kind of ground connection. Nor has NEMA issued any separate "ground pad standard." The intended reference is probably to NEMA CC1-1981, "Electric Power Connectors"—which includes several different arrangements of number and size of bolt holes for terminal lug pads. Which one is desired is seldom specified by the motor purchaser.

It's also common simply to call out a hole or stud size, like this: "All motors shall be provided with an earthing bolt of at least 6 mm diameter." That's easy enough to supply—but how much flat surface around it will be needed for adequate contact between the ground wire terminal lug and the motor frame?

Perhaps the most frequent, yet most useless, spec requirement is that grounding provision be made "in accordance with NEC Article 250-94." (That's for *system* grounding anyway, not *equipment* grounding; the latter is covered by Article 250-95.) Table 250-94, for system grounding conductors, bases wire size on the size of the largest service-entrance conductor—an unknown quantity to the motor manufacturer. Table 250-95, for equipment grounds, instead bases ground wire size entirely on "Rating or Setting of Automatic Over-current Device in Circuit Ahead of Equipment...." with a range from No. 14 up to 800 MCM. The motor manufacturer never knows that either, so how can he possibly take account of it in choosing a ground wire size?

What about other pertinent sections of the Code? Article 250-23(b) applies only to circuits below 1000 volts; it references Table 250-94 but does not require a ground wire bigger than the largest ungrounded service conductor. For motors, Article 250-2 references Article 430, which cites Article 250! Article 710, above 600 volts, lacks any specific guide to motor grounding.

Let's look at the other standards that mention equipment grounding.

(1) ANSI C2-1984, "National Electrical Safety Code."

(2) IEEE 142/ANSI C114.1-1982, "Safety Standard for Grounding of Industrial and Commercial Power Systems."

(3) NEMA MG2-1977, "Safety standard for construction and guide for selection, installation and use of electric motors and generators."

ANSI C2 says "A grounding conductor for an alternating-current system shall have a current-carrying capacity not less than one-fifth that of the conductor to which it is attached." But that provision is applicable to a system, not to a piece of equipment, so is of no help in dealing with motors.

Does IEEE 142, under "Equipment Grounding," recommend specific ground conductor size? No. It says merely that: "The use of a large-cross-section grounding conductor of itself is not enough... the installation must also provide a more favorable (lower impedance) fault return path than other possible paths which might have inadequate current-carrying capacity."

According to NEMA MG2-2.09: "... All exposed non-current-carrying metal parts which are likely to become energized under abnormal conditions shall make metal-to-metal contact or otherwise be electrically connected or bonded together...."

"When a motor or generator is provided with a grounding terminal, this terminal shall be on a part of the machine not normally disassembled during operation or servicing."

Where to connect?

In modern large motors there may be no parts "not normally disassembled during servicing." To what specific part, then, should a ground connection be made? The answer must take account of these facts:

(1) In many motors, no accessible exterior frame surface can be considered electrically bonded to the stator core itself. The entire enclosure may consist of bolted-together pieces of sheet metal. Any path between an external grounding point and the core itself may contain several separable joints of uncertain electrical quality.

(2) In many smaller machines having one-piece frames, attaching the ground connector desired may be impossible for two reasons: (a) Lack of a big enough flat surface. (b) Insufficient thickness of material for threads (or the drilling may be prohibited, for explosion proof motors).

(3) Connections within the terminal box may be needed. Yet some types of connection hardware aren't permitted for explosion proof motors. Or, not knowing how much space the user may need for the hookup, the manufacturer may not make the box large enough to contain it.

So in deciding where and how to make a proper motor ground connection, the size, type, and design of the machine must be carefully considered. Ideally, of course, the "frame ground" would attach directly to the stator core. Some users are calling for tapping a grounding bolt directly into the laminations, on motors as small as 60 hp. They may require the motor manufacturer to furnish green cable (leading from that point to the conduit box) as described in NEC Article 250-57 (green tape or adhesive markers may be used for the larger wire sizes).

However, threading a bolt directly into core laminations isn't recommended for any motor, regardless of size. Cutting threads into hard silicon steel sheets at right angles to the hole axis results in a joint that's poor to begin with. As the stator heats and cools throughout the life of the motor, that can only get worse. Small machines usually lack sufficient internal room to make such a connection anyway.

Small motors

Wherever the attachment may be, it shouldn't be assumed that small motors will present no problems because circuit (and grounding) conductors will be small. Consider this, from a chemical plant specification: "Terminal box shall be furnished with . . . one clamp terminal for grounding connection." An accompanying table calls for a 350 MCM ground cable for 150-hp motors. The smallest size compression lug for a copper cable that big would be about 3 inches long, with a tongue requiring a contact surface 1½ inches square, using a half-inch mounting bolt. Such a termination, allowing for the necessary bending or contouring of the cable itself, would never fit within the terminal box of so small a motor.

Nor is the manufacturer safe in allowing for only a single ground. One user asked for this: "Furnish . . . 3 Burndy clamp type Servit posts for 3 ground wires"—but still neglected to give the desired sizes or locations.

Furthermore, conduit box mounting (or grounding provision) isn't necessarily the motor manufacturer's sole concern. Once the motor reaches the job site, the installer should realize that box alterations, some test procedures, and future maintenance work all may affect ground integrity. Mistakes could result in severe fault damage, costly downtime, or danger to operating personnel.

Some users install their own conduit boxes. Perhaps the original box wasn't large enough. Whatever the reason, it gets reworked or replaced on the job site. At that time, the interconnecting joints, jumpers, etc., may be not be replaced correctly, or new bolted joints may be added to the original assembly. This may obstruct grounding paths.

Assembly bolts are normally steel, holding together steel parts that are prime-painted (if not fully finished) before assembly. In the shop or in the field, these components may not be thought of as current-carrying parts. Yet for the ground path typified by Figure 9-37 to be effective, at every bolted joint shown the steel plate surfaces should be carefully ground smooth and bright for good electrical contact before the joint is made up. Where holes are tapped, instead, there's less of a problem.

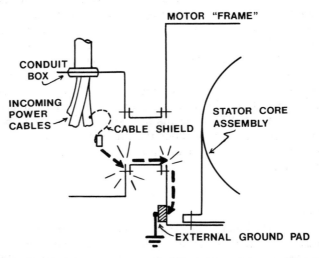

Figure 9-37: Some motor installations require conduit box internal grounding connections for cable shields, as shown here, but use an "earth" connection on the motor frame. The ground fault current path between these two locations involves several bolted joints of uncertain electrical quality (heavy arrows show the ground path). This is especially true of large motors, which are the most likely to use shielded cable (required by the NEC at 4 kv and up).

NEC Article 250-51 requires the path "to ground from ... equipment" to be "permanent and continuous," able to conduct safely "any fault current likely to be imposed on it" and with sufficiently low impedance "to limit the voltage to ground and to facilitate the operation of the circuit protective devices. ... " Adds Article 250-75, "Metal raceways, cable trays, cable armor, cable sheath, enclosures, frames, fittings and other metal noncurrent-carrying parts that are to serve as grounding conductors shall be effectively bonded where necessary to assure electrical continuity and the capacity to conduct safely any fault current likely. ... Any nonconductive paint, enamel, or similar coating shall be removed at threads, contact points, and contact surfaces. ... "

It's not enough to do all that at the factory. Even user specifications calling for "serrated washers" at conduit box mounting bolts, to improve electrical contact, can't overcome attack by environment or neglect in the field or in the repair shop.

Unfortunately, grounding connections normally carry no current. Line connections may loosen or corrode, too, but the damaging results soon become apparent—as overheating which can be spotted during maintenance, or as self-evident "blowout." But the quality of a ground path has no such self-checking property.

Proper earth resistance

How high a path resistance is acceptable? There are no standards. One suggested upper limit is 3 to 5 ohms from the motor ground point to "earth." But just where is that ground point? If resistance from a ground pad on the motor enclosure is half an ohm to earth, is the ground adequate if there are several more ohms between that pad and the stator core?

At least one installation spec has called for a maximum of 1/10 ohm, by earth megohmmeter test, between the motor and the plant grounding grid. Asked how that figure was arrived at, the user replied "We wanted it as low as possible, but didn't want to specify zero." If that seems an unscientific approach, remember that there are no rules to go by. The latest OSHA electrical safety standards, in effect since April 1981, specify no particular limiting value of grounding conductor impedance.

That impedance can't be accurately measured, simply because at any given time there are normally other low resistance paths in parallel with a grounding conductor. An equipment ground wire or bus could actually be open-circuited and a test may not disclose that fact. In recognizing this, OSHA says: "... It appears that visual inspection and periodic maintenance are the only practical approach to assure a safe condition of the equipment grounding conductor." Only a continuity test of that conductor is required by OSHA standard 1926.400(h) on ground fault protection.

Incidentally, both OSHA and NEC rules now consider connection to grounded metallic structural building frames as inadequate equipment grounding. Installations made after April 1981, says OSHA, must use a separate grounding conductor. The reason is that heavy ground currents flowing through unpredictable paths far from the circuit conductors can generate stray fields damaging to low voltage devices, as well as arcing—especially in piping systems. Figure 9-38 shows one example, taken from a major electrical breakdown which eventually resulted in a $9 million fire.

Even using the conduit itself as a ground may be dangerous. In high-current tests reported years ago in an AIEE paper, "normally tight" threads at conduit fittings emitted "showers" or "blowtorch streams" of sparks, burning out threads and creating a major fire hazard. In some of these tests the alternate ground path through steel building framing was measured at .0001 to .0006 ohm—yet was incapable of diverting ground return current from the conduit system containing the line conductors. Because of magnetic field interactio an effective ground conductor must be inside, not outside, the conduit itself. This is increasingly common, as shown in Figure 9-39—though it does not solve all the problems inside the motor.

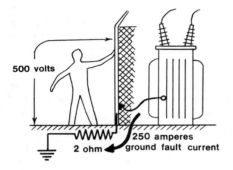

Figure 9-38: In instances where system design and relaying do not immediately trip out grounded circuits, a fault can result in this kind of hazard to personnel, despite what appears to be a low ground resistance.

Figure 9-39: This connection arrangement relies entirely on the grounding point within the conduit box. A stator winding ground fault then sets up the path shown by these arrows—again involving various bolted joints.

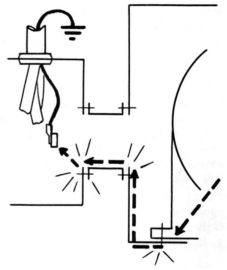

Surge protectors

Of particular concern is the grounding of surge protection equipment in a motor conduit box. Surge capacitors, often accompanied by lightning arresters, are often used on motors at 2300 volts and up. To properly protect motor windings, those devices must be solidly grounded to the stator core itself—usually via some kind of frame connection. Simply bolting the units to the conduit box structure isn't enough.

Capacitor and arrester cases have grounding terminals that can accept certain cable size ranges. Typically the one size that will fit both is No. 2. In any

Installing Motors Properly 479

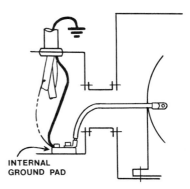

Figure 9-40: Here a single conduit box grounding pad will serve to ground all motor parts solidly. The bolted joints are bypassed by a cable directly linking the pad with the structural part of the stator core assembly. Any surge protection devices furnished in the box would be grounded to that same pad. Figures 9-42 and 9-43 show other arrangements that have been used to minimize unexpectedly high ground path impedance.

event, this cable should be carried back to the stator assembly as in Figure 9-40. The terminal box of Figure 9-41 includes such a connection.

Because the arresters and capacitors are solidly connected to the incoming power leads, and it is often difficult or expensive to disturb those connections, maintenance personnel may prefer to avoid unhooking the surge protection for periodic insulation tests on motor windings. But if that isn't done, as required by IEEE test standard 112, Para. 7.2.2, test voltages could damage the accessories.

Figure 9-41: Terminal box with connection described in Figure 9-40.

Therefore, some users now specify disconnection on the ground side for both capacitors and arresters. For that purpose the devices must be mounted on insulated supports, grounded to the conduit box structure only through removable cable jumpers. Because the ground will then be removed at least once a year for winding tests, special care must be exercised to get those joints clean and tight at re-assembly. Remember—there will be no warning of a poor connection. If it results in failure of the surge protection equipment to properly

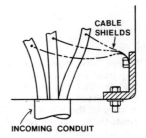

Figure 9-42: A way of allowing cable shield grounds to be connected outside the conduit box directly to a grounding bus.

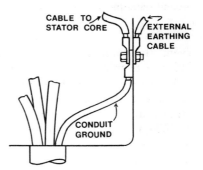

Figure 9-43: Similarly, this permits external connection to a stator core grounding cable.

drain off a damaging surge, however, a catastrophic winding failure may occur.

In summary: Though there's an understandable tendency to blame the equipment supplier when "things don't fit," the installer or operator should realize that actually the system designer or consultant may have generated the problem. When describing motor grounding requirements, spec writers tend to work under two handicaps:

(1) Lack of awareness of how motor standards deal with grounding.

(2) Unfamiliarity with motor construction and its effect on adequate grounding.

Because there are no standards for size, type, location, or number of motor grounding points, a user should not assume that all motors will lend themselves to his particular grounding practice. He and the manufacturer should agree beforehand on just what is to be furnished.

Then, sound installation, maintenance, and repair practices must be followed to make sure grounds will remain fully functional. (See Figure 9-44.) It's fairly obvious that motor stators repainted or revarnished during overhaul should have grounding pads or terminals carefully masked beforehand, or thoroughly cleaned afterward, for a good electrical connection. Look at other assembly joints, also, in light of Figures 9-37, 9-39 and 9-40, to see what else may interfere with good grounding. And keep that ground path in mind when a new ground connection is furnished.

Installing Motors Properly

Figure 9-44: A typical "frame ground" bus (arrow) bolted to housing of 700-hp motor. The joint must be kept tight and properly made up whenever the motor is reinstalled after being taken off its base for servicing. Use these bolt torque values unless others have been specified:

Bolt diameter	Bolt torque, lb-ft
1/4	8
5/16	14
3/8	20
1/2	40

Bibliography

Chapter 1

Arnold, R. E. "NEMA Suggested Standards for Future Design of AC Integral Horsepower Motors," *Transactions of the IEEE,* Vol. IGA-6, No. 2, Mar/Apr 1970, pp. 110-114.
Goldman, D. T. "The Metric System: Its Status and Future," *IEEE Spectrum,* April 1981, pp. 60-63
Mason, J. "Why Parts 'fit': The Role of IEC," *IEEE Spectrum,* June 1980, pp. 42-44.
Merrill, E. F. "Should I Select a Service Factor Motor?" *Transactions of the IEEE,* Vol. IA-17, No. 5, Sept/Oct 1981, pp. 458-462.
Panesar, K. S. "Match Motor to Driven Machine," *Hydrocarbon Processing,* August 1979, pp. 79-82.
Paulus, C. F. and D. R. Green. "Proper Motor Application Engineering Can Reduce Costs and Improve Reliability," *Transactions of the IEEE,* Vol. 81 Part III, 1962, pp. 886-891.
Saylor, C. "Why Motors Balk when the Power Isn't Right," *Machine Design,* May 16, 1974, pp. 139-141.
Szabo, L. "Is the Sun Setting for Voluntary Standardization?" *Power Transmission Design,* July 1979, pp. 71-76.
Woll, R. F. "Comparison of Application Capabilities of U and T Rated Motors," *Transactions of the IEEE,* Vol. IA-11, No. 1, Jan/Feb 1975, pp. 34-37.

Chapter 2

Aldworth, D. "Effect of Motor Efficiency on Energy Savings," *Electrical Construction and Maintenance,* September 1981, pp. 38-42, 47.
Andreas, J. C. "Energy-Efficient Electric Motors," (book), Marcel Dekker, Inc., New York, 1982.
Bergman, A. "How to Evaluate Savings from Power Factor Improvement Projects," *Plant Engineering,* April 1981, pp. 56-60.
Bonnett, A. H. "Understanding Efficiency in Squirrel-Cage Induction Motors,"*Transactions of the IEEE,* Vol. IA-16, No. 4, July/Aug 1980, pp. 476-483.
Bowers, W. D., P. G. Cummings, and W. J. Martiny. "Induction Motor Efficiency Test Methods," *Transactions of the IEEE,* Vol. IA-17, No. 3, May/June 1981, pp. 253-272.
Buschart, R. J. "Motor Efficiency," *Transactions of the IEEE,* Vol. IA-15, No. 5, Sept/Oct 1979, pp. 507-510.
Freund, A. "Reducing Motor Losses—the NASA Invention," *Electrical Construction and Maintenance,* March 1981, pp. 69-72; 168-169.
Krishnan, K. R. "High-Efficiency Motors Require Special Component Adjustments," *Pulp and Paper,* January 1982, pp. 134-135.
Krishnan, K. R. "Power Factor Control Shows Little Improvement in Motor Efficiency," *Pulp and Paper,* December 1981, pp. 164-167.
Lazar, I. "Why You Should Specify Energy-Saving Motors," *Specifying Engineer,* August 1981, pp. 80-84.

Marder, A. R. and E. T. Stephenson, eds. "Energy Efficient Electrical Steels," (book), Metallurgical Society AIME, P.O. Box 430, Warrendale, Pa., 1981.
Mongomery, D. C. "How to Specify and Evaluate Energy-Efficient Motors," Publication GEA-19051, General Electric Co., Schenectady, New York, 1981.
Palko, E. "Evaluating Energy-Efficient Motors," *Plant Engineering,* September 2, 1982, pp. 40-44.
Schwartz, K. K. "Survey of Basic Stray Losses in Squirrel-Cage Induction Motors," *Proceedings of the IEEE,* Vol. III, No. 9, September 1964, pp. 1565-1574.
Stebbins, W. L. "How a Big User of Energy Manages to Cut Waste," *Power Transmission Design,* January 1978, pp. 26-32, 52.
Toth, A. "Power Factor Controller Curtails Energy Waste in Motors," *Control Engineering,* August 1980, pp. 67-69.

Chapter 3

Alger, P. L. "Performance Calculations for Part-Winding Starting of 3-Phase Motors," *Transactions of the AIEE,* Vol. 75 Part III, 1956, pp. 1535-1543.
Babun, T. J. Jr. "Choosing a Reduced-Voltage Starter for Motors," *Contractors Electrical Equipment,* June 1979, pp. 24-29, 32.
Bak, D. J. "Solid-State Motor Control Gives Smooth Starting," *Design News,* April 6, 1981, pp. 184-185.
Carlson, F. and B. R. Parker. "Selecting Reduced-Voltage Motor Starters," *Plant Engineering,* August 7, 1980, pp. 76-81.
Ho, T. T. and H. S. Robinson. "System Design for Motor Starting," *Power Engineering,* July 1968, pp. 34-37; August 1968, pp. 44-46.
Mungenanst, J. "Design and Application of a Solid-state AC Motor Starter," *Transactions of the IEEE,* Vol. IA-12, No. 1, Jan/Feb 1976, pp. 39-42.
Ponstingl, J. C. and J. R. Harbaugh. "Using Capacitors for Starting Large Electric Motors," *Plant Engineering,* April 19, 1979, pp. 259-263; May 3, 1979, pp. 97-100.
Williams, A. J. Jr. and M. S. Griffith. "Evaluating the Effects of Motor Starting on Industrial and Commercial Power Systems," *Transactions of the IEEE,* Vol. IA-14, No. 4, July/Aug 1976, pp. 292-305.

Chapter 4

"Applied Protective Relaying," (book), Westinghouse Electric Corporation, Newark, New Jersey, Relay-Instrument Division, 1976.
Beeman, Donald, ed. "Industrial Power Systems Handbook," 1st Edition, McGraw-Hill Book Co., New York, 1955.
Boothman, D. R., E. C. Elgar, and others. "Thermal Tracking—A Rational Approach to Motor Protection," *Transactions of the IEEE,* Vol. PAS-93, No. 5, Sept/Oct 1974, pp. 1335-1344.
Bottrell, G. B. and L. Y. Yu. "Motor Behavior Through Power System Disturbances," *Transactions of the IEEE,* Vol. IA-16, No. 5, Sept/Oct 1980, pp. 600-605.
Cornick, K. J. and T. R. Thompson. "Steep-Fronted Switching Voltage Transients and Their Distribution in Motor Windings," Part I, *Proceedings of the IEEE,* Vol. 129 Part B, No. 2, March 1982, pp. 45-63.
Daugherty, R. H. "Analysis of Transient Electrical Torques and Shaft Torques in Induction Motors as a Result of Power Supply Disturbances," *Transactions of the IEEE,* Vol. PAS-101, No. 8, August 1982, pp. 2826-2836.
Gill, J. D. "Transfer of Motor Loads Between Out-of-Phase Sources," *Transactions of the IEEE,* Vol. IA-15, No. 4, July/Aug 1979, pp. 376-381.

Heidbreder, J. F. "Induction Motor Temperature Characteristics," *Transactions of the IEEE*, Vol. 77 Part III, 1958, pp. 800-804.

Huening, W. C. Jr. "Calculating Short-Circuit Currents with Contribution from Induction Motors," *Transactions of the IEEE*, Vol. IA-18, No. 2, March/April 1982, pp. 85-92.

Koerber, A. R. "Surge Protection for Rotating Electrical Equipment," *Plant Engineering*, September 7, 1972, pp. 92-94; October 19, 1972, pp. 86-87.

Linders, J. R. "Effects of Power Supply Variations on AC Motor Characteristics," *Transactions of the IEEE*, Vol. IA-8, July/Aug 1972, pp. 383-400.

Martiny, W. J., R. M. McCoy, and H. B. Margolis. "Thermal Relationships in an Induction Motor under Normal and Abnormal Operation," *Transactions of the AIEE*, Vol. 80 Part III, 1961, pp. 66-78.

Paape, K. L. and M. J. Manthe. "Trade-Offs in Overload Protection for Lower Voltage Motors," *Transactions of the IEEE*, Vol. IA-18, No. 2, March/April 1982, pp. 110-119.

Pettigrew, R. "Automatic Transfer: New Way to Switch Spinning Motors," *Power*, October 1981, pp. 85-87.

Picozzi, V. J. "Factors Influencing Starting Duty of Large Induction Motors," *Transactions of the AIEE*, Vol. 70 Part III A, 1959, pp. 401-407.

Ponstingl, J. C. "Why You Can't Afford Motor Voltage Unbalance," *Plant Engineering*, November 11, 1982, pp. 60-62.

Rieland, P. A. "Relating Motor Safe Stall and Accelerating Times," *Allis-Chalmers Engineering Review*, Vol. 36, No. 2, 1971, pp. 28-31.

Schatz, M. W. "Overload Protection of Motors—Four Common Questions," *Transactions of the IEEE*, Vol. IGA-7, No. 2, March/April 1971, pp. 196-207.

Shulman, J. M., W. A. Elmore, and K. D. Bailey. "Motor Starting Protection by Impedance Sensing," *Transactions of the IEEE*, Vol. PAS-97, No. 5, Sept/Oct 1978, pp. 1689-1695.

Wright, M. T. and K. McLeay. "Interturn Stator Voltage Distribution Due to Fast Transient Switching of Induction Motors," *Transactions of the IEEE*, Vol. IA-18, No. 4, July/Aug 1982, pp. 363-367.

Chapter 5

Angehern, J. "Corrosion Protection for Electrical Equipment," *Electrical Construction and Maintenance*, March 1969, pp. 142-144.

Bartheld, R. G. "Motor Surface Temperatures in Hazardous Areas," *Transactions of the IEEE*, Vol. IA-14, No. 3, May/June 1978, pp. 220-227.

Binns, V. R. and R. B. Cole. "Arctic Duty Motors," *Transactions of the IEEE*, Vol. IA-14, No. 1, Jan/Feb 1978, pp. 33-39.

Brozek, R. "Noise in High-Speed Motors," *Machine Design*, March 5, 1970, pp. 123-128.

Cashmore, D. H. "How to Estimate Induction Motor Noise Levels," *Sound and Vibration*, October 1972, pp. 14-17.

Constance, J. D. "How to Pressure-Ventilate Large Motors for Corrosion, Explosion and Moisture Protection," *Chemical Engineering*, February 27, 1978, pp. 113-116.

Cowern, E. "How To Select Motors for Hazardous Locations," *Plant Engineering*, October 28, 1976, pp. 99-101.

Ghirardi, G. and R. W. Mills. "Enclosures for Large AC Motors for the Petroleum and Chemical Industry," *Transactions of the IEEE*, Vol. IA-14, No. 1, Jan/Feb 1978, pp. 18-22.

Harris, C. M. ed. "Handbook of Noise Control," (book), 2nd Edition, McGraw-Hill Co., New York, 1979.

Judd, S. H and J. A. Spence. "Noise Control for Electric Motors," *Am. Ind. Hygiene Assoc. Journal,* Nov/Dec 1969, pp. 588-595.
Lance, H. "Reducing Motor Noise," *Electrical Construction and Maintenance,* December 1976, pp. 99-100.
Loewe, R. "Approaches Other Than Explosion-Proofing to Electrical Safety," *Transactions of the IEEE,* Vol. IA-11, No. 1, Jan/Feb 1979, pp. 62-83.

Chapter 6

"ABC of Motor and Generator Insulation," Publication No. 100-SYN073, Electric Machinery Division, Minneapolis, 1973.
Albrecht, W. M. "Applying Electrical Insulation Test Instruments," *Plant Engineering,* February 21, 1974, pp. 99-101.
Davis, R. R. "Surge Comparison Tests Detect Motor Faults," *Electrical Construction and Maintenance,* May 1981, pp. 34-35; December 1981, pp. 34, 100.
Fromm, M. M. "Tests for Turn Insulation in Large Machines," *Power Engineering,* November 1978, pp. 70-72.
Heyne, C. J. "Interpreting Temperature Measurements in Large A-C Machines," *Power Engineering,* September 1974, pp. 58-61.
"Impulse Voltage Strength of AC Rotating Machines," (Working Group Report), *Transactions of the IEEE,* Vol. PAS-100, No. 8, August 1981, pp. 4041-4053.
Jonsson, K. "MICAPACT II Coils for High-Voltage Rotating Machines," *ASEA Journal,* Vol. 54, No. 2, 1981, pp. 27-32.
Kelen, A. "The International Electrotechnical Commission and Insulation Classification," *Insulation/Circuits,* July 1972, pp. 32-40.
Miller, H. N. "Insulation Testing and High-Potential Testing: Advantages and Limitations," *Transactions of the IEEE,* Vol. IGA-5, No. 3, May/June 1969, pp. 326-332.
Reason, J. "Why Dielectric Failure Is the Major Cause of Electrical System Breakdown," *Power,* July 1981, pp. 35-40.
Rejda, L. J. and Kris Neville. "Industrial Motor Users' Handbook of Insulation for Rewinds," (book), Elsevier-North Holland, Inc., New York, 1977.
Richon, G. L. "New Water Based, Solvent Based, and Solventless Varnishes," *Insulation/Circuits,* August 1980, pp. 46-47.
Shaffer, K. W., D. L. Graham, and R. E. Obenhaus. "Application of Inherent Thermal Protection to Industrial Motor systems," *Transactions of the IEEE,* Vol. IA-11, No. 1, Jan/Feb 1975, pp. 14-33.
"A Stitch in Time," Manual 21-P-89, Biddle Instruments, Blue Bell, Pa. 1981.

Chapter 7

Anderson, E. and W. C. LaBahn. "Electric Motor Bearings," *Plant Engineering,* November 2, 1972, pp. 78-81.
"Bearing Technical Journal," Link-Belt Bearing Division, PT Components, Indianapolis, Ind., 1982.
Beercheck, R. C. "How dirt and Water Slash Bearing Life," *Machine Design,* July 6, 1978, pp. 68-73.
Block, H. P. "Dry Sump Oil Mist Lubrication for Electric Motors," *Hydrocarbon Processing,* March 1977, pp. 133-135.
Booser, E. R. "When to Grease Bearings," *Machine Design,* August 21, 1975, pp. 70-73.

Corman, R. and P. Ranade. "Check on Vertical-Pump Thrust for Maximum Efficiency," *Power,* December 1982, pp. 65-67.
Hafner, E. R. "Understanding Sleeve Bearings," *Plant Engineering,* February 21, 1980, pp. 79-83; April 3, 1980, pp. 111-115.
Ivanick, F. M. "Proper Lubrication Supports Energy Management," *Heating/Piping/Air Conditioning,* October 1979, pp. 171-174.
Jordan, G. R. "Some Straight Talk about 31 Common Lubrication Myths," *Plant Engineering,* May 17, 1979, pp. 157-159.
Lewis, K. G. "Selecting and Applying Greases," *Plant Engineering,* December 23, 1982, pp. 53-56.
Marinello, R. L. "Interchangeable Lubricants," *Plant Engineering,* June 12, 1980, pp. 100-113.
Stefanides, E. J. "Ring, Roller Changes Boost Thrust-Bearing Capacity," *Design News,* September 27, 1982, pp. 66-67.
Stover, J. D. and R. L. Widner. "New Technology Yields Longer Life for Bearings," *Machine Design,* November 25, 1982, pp. 54-58.
Weihsmann, P. R. and W. L. Subler. "Modern Lubrication Practices," *Transactions of the IEEE,* Vol. IA-16, No. 4, July/Aug 1980, pp. 484-489.
Wolfe, G. "Bearings," *Plant Engineering,* July 9, 1981, pp. 71-89.

Chapter 8

Dikinis, D. V. and M. H. Yuen. "Solid-State Control—Low Voltage Heating of Motors," *Transactions of the IEEE,* Vol. IA-11, No. 3, May/June 1975, pp. 287-290.
Keller, R. A. III. "Selecting Electric Disc Brakes for Hazardous Locations," *Power Transmission,* July 1980, pp. 44-47.
Nailen, R. L. "Connecting Large Motors," *Electrical Construction and Maintenance,* February 1971, pp. 80-82, 144.
Nailen, R. L. " 'Quick Disconnects' in Large Motor Circuits," *Transactions of the IEEE,* Vol. PAS-96, No. 1, Jan/Feb 1977, pp. 150-157.
Peterson, H. S., J. W. Moss, and R. W. Eisbrener, "Failsafe Brakes," *Machine Design,* April 26, 1973, pp. 170-175.
Sandford, J. "DC Brake Stops AC Motors Quickly," *Design News,* April 7, 1975.
Wilson, J. J. "The Ins and Outs of Electromagnetic Clutches and Brakes," *Plant Engineering,* September 20, 1979, pp. 151-157.
Wrensch, B. E. "Electric Clutches and Brakes," *Machine Design,* October 31, 1974, pp. 55-59.

Chapter 9

Brueggeman, L. T. "Motor Reed Frequency," *Allis-Chalmers Engineering Review,* Vol. 35, No. 1, 1970, pp. 16-19.
Dreymala, J. "Try Dial Indicators for Close Alignment of Coupling-Connected Machinery," *Power,* June 1971, pp. 96-98.
Durkin, T. "Aligning Shafts," *Plant Engineering,* January 11, 1979, pp. 86 ff; February 8, 1979, pp. 102-105.
Eckart, G. C. "How Ground Fault Systems Protect Motor Loads," *Electrical Construction and Maintenance,* March 1982, pp. 42, 47.
Gibbs, C. R. "Installing Machine Supports," *Allis-Chalmers Engineering Review,* Vol. 32, No. 2, 1967, pp. 20-23.

Goodspeed, I. T. and F. K. Landon Jr. "How to Select Limited-End-Float Gear Couplings," *Power Transmission Design,* April 1973, pp. 50-53.

Kaufmann, R. H. "Important Functions Performed by an Effective Equipment Grounding System," *Transactions of the IEEE,* Vol. IGA-6, No. 6, Nov/Dec 1970, pp. 545-552.

Krouse, J. K. "Soft-Start Couplings," *Machine Design,* December 8, 1977, pp. 171-175.

Meyer, R. J. "Solve Vertical Pump Vibration Problems," *Hydrocarbon Processing,* August 1977, pp. 145-149.

Murray, M. G. Jr. "How to Specify Better Pump Baseplates," *Hydrocarbon Processing,* September 1973, pp. 157-160.

Pearch, D. A. "Clutches That Save Energy," *Power Transmission Design,* July 1980, pp. 35-38.

Phelps, D. R. "Flange Loads in a Pumping System," *Chemical Engineering Progress,* May 1970, pp. 43-50.

"Report on Reliability Survey of Industrial Plants, Part III," (Committee report), *Transactions of the IEEE,* Vol. IA-10, No. 2, March/April 1974, pp. 242-252.

St. John, R. C. "Centrifugal Clutch Basics," *Power Transmission Design,* March 1979, pp. 52-55.

Symons, W. R. "Mechanical Installation of Electric Motors," *Plant Engineering,* October 31, 1979, pp. 51-53; November 29, 1979, pp. 71-73.

Index

Accelerating torque, 21,22,62,92;
  loss of, 144
Acceleration;
  energy required for, 453-454;
  heating during, 101-105,109-115;
    clutches to prevent, 445-448;
  inertia's effect on, 101;
  time required for, 106-108;
  two-step, 355;
  unwanted trips during, 356-359;
  see also Starting
Acceptance tests, 269-270
Accessories, motor, 351;
  brakes see Brakes;
  heaters see Space heaters;
  selection of;
    brakes, factors to consider 368-373;
      electrical characteristics, 372;
      enclosure type, 372-373;
      heating capacity, 371;
      mounting position, 372;
      required torque, 368-371;
    clutches see Clutches, selection of;
  speed sensors, 351-363;
    non-contact, 359-363;
    see also Speed sensors;
  terminal boxes, 406-418;
    see also Terminal boxes
Adding grease, 321-322
AFBMA see Anti-Friction Bearing Manufacturers Association
Air filters, 215-221
Alignment, 342,458-469;
  of driving and driven shafts, 458-459;
    checking of, 460-465;
    definition of, 459;
    dial indicators for, 460-463;
    in vertical motors, 467;
  of lubrication pump shafts, 342;
  see also End play
Ambient temperature, 156;
  measurement of, 278
American National Standards Institute (ANSI), 4;
  mailing address, 12;
  standard covering bearing life prediction, 318,319;
  standard covering grounding practices, 474;
  standard covering terminal box construction, 414
ANSI see American National Standards Institute
Anti-friction bearings, 309-310,319
Anti-Friction Bearing Manufacturers Association (AFBMA), 318
Anti-plugging, 354-356;
  field-effect transducers used for, 360;
  see also Plug stopping
Application engineering (definition), 1-2
Autotransformers, 125,135,177
Average life of bearings, 318

Bearings;
  anti-friction, 309-310;
    end play required by, 466;
    purging of, 325;
    versus sleeve, 310,332-333;
  radial, 310-311,333;
  sealed, 327-328;
  shielded, 327;
  currents, 347-349;
  Electric Motor Quality, 332;
  guide, 315;
  insulation of, 347-349;
  limits, speed and temperature, 329;
  nameplate code numbers for, 33-34;
  overheating of, 331-332,465;
  plate-type, 314;
  predicting life of, 316-319;
  sleeve, 309-310,332-335;
    end play, 465-466;
    magnetic center in, 468;
    pressure-lubrication of, 335,336;
    split, 332;
    versus anti-friction, 332-333;
    wear, 333;
  spherical roller, 311;
  standards covering, 309-310;
  temperature limits of, 330,336;
  thrust, 311-316;
    plate type, 314;
    spherical roller, 318

Brakes;
  combination a-c/d-c brake circuit, 389;
  control of, 374-377;
    a-c, 374-375;
    d-c, 376-377;
    series, 377;
    electronic controllers, 390;
  d-c dynamic, 387-390;
    amount of current required by, 391-392;
    cost of, per horsepower, 391;
    disadvantages of, 389;
    energy saved by, 388;
    holding torque, 389-390;
    torque curve of, 387,389;
    winding connections for, 393;
  dual-winding types, 383-384;
  electronic d-c brakes, 390-394;
  friction type, 363-368;
    common features, 381;
    disc, 364;
    mounting of disc brakes, 364;
    fail-safety of, 364;
    shoe, 364;
    direct-acting shoe brakes, 368;
    hydraulic (thrustor) shoe brakes, 368;
    shoe brakes in d-c motors, 376;
    mounting of shoes, 364;
  selection factors, 368-373;
    electrical characteristics, 372;
    enclosure type, 372-373;
    heating capacity, 371;
    mounting position, 372;
    required torque, 368-371;
  servicing of, 377-379;
    failure to release, 377-378;
    failure to set, 378;
    routine maintenance procedures, 380-381;
  see also Plug stopping
Breakaway torque, 92
Breakers, vacuum, 182;
  see also Fuses

Capacitance, 69,70
Capacitors, 41,75-87,145;
  in hipot tests, 270;
  surge protection, 250;
    grounding of surge protectors, 478-480
Centering thrust, 467
Centrifugal clutches see Clutches, centrifugal
Centrifugal pumps, 96-100
Centrifugal speed switches, 359
Clearance in terminal boxes, 414-418
Clutches, 445-458;
  centrifugal, 446;
    effect on acceleration time, 458;
    versus fluid couplings, 451;
  dry fluid types, 448;
  non-centrifugal, 449;
  see also Couplings, fluid

Coating, conductive see Conductive coating
Coils, 243-246;
  end coil corona, 264-265;
  form-wound, 243-246;
  random- (mush) wound, 243-246
Compressors, 116-117
Condensation;
  prevented by heating, 394
Conductive coating, 261-267;
  resistivity of, 263;
  see also Semicon paint
Conduit boxes, explosion, 204-205
Connections, 408-431
Conrad bearings, 310-311
Conveyor belts, 113-114;
  counter-torque generated by, 545
Cooling of oil, 342
Coplanar base (definition), 433
Copper loss see Loss, copper
Core loss see Loss, core
Corona, 260-261;
  in end coils, 264-268;
  protection against, 268;
  treatment of, 262-264
Cost-cutting, 178
Cost studies, 50;
  regarding lubricant, 326
Couplings;
  flexible, 458-459;
  limited end float, 469;
  unloading, 446,453-458;
    fluid, 451-453;
  see also Clutches
Critical speed, 440-441;
  first and second critical, 440
Crossover in slots, 244
Current transformers (CTs), 157-167;
  varying turns ratio of, 159;
  metering with, 161;
  operation of, 162;
  polarity of, 161;
  relaying, 161;
  secondary protection of, 164-165;
  window types, 157;
  wound types, 157

Deceleration, heating during, 352
Deep bar effect, 104-105,171
Derating, 173
Destructive tests, 270-271;
  see also Hipot tests
Detectors see Sensors
Dielectric absorption tests, 273-274
Direct-acting shoe brakes, 368
Direct winding heating, 402-406;
  black box, 402
Disc brakes, 364;
  mounting of, 364;
    on floors, 364;
  versus shoe brakes, 364;

(Disk brakes, Cont.)
  see also Brakes and Shoe brakes
Drives, variable speed, 48
Drip proof motors, NEMA definition, 189-190
Drying motors *see* Space heaters
Dust, 192,194;
  brakes for use near, 374;
  explosions caused by, 194;
  ventilation blocked by, 192
Dv/dt protectors, 145
Dynamometers, 57,58
Dynamic braking *see* Brakes, d-c dynamic

Economics, 44;
  depreciation, 44;
  payback period, 44,51;
  of bearings, 309;
  of buying lubricant, 326;
  of fluid couplings, 453-458;
  of high-rated explosion-proof motors, 200;
  world market, 11
Efficiency, 18,23-29,41-45,50-52,72,144,186;
  evaluation of, 45-46;
  nominal, 27;
  premium, 51;
  testing, 53-58;
  unloading couplings' effect on, 457;
  variations between motor manufacturers, 48-49
Electrocution, 164
Electronic d-c brakes, 390-394
Electronic speed sensors, 360
EMQ bearings *see* Bearings, Electric Motor Quality
Enclosures, motor, 189-194;
  totally enclosed water-to-air cooled (TEWAC), 222-227;
  tube-type totally enclosed, 221-222,224
End coils, 264-268
End play, 465-467
Energy crisis, 11,47
Energy efficient motors, 23,29
Energy saving, 185-188;
  of fluid couplings, 453-457
Environment *see* Motor environment
EP grease *see* Extreme pressure grease
External devices, 174-176;
  solid-state electronic relays, 175;
  voltage monitors, 175;
  winding temperature sensors, 176
Explosion-proof motors *see under* NEMA enclosure classifications
Explosions, 194;
  dust, 194,196-198;
  grain, 197;
  testing, 193
Extreme pressure (EP) grease, 326

Fail-safety;
  of friction brakes, 364
False tripping, 167
Fans, 100-101,200;
  non-sparking, 200;
  fan-cooled motors *see under* NEMA enclosure classifications
Faults during tests, 271
Field-effect transducers *see* Transducer-type motion detectors
Fires, 148,196-197;
  electrical, 148;
  hazards, 167,343
  see also Explosions
First critical, 440
Flexible couplings, causes of vibration and wear, 458-459
Flow rate;
  of lubrication pumps;
  calculation of, 340-342;
  shaft-driven, 342
Fluid couplings *see* Couplings, fluid
Forced lubrication *see* Lubrication, pressure
Force-ventilated motors, 200,212-215;
  NFPA Standard, 496
Form-wound coils, 243-246;
  stresses on, 247
Foundations, 68,433-440;
  bolts for, 444-445;
    torque limits, 444;
  coplanar (definition), 433;
  grouting, 434-435;
  resonance, 439-440;
    checking for, 441;
  shimming, 435-436;
  thermal expansion of, 438
Free baseplate, 443-444
Frequency of greasing, 321
Friction, 52,73;
  anti-friction bearings, 309-310;
  effect on lubrication, 329;
  loss *see under* Loss;
  sliding friction, 446;
  sparks, 200;
  see also Inertia *and* Lubrication
Friction brakes *see* Brakes, friction
Fuses, 147;
  see also Breakers

Gradient paint *see* Semicon paint
Grease, 320-326;
  extreme pressure, 326;
  hardening of, 321;
  NFPA standard covering, 321;
  preventing contamination of, 320;
  regreasing, 323-325;
  replacement of, 320-321;
  versus oil, 320;
  when to add, 321-322

Grease guns, 323
Grounded neutrals, 248,251
Grounding, 167,471-480;
  brushes, 349;
  connections, 474-475;
  circuits, 472-473;
  drilling, 473-474;
  of small motors, 475-477;
  of surge protectors, 478-480;
  resistance, 477
Grouting in foundations, 434-435
Guarded motors, NEMA definition, 190-191
Guide bearings, 315

Harmonics generated in SCR systems, 139-144
Head loss see Loss, head
Heaters, space see Space heaters
Heating;
  acceleration, 101-104,114,454;
  during plugging, 352,382;
  during starting, 112;
  end play required by, 465;
  [for information about heating idle motors, see
  Space heaters and Direct winding heating]
High torque loads, 113-117
Hipot tests, 180,268-269;
  a-c (destructive), 270;
  d-c, 271
Holding torque, provided by d-c dynamic braking, 389-390
Hydraulic (thrustor) brakes, 368;
  see also Brakes

Idle saturation, 54
IEC see International Electrotechnical Commission
IEEE see Institute of Electrical & Electronics Engineers
Inertia;
  effect on starting, 100-101;
  evaluations of, 118-119;
  see also Friction and High torque loads
Inductance, 69,70
Inrush current, 31-32
Insulation;
  between turns, 276;
    Rylander test, 276-277;
    see also Surge test;
  choice of material, 254-257;
  grading of, 248-251;
  heat-shrinkable tubing, 415-417;
  of bearings, 347-349;
  of mush windings, 246;
  variation of, with voltage class, 252;
  wall thickness of, 254;
  resistance, 272,275;
    test for, 271;
  see also Wrapper and Tape

Internal braking, 381-382;
  d-c dynamic, 387-390;
    disadvantages of, 389;
    energy saved by, 388;
    holding torque, 389-390;
    speed-torque curve of, 387;
    torque curve of, 389
International Electrotechnical Commission (IEC), 10,15
Institute of Electrical & Electronics Engineers (IEEE), 7,54,179,180;
  history of, 2-5,37;
  mailing address for, 12;
  standard covering grounding practices, 473,474;
  Standard No. 43, 271;
  Standard No. 112, 54,57;
  Standard No. 620, 152;
  standardization, 179;
  voltage classes defined by, 252
International Standards Organization (ISO), 5,12,14
Iron loss see Loss, core
ISO see International Standards Organization
I^2R Loss see Loss, I^2R;

Kelvin bridges, 66
Kingsbury bearings, 311

Life span of bearings, 316-318
Linear motion (Translation), 105-109
Load torque, 96-101
Locked rotor, 157;
  current, 31-32
Loss, 46,48,169,171;
  copper, 46,245,253;
  core (iron), 43,46,50,171,186;
  friction, 43;
    caused by clutches, 448;
  head, 52;
  iron see Loss, core;
  I^2R, 50,104,109,171,187;
  segregated, 62;
  starting, 456;
  stray load, 46,48,61;
  torque, 144,171
Lubrication;
  economics of, 326;
  forced see Lubrication, pressure;
  of shielded bearings, 327;
  of sleeve bearings, 333-335;
  over-lubrication, 323;
  pressure, 335-338;
    maintenance of, 343-346;
    oil-mist system of, 343-346;
  regreasing schedule, 321;
  speed limit, 329;

Index 493

(Lubrication, Cont.)
  under-lubrication, 323;
  see also Oil *and* Grease

Machine torque, 92
Magnetic center, 467
Magnetic flux, 374-375
Maintenance, 52,198;
  of brakes, 377-381;
  of explosion-proof motors, 198;
  of pressure lubrication systems, 343-346;
    oil-mist system, 343,346;
  of shaft alignment, 460-465;
  of terminal connections, 414;
  replacement of grease, 320-325;
  testing, 269
Mechanical center, 467
Metrication, 10-12;
  effect on power ratings and dimensions, 12-17;
  measurement of torque, 92
Misalignment *see* Alignment *and* End play
Mixers, 115
Mixing of greases, 326
Motor accessories *see* Accessories, motor
Motor construction, 196-197
Motor drying *see* Space heaters
Motor environment, 198-204;
  corrosive, effect on heaters, 399;
  influence over placement of temperature sensors, 331;
  influence over selection of brakes;
  influence on regreasing schedule, 321-322
Motor failure, 180
Motor installation *see* Installation
Motor losses *see* Losses
Motor performance (definition), 7;
  see also Efficiency
Motor plugging *see* Plug stopping *and* Antiplugging
Motor protection, 165,196-197;
  see also Protection, motor
Motor redesign, 53,73
Motor repair, 250
Motor safety *see* Safety
Motor servicing, 165,196-197
Motor specifications *see* Specifications
Motor terminal boxes *see* Terminal boxes
Mounting disk and shoe brakes, 364
Mush-wound coils, 243-246;
  stresses on, 247

Nameplate markings, 25-33
National Board of Fire Underwriters (NBFU), 4-5,194
National Electrical Code (NEC), 37,166,195;
  articles covering operating voltages in terminal boxes, 410;

(National Electrical Code, Cont.)
  articles related to grounding, 473-474,476;
  Article 300-3, 5,166;
  Article 300-20, 166;
  Article 500, 5,193;
  history of, 3,194-195;
  temperature code (Table 500-2b), 195-196
National Electrical Manufacturers Association, 15-16,50,156,166,172,173;
  derating factor, 173;
  enclosure classifications, 189-194;
    open, 189-191,199;
      drip proof, 189-190;
      guarded, 190-191;
      splash proof, 190;
      weather-protected, 191;
    totally-enclosed, 191-194;
      dust-ignition-proof, 194-204;
      explosion-proof, 193-200;
      fan-cooled (TEFC), 192,199,203;
      non-ventilated (TENV), 192;
      water-proof, 192-193;
  mailing address, 12;
  standard covering grounding practices, 474;
  standard covering terminal box clearance, 415;
  Standard MG1, 2,5-10,50;
  Standard MG1-14.34, 172;
  Standard MG1-14.36, 129;
  Standard MG1-20.42, 156;
  Standard MG1-20.62, 166;
  torque and inrush ranges, 7-10
National Fire Protection Agency (NFPA), 4-5,194-195;
  history of, 4;
  mailing address, 12;
  standard 325M;
  standard covering relubrication, 321
NBFU *see* National Board of Fire Underwriters
NEC *see* National Electrical Code
NEMA *see* National Electrical Manufacturer's Association
NFPA *see* National Fire Protection Agency
Noise;
  definition, 227;
  directionality, 231;
  frequency variations, 228-229;
  measurement, 54;
  methods of reducing, 237-241;
  origins, 235-237;
  OSHA rules covering, 232-234;
  standards governing, 232-234;
  testing, 54,234-235
Non-contact speed sensors, 359-363;
  electronic, 360;
  transducer type, 360
Nuisance tripping, 154;
  prevention of, 356-359

Ohm's Law, 163,271,275
Oil;
  cooling of, 336,342;
  extreme pressure, 326;
  flow, 338;
  life of, 326;
  mixing of, 326;
  oiler bottles, 335;
  selection of, 320-321,326;
  versus grease, 320;
  see also Grease and Lubrication
Oil-mist system of bearing lubrication, 343-346;
  applications, 346;
  maintenance of, 343,346
Optical speed switches, 359
Overgreasing see Over-lubrication
Overheating, 150,171,173;
  caused by loss of oil pressure, 335;
  during plugging, 352,382;
  of bearings, 331
Overloading;
  bearings 319
Overloads on stator, 151,300-302,303,304-307
Over-lubrication, 323,325
Ozone, effect on insulation, 260

Paint see Slot paint
Part winding starting, 130-134;
  connections, 127-130
Partial discharge see Corona
Payback period, 44,51
Performance, 245;
  see also Efficiency
Performance evaluation, 51
Performance testing, 26,54
Plant power factor (PPF), 72;
  see also Power factor
Plate-type bearing, 314
Plug stopping, 115,144,351-353,381-382;
  drawbacks of, 382;
  of large motors, 382;
  see also Anti-plugging
Plugging switches, 351-353
Polarity marks, 160
Polarization index, 273
Power factor, 18,24-29,52,163,169;
  controller, 87-90;
  misconceptions about, 68-73;
  plant power factor, 72
Power line pollution, 147
Power supply;
  tests, 58
Preferred numbers see Renard preferred numbers
Premium efficiency motors, 52
Pressure lubrication see Lubrication, pressure
Proof tests see Acceptance tests
Protection, motor, 173;
  against unwanted trips, 356;
  against plugging, 354;
  against surges see under Surges;
  from hostile surroundings see Chapter 5
Pulverizers, 114
Pumps;
  axial flow, 98-100;
  centrifugal, 96-100;
  lubrication of, 339;
  piston (plunger), 9;
  viscosity, 323-324
Purged (force-ventilated) motors, 200-2033;
  NFPA Standard, 496

Radial bearings, 310-315;
  see also Bearings
Radial loads, 310-311
Random-wound coils see Mush wound coils
Reclosure, 185
Redesign, 53,73
Reduced voltage starting, 130
Reed frequency vibration, 442
Renard preferred numbers, 12-16
Repair, 250
Rerating, 42,178;
  uprating, 53
Residual voltage, 182;
  defined, 363;
  effect on brakes, 376;
  speed sensors driven by, 362;
  see also Transient voltage
Resistance;
  contact, 48;
  grounding, 477
Resistance grounding, 251
Resonance in motor baseplates, 439-440;
  checking for, 441
Reverse rotation, 354-356;
  caused by plugging, 382
Ring expansion, 104
Rotation in machine drives, 106-108;
  effect on lubrication life, 329;
  measurement of, 351-353;
  reverse, prevention of, 355-356,382
Rotor centering, 467-469
Rotor heating, 104-105,144
Rylander tests, 276-277

Safety, 5,192;
  considered in terminal box design, 412;
  factor, 338;
  fan shrouds, 192;
  provided by speed switches, 358,360;
  warning signs, 35,403
SCRs see Semiconductors

Index   495

Secondary protectors, 164,165
Second critical, 440
Segregated loss *see* Loss, segregated
Selection of accessories;
   bearings; *see* Bearings
   brakes, factors to consider 368-373;
      electrical characteristics, 372;
      enclosure type, 372-373;
      heating capacity, 371;
      mounting position, 372;
      required torque, 368-371;
   clutches *see* Clutches, selection of;
Semiconducting paint *see* Semicon paint
Semiconductors (SCRs);
   harmonics generated by, 139-144;
   in solid-state starters, 136
Semicon paint, 266-268
Sensors;
   bearing temperature, 330-332;
   direction, 355;
   speed, 351,355;
   temperature *see* Temperature detectors;
   vibration, 351
Separable connectors, 417-422;
   disassembly of, 426-428;
   features of, 419;
   insulation for, 426-427;
   non-loadbreak types, 421
Service factor, 17-25,178
Servicing *see* Maintenance
Severe starts, 154;
   end play required during, 465;
   preventing unwanted trips during, 356-359;
   unloaded couplings used for, 458
Shaft misalignment *see* Alignment, checking of
Shaft voltage, 347-348;
   protection against, 348-349;
   see also Transient voltage *and* Shaft voltages
Shimming motor baseplates, 435-436,445
Shoe brakes, 364;
   mounting of, 364;
   hydraulic (thrustor), 368;
   direct-acting, 368;
   see also Brakes
Single-phasing, 168-171;
   as alternative braking scheme, 383;
   see also Voltage unbalance
Sleeve bearings, 309-310;
   see also Bearings, sleeve
Slip, 110
Slot discharge, 261
Slot paint, 261-264
Solid-state starters *see* Starters, solid-state
Space heaters, 351,394-401;
   elements of, 397;
      for explosive atmospheres, 398;
      location of, 398-399;
      tubular, 398;
   enclosures for, 399;

(Space heaters, Cont.)
   environmental considerations, 395;
   judging how much heat to supply, 396;
   reliability of, 399;
   sizing of, 396;
   thermostats for, 395;
   2DL rule applied to, 396;
   wrap-around type, 400;
   see also Direct winding heating
Specifications, 36-40;
   bearings, 314;
   writing of, 36-38;
   terminology, 38-40
Speed switches *see* Speed sensors
Speed limits;
   for bearings, 329;
      for sleeve bearings, 335;
   for lubrication, 329
Speed sensors, 155,351-353;
   belt driven, 359;
   centrifugal, 359;
   electronic, 360;
   non-contact, 359-363;
      electronic, 360;
         zero-motion detectors, 362;
      magnetic, 359;
      transducer-type, 360;
      used as tachometers, 361;
   optical, 359
Speed versus torque curve, 94-96;
   for fans, 100
Speed switches;
   centrifugal, 359;
   overriding relays with, 356-359
Spherical roller bearings, 311
Spikes, 145,180;
   see also Surges
Splash proof motors, NEMA definition, 190
Stacked bearings, 315
Standards, 2-12,203-204;
   American, 2-10,204;
   covering terminal boxes, 412;
   definition of, 3;
   forming of and compliance with, 3;
   European, 203-204;
   history of, 3-5;
   industry, 2-10;
   international, 10-11,248
Starters, solid-state, 136-146;
   advantages of, 136-138;
   drawbacks of, 138-145;
   servicing of, 145-146;
Starting, 119-127,154;
   capacitors for, 120;
   clutches for, 445-448;
   current, 31-32;

effect on bearings, 310;
effect on windings, 247;
frequency of, 118-119;
heating during, 112;
loss, 456;
part-winding, 125-132;
 of 2-pole motors, 131-134;
severe *see* Severe starts;
reduced voltage, 120,124,135;
three-step, 130-131;
unloaded, 114
Step-voltage tests, 274-275
Stray load loss, 61,73;
 see also Loss
Surface creepage;
 in mush coils, 244
Surge comparison tests, 276
Surges, 178;
 capacitors for, 184;
 origins of, 181-182;
 protection against, 179-186,351,429-431;
  grounding of surge protectors, 478-480;
 see also Spikes
Switches;
 plugging, 351-353;
 speed, 356-358
Symmetrical components, 169

Tachometers, non-contact speed sensors used as, 361
Tape insulation, 252-257;
 application of, 257-260;
  by machine, 257;
 versus heat-shrinkable tubing, 415-416
Temperature detectors, 351
Temperature limits;
 of bearings, 329
Temperature rise, 31,65,187;
 of bearings, 330-331;
  sensors for monitoring, 330-331;
 heat energy versus, 104-105;
 incremental, 156;
 winding, 253
Terminal boxes, 406-410;
 building your own, 430-431;
 clearance in, 414;
 corrosion of, 410;
 damage to during installation, 410;
 explosion of, 412;
 failure, causes of, 413-414;
 grounding of, 475;
 National Electrical Code rules governing operation of;
  voltages of, 410;
 size of feeder cable in, 407-209;
 size selection, 406-409;
 specifying, 406-409;
 standards covering, 412

Tests;
 acceptance 269-270;
 efficiency, 53-58;
 insulation, 268;
  d-c, 273-275;
 hipot, 180,268-271;
 Rylander, 276-277;
 surge comparison, 276;
 thermal endurance, 252;
 load, 62;
 locked, 60;
 maintenance, 269;
 noise, 54;
 performance, 26,54;
 power supply, 58-60;
 procedures, 53,68;
 shaft alignment, 460-464;
 temperature, 60,65-68;
 vibration, 54
Thermal expansion of foundations, 438
Thermal limits, 147-150;
 of centrifugal clutches, 450
Thermistors, 195,290-299
Thermostats, 144,195-196,289-290;
 for sensing bearing temperature, 330
Throttling, 52
Thrust bearings, 311,315-316;
 Kingsbury, 311;
 see also Bearings
Thrust loads, 310-316
Thrustor brakes, 368
Thyristors, 165
Torque, 182;
 accelerating, 21,22,62,92;
 breakaway, 92;
 breakdown, 21;
 holding, provided by d-c dynamic braking, 389-390;
 load *see* Load torque;
 locked, 21,62;
 loss *see* Loss, torque;
 measurement of, 62-65,92;
 transient, 182,185
Totally-enclosed motors;
 see under NEMA enclosure classifications
Totally-enclosed inert gas filled (TEIGF) motors, 203
Transducer-type (field effect) motion detectors, 360
Transfer point, as controlled by speed switches, 355
Transformers, 41,167
Transient surges, 147,178,179,248;
 between turns, 276;
 see also Transient voltages
Transient torques;
 during plugging, 382;
 in d-c dynamic braking, 390
Transient voltages, 178-182,250;
 lightning, 178,250;

Index

(Transient voltages, Cont.)
    lightning arresters, 184,429;
    spikes, 178;
    see also Residual voltage
Translation (linear motion), 105-109
2DL rule, 396;
    see also Space heating

UL *see* Underwriters Laboratories
Unbalance *see* Voltage unbalance
Under-lubrication, 323
Underwriters Laboratories (UL), 4-5,193,196,198-199;
    history of, 4;
    standard covering motor enclosures, 194;
    standard covering brakes in combustible atmospheres, 374;
    Standard 674A, 195;
    Standard 674B, 195
Unloaded starting *see* Starting, unloaded
Unloading couplings *see* Couplings, fluid
Uprating, 53;
    see also Rerating

Vacuum-pressure impregnation (VPI), 256-257,261-262,
Ventilation, 200;
    air filters, 215-221;
    cooling air requirements, 207-212;
    forced, 200
Variable speed drives, 48
Vibration, 66,310,422;
    caused by flexible couplings, 458-459;
    in motor foundations, 440-442;
        altering, 442-443;
        diagnosing, 434;
    loosening of bolts, 444-445
Viscosity pumps, 323-324
Voltage classes, IEEE, 252
Voltage unbalance, 50,147,168-173;
    causes of, 176-177;
    protection against, 174-176
Voltage ranges, standard, 251
Voltage sag, 50
VPI *see* Vacuum-pressure impregnation

Water-proof motors *see under* NEMA enclosure classifications
Water System Council, 171
Weather protected *see under* NEMA enclosure classifications
Winding connections, 127-130
Winding life, 52,185
Winding rise, 277-278
Winding temperature, 144,150,253,277-288;
    measuring, 278-280;
    see also Winding rise
Windmilling, prevented by d-c dynamic braking, 387

Wire-to-wire voltage, 244
Wrap-around heaters, 400
Wrapper, 252-260;
    application of, 257-260;
    selection of, 254-257

Zener diodes, 165
Zero motion detectors, 362
Zero speed switches *see* Plugging switches

Photo Credits

Figure	Page	Source
1-12	28	Louis Allis Division, MagneTek, Inc.
1-13	30	Gould, Inc.
1-14	30	Westinghouse, Inc.
1-16	34	Louis Allis Division, MagneTek, Inc.
2-7	56	Louis Allis Division, MagneTek, Inc.
2-8	56	Louis Allis Division, MagneTek, Inc.
2-28	77	Westinghouse, Inc.
2-33	83	Sprague Electric Co.
4-3	149	Louis Allis Division, MagneTek, Inc.
5-18	216	Louis Allis Division, MagneTek, Inc.
5-24	220	Louis Allis Division, MagneTek, Inc.
5-25	221	Louis Allis Division, MagneTek, Inc.
5-26	221	Louis Allis Division, MagneTek, Inc.
5-28	223	Louis Allis Division, MagneTek, Inc.
5-29	223	Louis Allis Division, MagneTek, Inc.
7-1	311	SKF Industries, Inc.
7-2	312	SKF Industries, Inc.
7-3	312	SKF Industries, Inc.
7-4	312	SKF Industries, Inc.
7-5	313	Kingsbury Machine Works
7-15	325	Louis Allis Division, MagneTek, Inc.
7-18	330	Louis Allis Division, MagneTek, Inc.
7-20	333	Louis Allis Division, MagneTek, Inc.
7-21	334	Westinghouse, Inc.
7-24	337	Louis Allis Division, MagneTek, Inc.
7-25	338	Oil-Rite Corp.
7-27	339	Louis Allis Division, MagneTek, Inc.
7-29	341	Oil-Rite Corp.
7-32	344	Eaton Corp., Fluid Power Division
8-6	355	Louis Allis Division, MagneTek, Inc.
8-7	356	Westinghouse, Inc.
8-13	360	Square D
8-14	361	Sigmation
8-20	366	Westinghouse, Inc.
8-21	366	Cutler Hammer
8-24	367	Louis Allis Division, MagneTek, Inc.
8-31	373	Louis Allis Division, MagneTek, Inc.
8-52	391	Ambi-Tech
8-60	398	Louis Allis Division, MagneTek, Inc.
8-61	400	Louis Allis Division, MagneTek, Inc.
8-79	421	Elastimold Division, Amerace Corp.
8-81	423	Elastimold Division, Amerace Corp.
8-88	428	Adalet-PLM Division
8-89	429	Joy Manufacturing
8-90	429	Cam-Lok Division